LIFE IN ANCIENT ICE

LIFE IN ANCIENT ICE

Edited by John D. Castello and Scott O. Rogers

PRINCETON UNIVERSITY PRESS   PRINCETON AND OXFORD

Library of Congress Cataloging-in-Publication Data

Life in ancient ice / edited by John D. Castello and Scott O. Rogers.
p.   cm
Includes bibliographical references and index.
ISBN 0-691-07475-5 (cloth : acid-free paper)
1. Ice—Microbiology. 2. Microbiology—Polar Regions. I. Castello,
John D., 1952– II. Rogers, Scott O., 1953–
QR100.9.L54 2005
579′.17586—dc22   2004050523

British Cataloging-in-Publication Data is available

This book has been composed in Times Roman with Helvetica Display

Printed on acid-free paper.∞

pup.princeton.edu

Printed in the United States of America

10 9 8 7 6 5 4 3 2 1

Contents

Figures

Tables

Contributors

VIRGINIA ABERDEEN, Biology Department, Syracuse University, Syracuse, NY 13244.

S.S. ABYZOV, Institute of Microbiology, Russian Academy of Sciences, Moscow, Russia.

EDWARD E. ADAMS, Department of Civil Engineering, Montana State University, Bozeman, MT 59717

RYAN BAY, Department of Physics, University of California, Berkeley, CA 94720.

ROBIN BELL, Lamont-Doherty Earth Observatory of Columbia University, Palisades, NY 10964.

A. BOLSHAKOVA, Chemistry Faculty, Moscow State University, Vorobyovy Gory, Moscow 119899, Russia.

NATHAN BRAMALL, Department of Physics, University of California, Berkeley, CA 94720.

LLOYD BURCKLE, Lamont-Doherty Earth Observatory of Columbia University, Palisades, NY 10964.

JOHN D. CASTELLO, Faculty of Environmental and Forest Biology, College of Environmental Science and Forestry, State University of New York, 1 Forestry Drive, Syracuse, NY 13210.

CATHERINE M. CATRANIS, Faculty of Environmental and Forest Biology, College of Environmental Science and Forestry, State University of New York, 1 Forestry Drive, Syracuse, NY 13210.

BRENT C. CHRISTNER, Department of Land Resources and Environmental Science, Montana State University, Bozeman, MT 59717.

JOHN E. DORE, SOEST, Department of Oceanography, University of Hawaii, 1000 Pope Road, Honolulu, HI 96822.

VITALYI I. DUDA, Institute of Biochemistry and Physiology of Microorganisms, Russian Academy of Sciences, Pushchino, Moscow Region, 142290, Russia.

LUDMILA G. EROKHINA, Institute of Basic Biological Problems, Russian Academy of Sciences, Pushchino, Moscow Region, 142290, Russia.

RUSHANIYA N. FAIZUTDINOVA, Institute for Physicochemical and Biological Problems of Soil Science, Russian Academy of Sciences, Pushchino, Moscow Region, 142290, Russia.

JACK W. FELL, Rosenstiel School of Marine and Atmospheric Science, University of Miami, 4600 Rickenbacker Causeway, Key Biscayne, FL 33149.

CHRISTIAN H. FRITSEN, Division of Earth and Ecosystem Science, Desert Research Institute, 2215 Raggio Parkway, Reno, NV 89512.

DAVID A. GILICHINSKY, Institute of Physicochemical and Biological Problems in Soil Science, Russian Academy of Sciences, Pushchino, Moscow Region, 142290, Russia.

M.V. IVANOV, Institute of Microbiology, Russian Academy of Sciences, Moscow, Russia.

NATALIYA E. IVANUSHKINA, Skryabin Institute of Biochemistry and Physiology of Microorganisms, Russian Academy of Sciences, pr. Nauki 5, Pushchino, Moscow Region, 142290 Russia.

DAVIDA E. KELLOGG, Climate Change Institute, University of Maine, Orono, ME 04469.

THOMAS B. KELLOGG, Department of Geological Sciences, University of Maine, Orono, ME 04469.

GALINA A. KOCHKINA, Skryabin Institute of Biochemistry and Physiology of Microorganisms, Russian Academy of Sciences, pr. Nauki 5, Pushchino, Moscow Region, 142290 Russia.

KAYASTAS LAURINAVICHYUS, Institute of Biochemistry and Physiology of Microorganisms, Russian Academy of Sciences, Pushchino, Moscow Region, 142290, Russia.

JOHN T. LISLE, U.S. Geological Survey, Center for Coastal & Regional Marine Studies, 600 4th Street-South, St. Petersburg, FL 33701.

LI-JUN MA, Whitehead Institute, Massachusetts Institute of Technology, Center for Genome Research, 320 Charles St., Cambridge MA 02141–2023.

A.G. MAMUKELASHVILI, Department of Soil Science, Soil Science Faculty, Moscow State University, Vorobyovy Gory, Moscow, 119899, Russia.

JILL A. MIKUCKI, Department of Land Resources and Environmental Sciences, Montana State University, Bozeman, MT 59717.

J.N. MITSKEVICH, Institute of Microbiology, Russian Academy of Sciences, Moscow, Russia.

ELLEN MOSLEY-THOMPSON, Department of Geography, and Byrd Polar Research Center, Ohio State University, Columbus, OH 43210–1292.

A.L. MULYUKIN, Institute for Microbiology RAS, Pr. 60-letiya Oktyabbrya, 7/2, Moscow, 117811, Russia.

DAVID S. NICHOLS, School of Agricultural Science and Antarctic CRC, University of Tasmania, Private Bag 54, Hobart, Tasmania 7001.

SVETLANA M. OZERSKAYA, Skryabin Institute of Biochemistry and Physiology of Microorganisms, Russian Academy of Sciences, pr. Nauki 5, Pushchino, Moscow Region, 142290, Russia.

HANS W. PAERL, Institute of Marine Sciences, University of North Carolina at Chapel Hill, 3431 Arendell Street, Morehead City, NC 28557.

LADA E. PETROVSKAYA, Shemyakin & Ovchinnikov Institute of Bioorganic Chemistry, Russian Academy of Sciences, Miklukho-Maklaya Str., 16/10 GSP-7 Moscow, 117997, Russia.

M.N. POGLAZOVA, Institute of Microbiology, Russian Academy of Sciences, Moscow, Russia.

P. BUFORD PRICE, Department of Physics, University of California, Berkeley, CA 94720.

JOHN C. PRISCU, Department of Land Resources and Environmental Sciences, Montana State University, Bozeman, MT 59717.

JOHN N. REEVE, Department of Microbiology, Ohio State University, Columbus, OH 43210–1292.

ELIZAVETA RIVKINA, Institute of Physicochemical and Biological Problems in Soil Science, Russian Academy of Sciences, Pushchino, Moscow Region, 142290, Russia.

SCOTT O. ROGERS, Department of Biological Sciences, Bowling Green State University, Bowling Green, OH 43403.

RAYMOND SAMBROTTO, Lamont-Doherty Earth Observatory of Columbia University; Palisades, NY 10964.

ANASTASIA V. SHATILOVICH, Institute for Physicochemical and Biological Problems in Soil Science, Russian Academy of Sciences, Pushchino, Moscow Region, 142290, Russia.

SEUNG-GEUK SHIN, Department of Biological Sciences, Bowling Green State University, Bowling Green, OH 43403.

DANY SHOHAM, Begin-Sadat Center for Strategic Studies, Bar-Ilan University, Ramat-Gan, Israel.

JAMES E. SMITH, Faculty of Environmental and Forest Biology, College of Environmental Science and Forestry, State University of New York, 1 Forestry Drive, Syracuse, NY 13210.

V.S. SOINA, Department of Soil Science, Soil Science Faculty, Moscow State University, Vorobyovy Gory, Moscow, 119899, Russia.

ELENA V. SPIRINA, Institute for Physicochemical and Biological Problems in Soil Science, Russian Academy of Sciences, Pushchino, Moscow Region, 142290, Russia.

WILLIAM T. STARMER, Biology Department, Syracuse University, Syracuse, NY 13244.

MICHAEL STUDINGER, Lamont-Doherty Earth Observatory of Columbia University, Palisades, NY 10964.

NATALIYA E. SUZINA, Institute of Biochemistry and Physiology of Microorganisms, Russian Academy of Sciences, Pushchino, Moscow Region, 142290, Russia.

VINCENT THERAISNATHAN, Department of Biological Sciences, Bowling Green State University, Bowling Green, OH 43403.

LONNIE G. THOMPSON, Department of Geological Sciences and the Byrd Polar Research Center, Ohio State University, Columbus, OH 43210–1292.

ANAHITA TIKKU, Lamont-Doherty Earth Observatory of Columbia University, Palisades, NY 10964.

ALEXANDER I. TSAPIN, Jet Propulsion Laboratory, 4800 Oak Grove Drive, Pasadena, CA 91109.

TATIANA A. VISHNIVETSKAYA, Institute for Physicochemical and Biological Problems in Soil Science, Russian Academy of Sciences, Pushchino, Moscow Region, 142290, Russia.

ELENA A. VOROBYOVA, Department of Soil Science, Soil Science Faculty, Moscow State University, Vorobyovy Gory, Moscow, 119899, Russia.

CRAIG F. WOLF, Department of Land Resources and Environmental Sciences, Montana State University, Bozeman, MT 59717.

I.V. YAMINSKY, Chemistry Faculty, Moscow State University, Vorobyovy Gory, Moscow 119899, Russia.

GANG ZHANG, Department of Biological Sciences, Bowling Green State University, Bowling Green OH, 43403.

YINGHAO ZHAO, Department of Immunology and Microbiology, School of Medicine, Wayne State University, 540 E. Canfield, Detroit, MI 48201.

SHUANG ZHOU, Faculty of Environmental and Forest Biology, College of Environmental Science and Forestry, State University of New York, 1 Forestry Drive, Syracuse, NY 13210.

Acknowledgments ⸻⸻⸻⸻⸻⸻⸻⸻⸻

WE WISH TO THANK the National Science Foundation for its generous financial support (grant number 000384) of the 2001 Life in Ancient Ice Workshop, which was the foundation for this book, as well as all of the workshop participants for lively and constructive discussions. JDC thanks his parents, Kay and Dom; and his Cuddles for their moral support. SOR thanks his wife, Mary; children, Liz and Ben; and parents, Harold and June, for all of their moral support. It was all needed to complete this book. Finally, we thank the editors, reviewers, and others at Princeton University Press for all of their time, help, and useful comments in producing the final product.

LIFE IN ANCIENT ICE

1

Introduction

John D. Castello and Scott O. Rogers

LIFE FLOATS in the clouds and is carried on the winds to all corners of the globe. This life includes fungi, protists, bacteria, archaea, viruses, pollen grains, and other biotic particulates that originate from the oceans, lakes, deserts, tropics, temperate regions, cities, forests, agricultural fields, and other habitats all over the world. Eventually these particles fall from the sky affixed to raindrops or snowflakes. Some die during their airborne journey, but many survive. If they fall to the earth in the polar regions they may become immured within the vast Arctic and Antarctic ice sheets for tens, hundreds, or thousands of years, and possibly longer. But is this the end of the story? No, it is not. Many of these life forms die, some become dormant, and still others thrive under such harsh but stable conditions. Eventually, the ice in which they are entombed melts, and these organisms, including the dead, dormant, and living ones, are released into the contemporary biosphere. This book is a compilation of the research that describes the environs and organisms that have been found in ice and permafrost.

What happens after the organisms reenter the contemporary environment? We do not know the answer to this question, but research to address this question must be conducted because some of the effects could be profound. Do these organisms reproduce and spread throughout the biosphere once again, or do they die because they cannot adjust to a different environment? If they survive, do they reintroduce ancient genes back into the contemporary gene pools? If so, what are the consequences of this gene flow on evolutionary rates, adaptability, fitness, and so on.? What are the consequences if some of these organisms are pathogens? Could they possibly start epidemics or a pandemic? These are just a few of the many questions that can be raised. The answers to them and others are unknown at this time, although the authors of the chapters in this book are addressing many of these questions.

This scenario may sound like science fiction, but it is not. Within this book is the evidence to support the entire story. It is a fascinating story. It is a new story because research into this field of science is new. Serious investigation began only thirty years ago in Russia with the work of Sabit Abyzov. It also is a very old story because it is likely to have been repeated annually since the vast ice sheets formed millions of years ago. As you will see, some problems were solved, but others are yet to be solved. There were successes and failures; some questions were answered, but many more were raised.

This book had its origins in a workshop sponsored by the National Science Foundation and organized by the editors that was held at the Westin Salishan Lodge, Gleneden Beach, Oregon, from 30 June to 3 July 2001. The intensity and excitement were evident, indicating that additional collaborations and enhanced research activity should occur. The purpose of the workshop was to bring together experts studying ice, permafrost, ancient life, biological preservation, evolution, and astrobiology to assess current and future research that would extend our knowledge of life in ancient frozen matrices. Thirty-five internationally renowned scientists from Australia, Canada, Denmark, Germany, Israel, Russia, and the United States attended the workshop, and their presentations are the basis for the book. Many additional coauthors contributed to the chapters. The interactions among the participants were stimulating, and we hope that the workshop will lead to future meetings, collaborations, and research.

There were several objectives of the workshop, among which were

- providing the evidence for the presence of life in glacial ice and permafrost
- evaluating the reliability of that evidence
- discussing the significance and implications of the results to date
- identifying the more important and immediate research needs, and our recommendations for future research

The workshop and the book were organized according to scientific discipline. Chapter 2 introduces the many protocols that have been used when studying organisms and biological molecules entrapped in ice. It discusses some of the central problems, issues, and challenges in this research. One of the major challenges is to exclude, or somehow separate, external contaminants from the analyses of the internal ancient organisms and biological molecules. It recommends a set of goals to assure that only authentic entrapped organisms are identified as such. Chapters 3 through 6 outline the potential sources of microbes in polar environments, as well as describing microbes that each research group has found in ice, water, and the atmosphere. They state the quantities and types of microbes found in polar oceans, lakes, and the atmosphere, and describe some of the ways in which microbes can be transported to and deposited in glacial ice. In particular, waterborne and windborne (aeolian) transport is discussed in detail.

In chapter 7, Rivkina et al describe their evidence for microbial metabolic activity at temperatures below freezing. The authors state that the microbes do this by surrounding themselves with supercooled liquid water. This introduces the possibility that the microbes are not in states of suspended animation but, rather, may be metabolically active, although at low levels. This is an important point, because if the organisms are able to survive for millennia or longer in the ice, they would probably need to continue to repair their DNA, since some damage to their chromosomes would be expected over such long spans of time.

The next several chapters (chapters 8 through 16) detail the microbes that have been found in permafrost (chapters 8 through 10) and glacial ice (chapters 11 through 16). Several describe fungi (chapters 8, 9, 11, and 12), many of which are common in cold climates; however, many others appear to have blown in from temperate and tropical zones. Similarly, many of the bacteria described (chapters 10, 15, and 16) appear to have been transported by wind from warmer climates. This is the case whether the glacier is near temperate zones (chapter 15) or very far from any temperate zone (chapter 16). Chapter 10 describes cyanobacteria that are able to survive and metabolize in cold and dark permafrost, which emphasizes that most of the viable microbes in permafrost and ice are able to survive in a range of environments, some of which are extreme.

Viruses are discussed in two of the chapters. In chapter 13, bacteriophage are described that were isolated from bacteria that had been isolated from polar ice cores. In a previous paper, the same authors described plant viruses isolated from ancient glacial ice. Together, these findings indicate two possible modes of preservation of viruses in glacial ice, the first being as intact virus particles (or fragments of particles), and the second is through integration into bacterial chromosomes (becoming prophage). In chapter 14, the significance of the disappearance and reappearance of viruses through time is discussed. The importance of glacial ice as a reservoir for these viruses also is discussed. In the future, this topic will probably be the focus of many research studies. The diversity of organisms described in these chapters, the extreme environments where they are found, the modes of movement, and the implications of their reentry into a more recent environment cause us to reexamine our views of where and how microbes can survive.

The final chapters describe several areas of research that have been developed over the past few years. The first of these chapters (chapter 17) presents details about Lake Vostok, a subglacial lake about the same size as Lake Ontario. The lake is buried by more than 3500 m of glacial ice and has been covered with ice for millions of years. Although Lake Vostok is the largest such lake found to date, there are hundreds of other subglacial lakes. No one knows whether life exists in Lake Vostok. It has been the subject of a great deal of speculation and will be the focus of research for many years to come. In the next chapter (chapter 18) a unique robot is described. This robot can be lowered into ice core boreholes to detect the presence of microbes. It can log data and send it to computers on the surface. It is possible that robots can be used more extensively to study other boreholes, as well as to examine subglacial lakes such as Lake Vostok. The following chapter (chapter 19) discusses the use of icy environments on Earth as model systems for the study of similar environments on planets and moons. Ice has been found on our moon, Mars, and Europa (one of Jupiter's moons). Ice, permafrost, and ice-covered bodies of water on Earth are being used to test methods for detecting life on other bodies in the solar system. This chapter indicates one of the very practical aspects of the study of life in ancient ice. The final chapter (chapter 20) summarizes the most significant and important points in the book. It details some of the most key discoveries and outlines the areas that will be of importance in the future.

This book presents many of the facets of the study of organisms entrapped in ice. One can consider this a summary of the current state of research in the study of life, both contemporary and ancient, in ice. We anticipate that this will change rapidly in the near future. From the pioneering work of Sabit Abyzov and co-workers to the most recent publications, it is clear that ice and permafrost represent unique ecosystems and preservation matrices that can yield valuable information on microbial longevity, biological molecule preservation, past climates, recent climate change, evolutionary processes, epidemics, origins of life, life on other planets/moons, and other fields of study. It is true that with the scores of ice and permafrost cores that have been collected over the past several decades, we have gained a deep understanding of the potential benefits for the study of the ancient cryosphere. However, it also is clear that we have only begun to scratch the surface. At the time of printing, this is the only book of its kind, but we are hopeful that many more books on this subject will follow.

2

Recommendations for Elimination
of Contaminants and Authentication
of Isolates in Ancient Ice Cores

Scott O. Rogers, Li-Jun Ma, Yinghao Zhao, Vincent Theraisnathan,

Seung-Geuk Shin, Gang Zhang, Catherine M. Catranis,

William T. Starmer, and John D. Castello

VIABLE FUNGI, BACTERIA, AND VIRUSES have been isolated from glacial ice up to 400,000 years old and permafrost up to 3 million years old, originating from many geographic locales (1, 17–20, 25, 33, 40, 42). While many bodies of water contain higher concentrations of viable microbes than do deposits of ice, microbes encased in ice are usually in states of suspended animation or their growth is limited. This makes ice a unique repository of microbes frozen at various points in the past. Permanent ice covers approximately 10% (about 15 million square km) of the land area on Earth. Over 84% of this ice covers Antarctica. Another 14% covers land areas in the Arctic. The remaining 2% is in the mountains of Asia, North America, South America, Europe, New Zealand, and Africa (in descending order of percentage). Worldwide, the number of microbes entrapped in this ice is enormous. Recent microbiological studies of ice cores indicate that there are from 10^3 to 10^7 viable microbes entrapped l^{-1} of glacial meltwater (8, 18, 20, 31, 39, 40, 53, 56). At current rates of melting, this translates into an annual release of approximately 10^{17} to 10^{21} viable organisms worldwide. Spread evenly over the Earth, this represents approximately 10^8 to 10^{12} propagules $\cdot$ m^{-2}, or about 10^2 to 10^6 propagules $\cdot$ m^{-2} per species. Although these are estimates extrapolated from a handful of studies, they are indicative of the enormity of the releases of microbes from glaciers. Any global warming events would increase these numbers further. We have identified plant pathogens in the ice cores (17), but there are likely to be many other types of pathogens, including those that infect humans. While the number of organisms in ice worldwide is high and they are likely to be important ecologically, evolutionarily, and epidemiologically, extracting them from the ice for study is problematic.

Preservation of Ancient Specimens and Nucleic Acids

The primary reason that ancient microbes and nucleic acids are difficult to study is that they are always surrounded by recent contaminating microbes. Also, the ancient organisms are in relatively low concentrations, and they may have become damaged during their encasement in the ice. All of these factors make experimental manipulation difficult and replication impractical, in most cases. Nonetheless, there is an inherent interest in knowing how long microorganisms can remain viable, how long nucleic acids can survive, and what conditions are ideal for preservation. Theoretical expectations once had placed the limit for recovering viable bacterial spores at about 200,000 years (28, 57). Empirically based mutation rates that result in detrimental effects on spore viability have been used to estimate the half-life of *Bacillus subtilis* at 7000 years. Assuming an exponential death rate, a large population of viable spores would be detectable after several hundred thousand and possibly for several million years (28). These expectations assume that the immediate environment protects the spores from UV irradiation, oxidation, and other chemical damage. Studies on deep mud cores provide evidence for survival of thermoactinomycete endospores in excess of 1000 years (23). Viable microbes have been isolated from 320-year-old herbarium specimens (57). Thermophilic bacteria have been isolated from ocean basin sediment cores estimated to be 5800 years old (12). Bacteria also have been recovered from Roman archaeological sites that are 1900 years old (52), from Siberia up to 8000 years old (54), and from Vostok Station (Antarctica) ice cores, some of which are more than 400,000 years old (2–7, 17, 18, 33, 34, 37–40, 42, 48). Reports of the isolation, growth, and DNA characterization of bacteria from amber and halide crystals have pushed the theoretical limit to several hundreds of millions of years (15, 24, 58). However, this estimate has yet to be confirmed, because the isolations from the oldest specimens have not been independently replicated. Authenticity has been challenged in studies with amber, halide crystals, fossils, and others (10, 22, 36, Willerslev, Hansen, and Poinar, pers. comm.).

Currently, the most conservative estimates on the limits of DNA integrity range from 10^3 to 10^4 years for hydrated DNA within matrices at or near physiological temperatures (36, Willerslev, Hansen, and Poinar, pers. comm.). DNA is subject to chemical and physical damage (Table 2.1) by a variety of agents (35). In general, DNA is most vulnerable to degradation when hydrated, at low pH, and at elevated temperatures. Under these conditions, the purine and pyrimidine bases frequently are cleaved off the DNA, creating AP (apurinic and apyrimidinic) sites. Deamination of guanine, adenine, and cytosine occurs often at low pH (although deamination can also occur at high or neutral pH) and/or high temperatures. The sugar phosphate backbone can be broken by a

TABLE 2.1.
The most frequent damaging reactions to DNA

Original	Reaction	Result	Pairs with	in vivo Repair	PCR / Sequencing Errors
C	Deamination	U	A	+	+
5meC	Deamination	T	A	−	+
A	Deamination	Hypoxanthine	C	+	+
G	Deamination	Xanthine	C	+	−
A	Depurination	No base	N	+	+
G	Depurination	No base	N	+	+
G	Oxidation	8-OH-G	A	−	+
A	Cyclic oxidation	8,5′ cyclic A	N	−	+
Py/Py	Dimerization	Dimers	None	+	+
N	Hydroxymethylation	H-N-CH$_2$OH	N	N	+

Notes: The result of each chemical reaction is indicated, as well as the change in pairing during replication, PCR amplification, and sequencing reactions; and whether the mutation causes a change in the resulting sequences. 5meC indicates 5-methyl-cytosine, a variant of cytosine, which when deaminated becomes thymine. Py/Py indicates regions with adjacent pyrimidine residues (cytosine and thymine). The most frequent pyrimidine dimers that are formed are T-T, followed by C-T, T-C, with C-C being the least frequent. N indicates any nucleotide, A, G, C, or T.

number of agents, including ionizing radiation and alkalai (27). Oxidative damage can occur in a number of sites on nucleic acids. Aside from these major types of damage, other damage is caused by UV irradiation, as well as enzymatic and nonenzymatic attack. Thus, DNA will undergo less degradation under the following conditions: limitation of free water, neutral pH, limitation of oxygen, limitation of Mg^{2+} (a cofactor for many nucleases), and low temperatures. These conditions are present in many ice deposits.

Dry and cold conditions may extend the longevity of DNA to over 10^6 years (Figure 2.1). Previously, we reported the isolation of DNA from plant seeds up to 100 years old and from mummified specimens from 500 to over 45,000 years old (46, 50), in addition to viable bacteria and fungi, as well as fungal, bacterial, and viral nucleic acids, from ice cores 500 to 420,000 years old (17, 18, 37–40, 48). The average size of DNA from some 45,000-year-old mummified plant seeds was up to 3 kilobases, with some fragments up to 10 kilobases. This indicates that the limit of DNA longevity is well over 4.5 × 10^4, and could approach 10^6 years for desiccated tissues buried in soil and rock. Because viable organisms have been isolated from 420,000-year-old ice, longevity of DNA is expected to be greater for tissues embedded in these cold matrices, especially when enclosed within living organisms. When tempera-

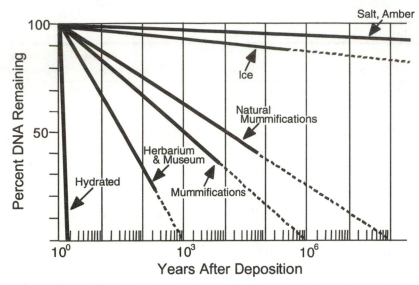

FIGURE 2.1. Representation of the amount of DNA remaining in ancient samples.
The percentages are based on average molecular weights. Solid lines represent
actual data, while dashed lines are extrapolations. Lines for ice, salt, and amber are
based on isolations of living organisms (assuming that at least 90% of the genome
is intact). Data are from the following sources: herbarium/museum—32, 45–47, and
Rogers unpublished data; mummifications—41, 46, 51; ice—19, 20, 39, 40, 42,
59; amber/salt—15, 21, 24, 58. The results from salt, amber, and ice have yet to be
independently confirmed by other laboratories.

tures are permanently below 0 °C, degradation slows down, because all of the
degradative reactions (outlined above) are slower. Therefore, preservation of
DNA in ice might exceed 10^6 to 10^7 years. Ice provides an environment that is
cold as well as dry, in that free water is unavailable.

For any preserved specimens, the initial conditions are important (49). If
the organisms or nucleic acids are deposited on the glacier already partially
degraded, encasement in ice can slow further damage, but cannot reverse any
prior damage. If the initial conditions are conducive to preservation of the
organisms or nucleic acids, then there is a possibility that organisms and/or
their nucleic acids can be recovered at a later time. It is clear that many types
of organisms remain viable after being frozen for relatively short lengths of
time. However, as indicated above, the upper limits of time for survival of
the organisms and the macromolecules are currently imprecisely known. With
larger organisms, the question is clearer than for microbes, because if a substantial number of vital cells die, the entire organism perishes. With microbes
and nucleic acids the situation differs, because a part of the longevity is de-

pendent on population size. If the population size is large, then there is a greater probability that some of the members of the population will survive for extended periods of time. Limiting DNA degradation is vital to the survival of the organisms, but it is also important to researchers when attempting to obtain accurate sequences from DNA isolated from ancient matrices, such as glacial ice.

Methods to Study Microbes in Ice

Many protocols have been used to study and characterize microbes from ice cores (Figure 2.2). Light and electron microscopy have been used extensively to examine the meltwater for plants and animals (or parts thereof), as well as for microorganisms. However, classification to species is seldom possible with these methods. Culture methods also are widely employed, and have yielded a large number of isolates, many of which closely resemble described taxa. However, a large number may be new species. While culture methods are very useful, many microbes will not grow in culture. Estimates range from 0.1 to 17% of taxa that can be cultured. Therefore, culturing results provide a gross underestimate of the diversity of viable microorganisms in the ice.

Molecular characterizations are being utilized more frequently in studies of microbes from ancient ice. For cellular organisms, ribosomal DNA sequences are used to determine both taxonomic and genealogical affiliations. For viruses, a number of gene regions have been used. The DNA (or RNA for RNA viruses) is amplified using polymerase chain reaction (PCR). Then, sequencing is performed using the amplified product as the template. This process is used on the cultured organisms as well as on the meltwater, separately. The resulting sequences can be compared to sequences from contemporary specimens using various methods, including phylogenetic analyses. A major consideration with all of the methods, especially those that are very sensitive, is the avoidance of contamination.

Contamination and Decontamination

The elimination of all contaminating microbes and biomolecules is of primary importance in specimen authentication. Effective decontamination depends on the sterilization methods, sensitivity of assays, controls, replication of assays by independent researchers, and use of indicator organisms and nucleic acids. Contamination is always present on the outside of the core sections. The contaminants originate from the drill, drilling fluids, handling, transport, storage, and final processing of the ice core sections. Attaining sterility at the field coring sites, and during subsequent transport to the storage and research facili-

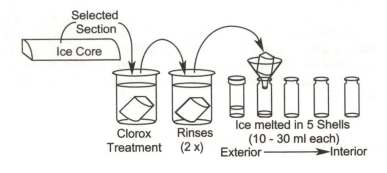

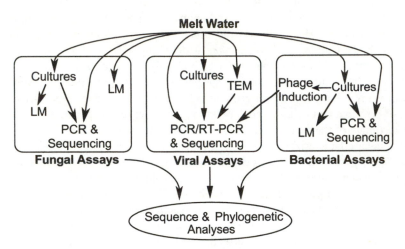

FIGURE 2.2. Flow diagram of decontamination, melting, and assay protocols. Upper portion shows the decontamination and melting protocol. Lower portion indicates the various assays for viable microbes and nucleic acids. TEM = transmission electron microscopy, LM = light microscopy, PCR = polymerase chain reaction, RT-PCR = reverse transcription-PCR.

ties, is impossible. Once the core is transported to the laboratory for study, sterility is possible, but the outer surfaces of the core section must be carefully and effectively decontaminated and/or removed. During culturing and molecular biology procedures, sources of contamination may include ventilation systems, work surfaces, culture media, reagents, researchers (hands, hair, breath, etc.), and others. In addition, contamination may have entered the core sections through minute cracks and breaks in the ice. Once in the laboratory, all sources of contamination must be eliminated, and the core must be decontaminated prior to assays for microbes and/or biomolecules.

Comparisons of Decontamination Protocols

There are two major objectives of any decontamination protocol. The first is to ensure that contaminants are completely eliminated. The second is to protect rare and/or sensitive organisms and nucleic acids inside the ice cores during decontamination and melting of the ice. Some of our research efforts have dealt with solving these important problems. Synthetic, or "sham," ice core sections (a cylinder 5 cm in diameter and 15 cm in length) have been produced in our laboratory, consisting of frozen suspensions of specific indicators on the interior (*Aspergillus tereus* spores and DNA, *Aureobasidium pullulans* spores, *E. coli* cells, and tomato mosaic tobamovirus [ToMV] virions), and other indicators spread onto the exterior (*Ulocladium atrum* spores and DNA; *Bacillus subtilis* spores; and lambda phage DNA). We evaluated UV irradiation, liquid sterilants (solutions of sodium hypochlorite, ethanol, hydrogen peroxide, sodium hydroxide, and hydrochloric acid), as well as ablation (removal of outer layers with sterile distilled water or sterilized razor blades) and mechanical methods (sterilized drills, saws, and heated probes) to extract inner core regions (Tables 2.2 and 2.3). Then, the treated cores were melted in shells, where shell 1 was the outermost part of the core and shell 5 was the innermost portion (Figure 2.2, upper). Each shell was approximately 5 to 10 mm in thickness, yielding approximately 20 ml of meltwater each. The treatment that assured complete elimination of all outer organisms and nucleic acids and preserved the largest proportion of the inner core regions was a 10 second exposure to a solution of 5% sodium hypochlorite (Table 2.3).

While UV irradiation was initially promising, we discovered that some microbes (most notably, *Ulocladium atrum*) are extremely resistant to treatment by this method. For example, 100% of the cultures of *Penicillium commune* were killed by a dose of 27, 540 $J \cdot m^{-2}$ of 265 nm UV irradiation (germicidal lamp placed 2 inches from the culture for 5 minutes). However, cultures of *U. atrum* withstood twelve times this dose with no notable effects (100% of the cultures survived with no effects on numbers or sizes of colonies). Three protocols (95% ethanol, heated probe, and drill extraction) all produced similar results. External contaminants were found through shell 2, and there was a reduction in the number of inner microbes and nucleic acids in shells 1 through 3. All three methods utilized a treatment of 95% ethanol, which might partially explain the similarity of results. While ethanol will kill many microbes, it is ineffective toward others and does not destroy nucleic acids. Thus, it was relatively ineffective. Treatment with hydrogen peroxide (6% or 35%) and sodium hydroxide/hydrochloric chloride (1N or 10N each) produced similar results. In each case, external contaminants were evident in shells 2 and 3, respectively. Additionally, the quantities of internal microbes and nucleic acids were reduced all the way into the innermost shell. With

TABLE 2.2.
Results of Clorox (5.25% sodium hypochlorite solution) decontamination tests of sham ice cores seeded with various organisms

| | | Growth[a] | | | | | | PCR[b] | | | | | | |
| | Number Propagules before Clorox | Before Clorox | After Clorox (shell number) | | | | | Nucleic Acid Concentration before Clorox | Before Clorox | After Clorox (shell number) | | | | |
Organism		0	1	2	3	4	5		0	1	2	3	4	5
Viruses[c]	NA	NA	NA	NA	NA	NA	NA	50 fg/µl	++	−	−	+	++	++
Bacteria[d]	300 /ml	++	−	−	+	+	++	1 pg/µl	++	−	+	++	++	++
Bacteria[d]	10⁷/ml	++	−	+	++	++	++							
Fungi[e]	1000/ml	++	−	+	++	++	++	1 pg/µl	++	−	+	++	++	++

NA = Not applicable.

[a] A volume of 100 µl of the suspension was spread onto nutrient agar plates, followed by incubation at 37 °C for bacteria or 22 °C for fungi. A minus (−) indicates no growth. A single plus (+) indicates a few colonies per plate. A double plus (++) indicates a large number of colonies per plate.

[b] A 10 µl aliquot of the icemelt was used to perform PCR or RT-PCR (in the case of RNA viruses) in a total volume of 25 µl. Prior to removal of the aliqot for the fungal and bacterial assays, the icemelt was frozen and then boiled to break open the cells. In some cases 1% Triton was added. The concentration of 50 fg/µl of ToMV RNA is equivalent to 15,000 virions, or 1000 fg/µl of whole virus. A minus (−) indicates no evident amplification (on ethidium bromide-stained gels). A single plus (+) indicates a faint amplification band. A double plus (++) indicates a heavy amplification band.

[c] ToMV was the first virus tested. Others will be tested later.

[d] *Escherichia coli* and *Bacillus subtilis* cells, as well as *B. subtilis* endospores were used in separate assays, with similar results.

[e] *Aspergillus terreus* was used to seed the sham ice cores on the inside. *Ulocladium atrum* was spread onto the outer surface of the ice core prior to Clorox treatment. In culture assays, no *U. atrum* was detected in any shell. *U. atrum* is resistant to desiccation, UV irradiation, and other types of decontamination methods, and was chosen because of its hardiness. *Aspergillus terreus* is a relatively labile fungus. All growth and PCR amplification was exclusively from *A. terreus*.

TABLE 2.3.
Microbial growth of inner and outer (surrogate contaminants) organisms from decontamination tests of sham ice cores.

Treatment	Shells[a]				
	1	2	3	4	5
6.0% NaOCl[b]	0/+[2]	0/+	0/+	0/++	0/++
5.25% NaOCl[c]	0/+	0/+	0/+	0/++	0/++
5.0% NaOCl[d,e]	0/+	0/+	0/++	0/++	0/++
4.0% NaOCl[d]	+/+	0/++	0/++	0/++	0/++ .
3.0% NaOCl[d]	+/+	+/++	0/++	0/++	0/++
2.0% NaOCl[d]	+/+	+/++	0/++	0/++	0/++
1.0% NaOCl[d]	+/+	+/++	0/++	0/++	0/++
95% Ethanol	+/+	+/+	0/+	0/++	0/++
Heated probe	+/+	+/+	0/+	0/++	0/++
drill	+/+	+/+	0/+	0/++	0/++
35% Hydrogen peroxide	+/+	+/+	0/+	0/+	0/+
6% Hydrogen peroxide	+/+	+/+	0/+	0/+	0/+
10N NaOH / 10N HCl	+/+	+/+	+/+	0/+	0/+
1N NaOH / 1 N HCl	+/+	+/+	+/+	0/+	0/+
Water ablation	+/+	+/++	+/++	+/++	0/++
Razor blade ablation	+/++	+/++	+/++	+/++	+/++
UV irradiation	+/+	+/+	+/+	+/++	+/++

Note: All treatments with solutions were for 10 seconds each.

[a]Symbols to the left of the slash indicate exterior microbes (*U. atrum* and/or *B. subtilis*). Symbols to the right of the slash indicate interior microbes (*Aspergillus terrreus, Aureobasidium pullulans,* and/or *E. coli*). 0 = absent; + = present, but at low levels; ++ = present at levels equal to those of controls.

[b] Undiluted Ultra Clorox = 6% NaOCl (sodium hypochlorite).

[c] Undiluted Clorox = 5.25% NaOCl.

[d] Dilution using Ultra Clorox.

[e] Designated as the best treatment, based on elimination of contaminants, while protecting the largest proportion of the ice core.

ablation methods, external contaminants were found through shell 4, leaving only shell 5 as usable for study of interior organisms and nucleic acids. However, there is a statistical likelihood that occasionally some outer contaminants would be found in shell 5.

Testing of various concentrations of sodium hypochlorite led to the conclusion that a 5% solution is ideal for decontamination of the external surfaces of the ice cores, since it eliminated all external contamination and high levels of internal organisms and nucleic acids were found consistently from shell 3 inward. Concentrations above 5% decontaminated the outer surfaces, but led to a reduction of internal microbes in shells 1 through 3. Concentrations below 5% failed to assure complete removal of all external contaminating microbes and biomolecules.

Chlorine is a strong oxidizing agent that is an effective sterilant, able to kill virtually all fungi, bacteria, and viruses shortly after contact. Initially, one concern was that the chlorine would penetrate the ice, killing the organisms and nucleic acids within the ice core in addition to those on the outer surfaces. This concern was unfounded, and we demonstrated that the effects of the chlorine disappeared from 2 to 10 mm into the ice core section, in that no inhibition of growth of microorganisms and of *Taq* DNA polymerase in PCR amplifications was detected. This was true for both high and low concentrations of microbes and nucleic acids.

Recommended Protocol for Aseptic Isolation of Microorganisms and Nucleic Acids from Ice

The protocol that we developed for aseptic isolation of microorganisms from ice is shown in Figure 2.2 (upper part), and is based on repeated testing of the protocol, as well as on the results from our extensive study of decontamination protocols (above). Prior to melting, the ice subcore sections are removed from the -80 °C freezer and placed in a sterile beaker at -20 °C for at least 24 hours to minimize cracking. The core sections are then placed in a new sterile beaker in a sterile room at room temperature. They are allowed to equilibrate to this temperature for 10 to 30 minutes. A 5% solution of cold (4 °C) sodium hypochlorite (800 ml) is poured into the beaker such that the ice core is completely covered. It is allowed to soak in the Clorox for exactly 10 seconds, after which time it is placed in another beaker containing 800 ml of cold (4 °C) sterile distilled water (18.2 MΩ, < 1 ppb TOC, DNase-free, and RNase-free) sufficient to cover the ice completely. This rinsing step is repeated twice. Next, the ice core is placed in a sterile funnel with a sterile collection jar below. Melting is at room temperature in a sterile hood. When approximately 20 ml have melted in the collection jar, the jar is sealed and labeled. The first meltwater aliquot is denoted "shell 1." A new collection jar is placed below a new sterile funnel and the process is repeated. Eventually, four (or more) additional shells are collected (denoted shells 2, 3, 4, etc.). These meltwaters are assayed immediately by culturing and PCR or reverse transcription PCR (RT-PCR for RNA viruses) amplification. The remaining portions are stored at −80 °C, and later assayed by culturing, microscopy, PCR, and sequencing (Figure 2.2, lower).

Results Using the Sodium Hypochlorite Decontamination Protocol

Our studies have employed a number of methods, including microscopy, culturing with sequencing, and sequencing directly from ice meltwater (17, 18, 39, 40, 48). We currently use the 5% sodium hypochlorite decontamination protocol (described above and in Figure 2.2) to decontaminate each ice core section in our laboratory. Also, we periodically test the cultures for susceptibility to sodium hypochlorite. Thus, we are certain that our cultures represent microbes that were entrapped within the ice cores. For culturing, we utilized four media for bacteria and seven for fungi, with several temperature regimes (4 °C, 8 °C, 15 °C, 22 °C, 37 °C, and combinations of these). More than 300 bacterial isolates and over 300 fungal isolates were obtained from twenty-one Greenland and six Antarctica ice cores ranging in age from less than 500 years before present (ybp) to over 400,000 ybp. Many of these have been identified to genus and species using cultural, molecular, morphological, and/or metabolic methods. Species of fungi were isolated within the genera *Alternaria, Aspergillus, Cladosporium, Cryptococcus, Emericella, Exophiala, Fusarium, Geotrichum, Penicillium, Phoma, Pleurotus, Rhodotorula,* and *Tricholoma.* Other fungi could only be classified at higher taxonomic levels. We have subjected many to DNA sequence and phylogenetic analyses (see refs. 39 and 41 and chapter 11, this volume, for details). In addition, fungal DNA sequences have been determined directly from the ice meltwater. Some sequences from ice are distant from all sequences included in national and international databases, while others closely resemble sequences from contemporary taxa. Bacterial isolates in the genera *Bacillus, Paenibacillus, Tatumella, Rhodococcus,* and others were isolated from the ice cores (see refs. 18 and 59 and chapter 13, this volume, for details). Three species of sheathed bacteria (*Leptothrix lopholea, Haliscomenobacter hydrosis,* and *Phragmidiothrix multiseptata*) appeared in the original slide preparations of the glacial meltwater (before filtrations) but were not recovered in pure culture. Some of the fungal and bacterial taxa were recovered from more than one section and depth; thus multiple isolations of the same taxon from progressively older ages is feasible.

We recovered few obligate psychrophiles, possibly due to the use of culture methods that are not optimal for these taxa. However, a variety of microorganisms were isolated or detected that were tolerant of low temperatures (*Cladosporium sphaerospermum*, *C. herbarum*, *Penicillium variabile*, *Micrococcus* spp., *Sarcina* spp., and *Chrysophyta* spp.), low oxygen tensions (*Leptothrix lopholea*), UV irradiation (*Ulocladium atrum*), and prolonged aerial transport (*Cladosporium* spp. *Penicillium* spp., *Phoma hibernica*, *Micrococcus* sp., and *Chrysophyta* spp.). Atmospheric circulation over Antarctica allows air exchange with lower latitudes (see ref. 1 and chapter 5, this volume). Therefore, microbes transported on dust and precipitation become embedded in ice that forms from falling snow. Most of the isolates may have originated from lower latitudes. Glaciers have entrapped large quantities of aerosolized microorganisms and viruses from ancient atmospheres, and thus ice is a natural air sampling matrix and repository from which to assess ancient microbial diversity on a global scale.

To date we have examined over thirty Greenland ice core meltwater samples for ToMV. We detected ToMV RNA in seventeen of the samples (see ref. 21 and chapter 13, this volume). Phylogenetic analyses indicated that the virus is genetically variable, and that a mixture of ToMV RNAs of varying antiquities exists in the ice cores. This concept of genome recycling (i.e., temporal gene flow) is currently being studied in another research project. Bacteriophage have been recovered from several *Bacillus subtilis* cultures (isolated from Greenland ice cores) by various induction methods (chapter 13). They are all most probably members of the PBSY group of bacteriophage, common in *Bacillus* species. We have also begun molecular assays for Coxsackie virus, parvovirus, polio virus, influenza virus, tomato bushy stunt virus, potato virus X, and pox viruses. We have assayed extensively for two of these, influenza and pox viruses. To date, influenza virus has been detected in one glacial ice core.

Many enteric viruses, such as parvovirus, foot-and-mouth disease virus, polio virus, influenza virus, Coxsackie virus, echovirus, and hepatitis A virus, are present in soil, water, and air in which they can spread and initiate disease outbreaks (43). Viruses surround us in the soil, air, and water. The dynamics of aerial transport and survival of human and plant pathogens in water droplets have been studied extensively from the earliest days of microbiology (for reviews, see 9, 26, 44). These studies, and our detection of ToMV in clouds (16), suggest that glacial ice viruses originate from frequent and varied aerosolizations. Sources of airborne viruses may include waterfall sprays, raindrops, sea spray and surf (13), hurricanes, tornadoes, volcanic activity, dust storms, soil (14), waste discharge and sanitary treatment (55), infected animals and their feces (29, 30), windblown plant debris (11), and agricultural activities. Although individual virions are small enough (10–450 nm) to remain airborne nearly indefinitely, they are not likely to occur free of other materials. For example, when ejecta droplets evaporate, the viruses will remain associated and stabilized on nonspecific colloidal organic and mineral debris. Therefore, they ultimately will settle out or become incorporated into snowflakes or raindrops. There is likely to be a range of viruses in ancient ice cores, but this area of research is in its early infancy.

Recommendations

Although a great deal of interesting and important information has been gleaned from studies of microbes in ice cores, standardization of methods has been elusive. Nonetheless, it will be necessary in the future to reach some consensus on the best ways to handle and research each of these valuable ice cores. While each ice core section is somewhat unique in its composition and inclusions, there should be a set of standard procedures in each research laboratory to ensure that only the microbes and nucleic acids originally entrapped in the environmental ice are isolated and/or characterized. This has the explicit requirement that the outer contemporary organisms and nucleic acids, which were introduced during drilling and afterward, be eliminated from subsequent processes. We suggest a set of goals to be attained regardless of the ultimate protocol to be used. They are:

1. Limitation or avoidance of drilling fluids.

2. Careful examination and selection of ice, to avoid cracks and fissures, and maintenance of ice core integrity throughout the process.

3. Rigorous decontamination of outer core surfaces (without damaging organisms or nucleic acids inside the core section). A 5% sodium hypochlorite solution was determined to be the most effective in our laboratory testing.

4. Maintenance of rigorous sterile conditions (including all reagents, solutions, tools, rooms, etc.)

5. Replication of assays on each sample.

6. For PCR and RT-PCR, repeated amplifications on the same sample. Cloning should be utilized whenever possible.

7. Inclusion of multiple controls in each protocol of the process.

8. Testing of each isolate with the protocol used for decontaminating the ice core from which it was isolated.

9. Consistency of methods and personnel.

10. Replication and confirmation by repetition in more than one laboratory. [*Note:* This is difficult because of the relatively low concentrations of organisms and nucleic acids in the ice. However, it is desirable to attempt this type of replication.]

It is recommended that effective sterilants be used for recommendation 3 (above) to ensure that the external contemporary organisms and nucleic acids are eliminated. Oxidizing agents have been used successfully, as well as commercial sterilizing agents. Additionally, chemicals that specifically degrade nucleic acids may be added to the decontaminating cocktail. UV irradiation and ethanol are ineffective in killing some microbes, and water only acts as a diluent of microbes and nucleic acids.

It is desirable to develop a set of standards and standard protocols for isolating viable microbes and nucleic acids (as well as other biological materials) from ancient ice core samples that could be utilized by the entire community of biologists studying microbes in ancient ice.

Literature Cited

1. Abyzov, S.S. 1993. Microorganisms in the Antarctic ice. pp. 265–295. In E.I. Friedman, (ed.), Antarctic Microbiology. Wiley, New York.
2. Abyzov, S.S., I.P. Babeva, V.I. Biryuzova, N.A. Kostrikina, and E.E. Azieva. 1983. Peculiarities of the ultrastructural organization of yeast cells from the interior of an Antarctic glacier. Biol. Bull. Acad. Sci. USSR (Eng. trans. IZV AKAD nauk SSSR Ser. Biol.) 10: 539–545.
3. Abyzov, S.S., N.E. Bobin, and B.B. Kudriashov. 1979. Microbiological analysis of glacial series of central Antarctica. Akademiia nauk SSSR Izvestia Seriia biologicheskaia 6: 828–836.
4. Abyzov, S.S., N.E. Bobin, and B.B. Kudriashov. 1982. Quantitative analysis of microorganisms during the microbiological investigation of Antarctic glaciers. Biol. Bull. Acad. Sci. USSR (Eng. trans. IZV AKAD nauk SSSR Ser. Biol.) 9: 558–564.
5. Abyzov, S.S., N.E. Bobin, and B.B. Kudriashev. 1982. Quantitative assessment of microorganisms in microbiological studies of Antarctic glaciers. Akademia nauk SSSR Investia Seriia biologicheskaia 6: 897–905.
6. Abyzov, S.S., N.E. Bobin, and B.B. Kudriashov. 1986. Central Antarctic glacier as an object of investigations of prolonged anabiosis of microorganisms in nature. Antarktiki; doklady komissii 25: 202–208.
7. Abyzov, S.S., V.Y. Lipenkov, N.E. Bobin, and B.B. Kudriashov. 1982. Microflora of the central Antarctic glaciers and control methods of sterile isolation of the ice core for microbiological analyses. Akademia nauk microbiologicheskikh Seriia biologicheskaia 4: 537–548.
8. Abyzov, S.S., I.N. Mitskevich, and M.N. Poglazova. 1998. Microflora of the deep glacier horizons of central Antarctica. Microbiology 67: 451–458.
9. Atlas, R.M., M.E. di Menna, and R.E. Cameron. 1978. Ecological investigations of yeasts in Antarctic soils. Terrestrial Biology III. Paper 2 in Antarctic Res. Ser. 30: 27–34.
10. Austin, J.J., A.J. Ross, A.B. Smith, R.A. Fortey, and R.H. Thomas. 1997. Problems of reproducibility—does geologically ancient DNA survive in amber-preserved insects. Proc. R. Soc. London B. 264: 467–474.
11. Banttari, E.E., and J.R. Venette. 1980. Aerosol spread of plant viruses: Potential role in disease outbreaks. Ann. New York Acad. Sci. 353: 167–173.
12. Bartholomew, J.W., and G. Paik. 1966. Isolation and identification of obligate thermophilic spore forming bacilli from ocean basin cores. J. Bacteriol. 92: 635–638.

13. Baylor, E.R., M.B. Baylor, D.C. Blanchard, L.D. Syzdek, and C. Appel. 1977. Virus transfer from surf to wind. Science 198: 575–580.

14. Büttner, C., and F. Nienhaus. 1989. Virus contamination of soils in forest ecosystems of the Federal Republic of Germany. Eur. J. For. Pathol. 19: 47–53.

15. Cano R.J., and M.K. Borucki. 1995. Revival and identification of bacterial spores in 25- to 40-million-year-old Dominican amber. Science 268: 1060–1064.

16. Castello, J.D., D.K. Lakshman, S.M. Tavantzis, S.O. Rogers, G.D. Bachand, R. Jagels, J. Carlisle, and Y. Liu. 1995. Detection of infectious tomato mosaic tobamovirus in fog and clouds. Phytopathology 85: 1409–1412.

17. Castello, J.D., S.O. Rogers, W.T. Starmer, C. Catranis, L. Ma, G. Bachand, Y. Zhao, and J.E. Smith. 1999. Detection of tomato mosaic tobamovirus RNA in ancient glacial ice. Polar Biol. 22: 207–212.

18. Catranis, C., and W.T. Starmer. 1991. Microorganisms entrapped in glacial ice. Ant. J. U.S. 26: 234–236.

19. Christner, B.C., E. Mosley-Thompson, L.G. Thompson, and J.N. Reeve. 2001. Isolation of bacteria and 16S rDNAs from Lake Vostok accretion ice. Environ. Microbiol. 3: 570–577.

20. Christner, B.C., E. Mosley-Thompson, L.G. Thompson, V. Zagorodnov. K. Sandman, and J.N. Reeve. 2000. Recovery and identification of viable bacteria immured in glacial ice. Icarus 144: 479–485.

21. Cooper, A. 1994. DNA from museum specimens. pp. 149–165. In B. Herrmann and S. Hummel (eds.), Ancient DNA. Springer-Verlag, New York.

22. Cooper, A., and H.N. Poinar. 2000. Ancient DNA: Do it right or not at all. Science 289: 18.

23. Cross, T., and R.W. Atwell. 1974. Recovery of viable thermoactinomycete endospores from deep mud cores. pp. 11–20. In A.N. Barker, G. W. Gould, and J. Wold (eds.), Spore Research. Academic Press, London.

24. DeSalle, R., J. Gatesy, W. Wheeler, and D. Grimaldi. 1992. DNA sequences from a fossil termite in Olio-Miocene amber and phylogenetic implications. Science 257: 1880–1882.

25. Dmitriev, V.V., D.A. Gilichinskii, R.N. Faizutdinova, I.N. Shershunov, W.I. Golubev, and V.I. Duda. 1997. Occurrence of viable yeasts in 3-million-year-old permafrost in Siberia. Mikrobiologiya 66: 655–660.

26. Fitt, B.D.L., H.A. McCartney, and P.J. Walkiate. 1989. The role of rain in dispersal of pathogen inoculum. Ann. Rev. Pathol. 27: 241–270.

27. Friedberg, E.C., G.C. Walker, and W. Siede. 1995. DNA Repair and Mutagenesis. American Society for Microbiology Press, Washington, D.C.

28. Gest, H., and J. Mandelstam. 1987. Longevity of microorganisms in natural environments. Microbiological Sciences 4: 69–71.

29. Gloster, J., R.F. Sellers, and A.I. Donaldson. 1982. Long distance transport of foot-and-mouth disease virus over the sea. Veterinary Record 110: 47–52.

30. Grant, R.H., A.B. Scheidt, and L.R. Rueff. 1994. Aerosol transmission of a viable virus affecting swine: Explanation of an epizootic of pseudorabies. Int. J. Biometeorol. 38: 33–39.

31. Gruber, S., and R. Jaenicke. 2001. Biological microparticles in the Hans Tausen Ice Cap, North Greenland, Meddelelser om Grønland. Geoscience 39: 161–163.

32. Higuchi, R., B. Bowman, M. Freiberger, O.A. Ryder, and A.C. Wilson. 1985. DNA-sequences from the Quagga, an extinct member of the horse family. Nature 312: 282–284.

33. Karl, D.M., D.F. Bird, K. Björkman, T. Houlihan, R. Shackelford, and L. Tupas. 1999. Microorganisms in the accreted ice of Lake Vostok, Antarctica. Science 286: 2144–2147.

34. Kudriashov, B.B., S.S. Abyzov, N.E. Bobin, and G.A. Kazakov. 1977. Elaboration of the technical means of selection of a sample of ice for microbiological analysis in Antarctica. Antarktika, doklady Komissii 16: 154–160.

35. Lindahl, T. 1993. Instability and decay of the primary structure of DNA. Nature 362: 709–715.

36. Lindahl, T. 1997. Facts and artifacts of ancient DNA. Cell 90: 1–3.

37. Ma, L.J., H. Fan, C. Catranis, S.O. Rogers, and W.T. Starmer. 1997. Isolation and characterization of fungi entrapped in glacial ice. Inoculum 48: MSA Abstracts, p. 23.

38. Ma, L., C. Catranis, W.T. Starmer, and S.O. Rogers. 1998. Study of glacial ice—a source of ancient fungi. Inoculum 49: MSA Abstracts, p. 34.

39. Ma, L., C. Catranis, W.T. Starmer, and S.O. Rogers. 1999. Revival and characterization of fungi from ancient polar ice. Mycologist 13: 70–73.

40. Ma, L., S.O. Rogers, C. Catranis, and W.T. Starmer. 2000. Detection and characterization of ancient fungi entrapped in glacial ice. Mycologia 92: 286–295.

41. Pääbo, S. 1985. Molecular cloning of ancient Egyptian mummy DNA. Nature 314: 644–645.

42. Priscu, J.C., E.E. Adams, W.B. Lyons, M.A. Voytek, D.W. Mogk, R.L. Brown, C.P. McKay, C.D. Takacs, K.A. Welch, C.F. Wolf, J.D. Kirchtein, and R. Avci. 1999. Geomicrobiology of subglacial ice above Lake Vostok, Antarctica Science 286: 2141–2144.

43. Rao, V.C., and J.L. Melnick. 1986. Environmental Virology. American Society of Microbiology, Washington, D.C.

44. Riley, R.L., and F. O'Grady. 1961. Airborne Infection: Transmission and Control. Macmillan, New York.

45. Rogers, S.O. 1994. Phylogenetic and taxonomic information from herbarium and mummified DNA, pp. 47–67. In R.P. Adams, J. Miller, E. Golenberg, and J.E. Adams (eds.), DNA Utilization, Intellectual Property and Fossil DNA. Missouri Botanical Gardens Press, St. Louis.

46. Rogers, S.O, and A.J. Bendich. 1985. Extraction of DNA from milligram amounts of fresh, herbarium and mummified plant tissues. Plant Mol. Biol. 5: 69–76.

47. Rogers S.O, and A.J. Bendich. 1994. Extraction of Total Cellular DNA from Plants, Algae and Fungi. D1: 1–8. In S.B. Gelvin and R.A. Schilperoort (eds.), Plant Molecular Biology Manual, 3rd ed., Kluwer Academic Press, Dordrecht, The Netherlands.

48. Rogers, S.O., L. Ma, J.D. Castello, Y. Zhao, J.E. Smith, and W.T. Starmer. 1999. Dilemmas and enigmas encased in ancient ice. XVI International Botanical Congress Abstracts, p. 452, no. 960.

49. Rogers, S.O., K. Langenegger, and O. Holdenrieder. 2000. DNA changes in tissues entrapped in plant resins (the precursors of amber). Naturwissenschaften 87: 70–75.

50. Rogers, S.O., S. Rehner, C. Bledsoe, G.J. Mueller, and J.F. Ammirati. 1989. Extraction of DNA from Basidiomycetes for ribosomal DNA hybridizations. Can. J. Bot. 67: 1235–1243.

51. Rollo, F.U., W. Asci, and S. Sassoroli. 1994. Assessing the genetic variation in pre-Columbian maize at the molecular level. pp 27–35. In R.P. Adams, J. Miller, E. Golenberg, and J.E. Adams (eds.), DNA Utilization, Intellectual Property and Fossil DNA. Missouri Botanical Gardens Press, St. Louis.

52. Seaward, M.R.D., T. Cross, and B.A. Unsworth. 1976. Viable bacterial spores recovered from an archaeological excavation. Nature 261: 407–408.

53. Sharp, M., J. Parkes, B. Cragg, I.J. Fairchild, H. Lamb, and M. Tranter. 1999. Widespread bacterial populations at glacier beds and their relationship to rock weathering and carbon cycling. Geology 27: 107–110.

54. Shi, T., R.H. Reeves, D.A. Gilichinsky, and E.I. Friedmann. 1997. Characterization of viable bacteria from Siberian permafrost by 16S rDNA sequencing. Microbial Ecol. 33: 169–179.

55. Shuval, H.I., N. Guttman-Bass, J. Applebaum, and B. Fattal. 1989. Aerosolized enteric bacteria and viruses generated by spray irrigation of waste water. Water Sci. and Technol. 21: 131–135.

56. Skidmore, M.L., J.M. Foght, and M.J. Sharp. 2000. Microbial life beneath a high Arctic glacier. Appl. Environ. Microbiol. 66: 3214–3220.

57. Sneath, P.H.A., 1962. Longevity of micro-organisms. Nature 195: 643–646.

58. Vreeland, R.H., W.D. Rosenzweig, and D.W. Powers. 2000. Isolation of a 250 million-year old halotolerant bacterium from a primary salt crystal. Nature 407: 897–900.

59. Zhao, Y.H. 2001. Identification of Selected Bacteria and Bacteriophage from Greenland Glacial Ice. MS Thesis, State University of New York, College of Environmental Science and Forestry, Syracuse, New York.

3

Perennial Antarctic Lake Ice: A Refuge for Cyanobacteria in an Extreme Environment

John C. Priscu, Edward E. Adams, Hans W. Paerl, Christian H. Fritsen,

John E. Dore, John T. Lisle, Craig F. Wolf and Jill A. Mikucki

THE POLAR DESERTS of Antarctica form one of the driest and coldest ecosystems known, and were originally thought to harbor little life (70). The diaries of Captain Robert Falcon Scott, one of the earliest explorers of the McMurdo and South Polar regions of Antarctica, referred to the McMurdo Dry Valleys as the "Valley of the Dead" during his first visit in 1903 (82). Recent studies in Antarctica have now revealed new information on the presence of microbial life in environments such as surface snow near the South Pole (9), 3.5 km deep Vostok ice (44, 71, 83), exposed soils (97), sandstones (23), meltwater ponds (94), glacial cryoconites (13), the liquid water column of permanently ice-covered lakes (77), and the ice covers of permanent lake ice (74, 78). Most of the microbes found in these habitats are prokaryotic (6, 31, 92). Except for the South Pole and Vostok sites, a large portion of these prokaryotes are photosynthetically active cyanobacteria that can provide new carbon and nitrogen to the ecosystem. As in all desert ecosystems, the production of new carbon and nitrogen compounds is critical given the general lack of liquid water and the dearth of allochthonous production. Without new carbon and new nitrogen entering the system, the presence of higher trophic levels and biodiversity in general is compromised.

Numerous studies on cyanobacteria-dominated systems have occurred in the McMurdo Dry Valleys. These valleys form the largest (ca. 4000 km^2) ice-free expanse of land in Antarctica, and are the centerpiece of a National Science Foundation (NSF)-funded Long Term Ecological Project aimed at integrating the understanding of meteorological, geophysical, hydrological, and biogeochemical processes in the area (49, 52). Meteorological conditions in the Dry Valleys reveal the extreme conditions that organisms must overcome to survive (18). Surface air temperatures average −27.6 °C, but range widely from 10 °C to −65.7 °C. Air temperatures above freezing are rare, averaging 6.2 degree days per year. Soil temperatures show similar averages and minimums but, owing to direct solar heating, reach almost 23 °C during cloud-free summer days. Given the persistent (avg = 4.1 m · s^{-1}) and often high winds (max = 37.8

m · s^{-1}) characteristic of the Dry Valleys in concert with low and seasonal stream flow, most organic matter is mobilized throughout the region by wind.

Perhaps the first description of cyanobacterially dominated assemblages in the McMurdo Dry Valleys was by Griffith Taylor during the 1911 British expedition to southern Victoria Land. Taylor noted the existence of "water plants" through the clear moat ice on a McMurdo Dry Valley lake (91). Since Taylor's early observations, little research on this system in the McMurdo Dry Valleys occurred until the 1960s, when Wilson (103) studied the cyanobacterial mats associated with lake and pond ice. Wilson observed pieces of cyanobacterial mat consisting of *Oscillatoria* and *Nostoc*- moving upward through the ice in the lakes and ponds of Wright and Taylor Valleys in the McMurdo Dry Valleys. These mats were 1 to 5 cm^2 in area and each was associated with a gas bubble. The phenomenon was observed only in small ponds and the shallow parts of lakes. Wilson used a thermodynamic model to show that the upward movement was a function of gas-induced buoyancy coupled with absorption of solar radiation by the mat itself, which melts the overlying ice. Following these initial reports, a considerable amount of research was conducted on benthic mats within the lakes (e.g., 62, 84, 99), streams (e.g., 41, 93), and ponds (e.g., 35, 36, 40). Except for the observation of pieces of mat within the lake ice of Lake Hoare by Squyres et al. (87), cyanobacteria within lake ice received little direct attention since its early description by Wilson. During studies on the biogeochemistry of nitrous oxide in McMurdo Dry Valley lakes, Priscu and co-workers observed a peak in nitrous oxide associated with a sediment layer 2 m beneath the surface of the 4 m thick ice cover of Lake Bonney (68, 73, 74). This observation together with elevated levels of chlorophyll -a (chla), particulate organic carbon, particulate organic nitrogen, ammonium, and dissolved organic carbon (105) led to the hypothesis that the cyanobacteria within the lake ice were not passive but grew actively within the ice cover. Subsequent research by Priscu's group showed that adequate liquid water was present (2, 24) to support an active prokaryotic ecosystem within the ice consisting of cyanobacteria and a diversity of bacterial species (e.g., 31, 60, 66, 74). DNA hybridization studies between cyanobacteria and other prokaryotes found in the permanent lake ice in the McMurdo Dry Valleys have revealed that cyanobacterial mats in ephemeral streams provide the biological seed for the lake ice cyanobacteria (30, 31, 74). It is clear that cyanobacterially colonized lakes and streams are the "life support system" of Antarctic polar deserts, providing organic carbon to the surrounding areas and seeding numerous habitats with microbes. No cyanobacteria have been observed in ice cores from the Antarctic ice sheet (44, 71) implying that dispersion is restricted to an area in close spatial proximity to the Dry Valleys.

This chapter will focus on the organisms and ecosystem processes associated with a novel habitat that exists within the permanent ice covers of lakes within the McMurdo Dry Valleys of Antarctica. Our specific objectives are to: (i) describe the microbes that exist within the lake ice; (ii) present biogeochemical data to reveal the structure and function of this unique ecosystem; and (iii) define the conditions that produce liquid water to support microbial growth and associated biogeochemical processes.

Study Area

The McMurdo Dry Valleys, southern Victoria Land (76°30′ to 78°30′S; 160° to 164°E), consist of a number of valleys that extend from the Polar Plateau to McMurdo Sound. The valleys harbor some of the only permanently ice-covered lakes on Earth (e.g., 32, 69). The lakes occupy closed basins and vary in surface area (1–ca. 6 km^2), depth (20–85 m), and ice cover thickness (3–5 m) (86). The permanent ice covers greatly reduce wind-driven mixing in the liquid water column (85, 86), gas exchange between liquid water and the atmosphere (68, 73, 101), light penetration (28, 48, 61), and sediment deposition into the water column (3, 56). Consequently, ecosystem properties in the water columns of the lakes are largely controlled by the presence of the perennial ice cover (26, 100). A majority of the research presented in this chapter was derived from three lakes in the Taylor Valley, particularly Lake Bonney.

The Dry Valley lakes are perennially ice-covered due to a combination of low mean annual air temperatures and seasonal glacial melt that flows into the lakes during the summer (51). McKay's model shows that the latent heat entering the lake in the meltwater plays a key role in the maintenance of a liquid water body beneath the permanent ice cover. Hence, the presence and thickness of the ice cover depends to a large extent on air temperature, which controls both freezing and melting processes. The ice thickness of most of the lakes that contain substantial liquid water bodies ranges from 3 to 5 m and varies with air temperature and the degree of mixing within the lakes (18, 86, 100). The thickness of the lake ice is ultimately governed by the dynamic equilibrium between lake water accretion (freeze) at the bottom of the ice and ablation at the surface. Hence, the ice can be depicted in a gross sense as an upward-moving conveyor belt carrying newly frozen ice from the bottom to the top, where it is lost through ablation (16). Sediment layers exist in most of the ice covers (2, 3, 24, 74, 87) as a result of aeolian deposition followed by summer melting. The sediments melt to a depth where solar radiation can no longer supply adequate energy to generate melt, that is, the downward melting is balanced by the upward conveyer motion of the ice described above.

Why Study Cyanobacteria in Ice?

Earth's biosphere is cold, with 14% being polar and 90% (by volume) cold ocean < 5 °C. More than 70% of Earth's freshwater occurs as ice, and a large portion of the soil ecosystem exists as permafrost. Recent information has revealed that the prokaryotic biomass in the Antarctic ice sheet and subglacial lakes approaches that in our planet's surface freshwaters (72). Hence, information on microbial dynamics in both young and ancient ice is important for understanding the dynamics of the global ecosystem. Expectations of commercial applications and interest in the early evolution of life have led many researchers to examine microbes in thermal systems. Based on the occurrence of evolutionarily deeply rooted microbes in extreme thermal systems, in concert with extensive geothermal activity during the early evolution of our planet, it is generally thought that life on Earth evolved in hot environments (43, 63). Recent considerations about the evolution of life, however, have suggested that a "hot start" was probably not the only alternative for the origin of life. Though there are strong arguments for a thermal origin of life based on small subunit ribosomal RNA (16S rRNA) phylogenetic relationships, the validity of this relationship is questioned by researchers who believe that phylogenies are strongly biased by the use of just a single gene for the construction of the tree of life. If lateral gene transfer is common among all prokaryotic organisms (57), then a hierarchical universal classification is difficult or impossible (64, 65), and evolutionary patterns must be reassessed (4, 17, 79). A hot origin of life also is not supported by new results from phylogenetic trees based on genes that do not code for ribosomal RNA, chemical experiments with alternative structure for the nucleic acid backbone (20), considerations about the thermal stability of basic molecules found in all organisms, and statistical analysis of the GC content of DNA (29). Adaptation to life in hot environments may even be a late adaptation (5).

Though much more research is required to determine whether life originated in hot or cold environments, it is highly probable that cold environments have acted as a refuge for life during major glaciations. Recent evidence has implied that around 600 million years ago during the Neoproterozoic, microbial life endured an ice age of such intensity that even the tropics froze over (38, 39, 81). According to this so-called Snowball Earth Hypothesis, the Earth would have been completely ice-covered for 10 million years or more, with ice thickness exceeding 1 km. Only the deepest oceans would have contained liquid water. One of the primary criticisms of the Snowball Earth Hypothesis is that the thick ice cover over the world ocean would cut off the supply of sunlight to organisms in the seawater below and thereby eliminate photosynthesis and all life associated with photosynthetic carbon production. Others have concluded that global-scale freezing would extinguish all surface life (102). Only the hardiest of microbes would have survived this extreme environment. Hoffman and Schrag (39), Vincent and Howard-Williams (95), and Vincent et al. (94) suggest that photosynthetic cyanobacteria and bacteria, similar to those found in the permanent ice covers of contemporary polar systems, may have acted as an icy biotic refuge during this period. The resultant high concentration of microbes in these icy environments would favor intense chemical and biological interactions between species, which would force the development of symbiotic associations and eventually eukaryotic development through evolutionary time (4, 79, 94). Though this "density-speeds-evolution" theory has been considered primarily in the context of thermal microbial mat communities (50), it is highly probable that ice-bound habitats also provided opportunities for microbial evolution, inducing the radiation of the eukaryotic cell type at the onset of the Neoproterozoic (39, 46).

Studies of earthly ice-bound microbes, particularly those in the McMurdo Dry Valleys, also are relevant to the evolution and persistence of life on extraterrestrial bodies. During the transition from a clement to an inhospitable environment on Mars, liquid water may have progressed from a primarily liquid phase to a solid phase and the Martian surface would have eventually become ice-covered (98). Evidence from Martian Orbiter Laser Altimeter images has revealed that water ice exists at the poles of Mars. Habitats in polar ice may serve as a model for life on Mars (60, 74, 77) as it cooled and may assist us in our search for extinct or extant life on Mars today (98). Surface ice on Europa, one of the moons of Jupiter, appears to exist in contact with subsurface liquid water (8, 33). Solar heating of the subsurface could result in melt layers similar to those described in Earth's polar habitats (94). With reports of life almost 4 km beneath the Antarctic ice sheet and implications for life in Lake Vostok, an enormous lake ca. 1 km deep and about 14,000 km^2 in surface area (14, 44, 71, 88), it is clear that we must extend the bounds of what is currently considered the "Earth's biosphere" to include subsurface ice environments.

The Lake Ice Environment

Vertical Profiles

A majority of research on microbes in permanent Antarctic lake ice has focused on the east lobe of Lake Bonney, though studies have encompassed at least seven other lakes in the McMurdo Dry Valleys. Clean techniques were used in the collection and processing of all samples. Particulate organic carbon (POC) in the ice cover of the east lobe of Lake Bonney is concentrated at 2 m in association with maximum sediment levels (Figure 3.1). Chlorophyll *a*, primary productivity, bacterial density, and bacterial activity (uptake and metabolism of organic matter) also coincide with the depth of maximum sediment concentration indicating that viable phototrophs and heterotrophs are present in the ice and can become active after a short exposure to liquid water.

Dissolved organic carbon (DOC), dissolved inorganic carbon (DIC), and ammonium show maxima in the region of highest sediment and POC concentration, suggesting active biogeochemical cycling of these constituents. The

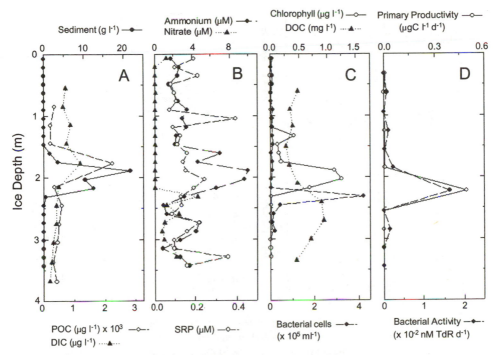

FIGURE 3.1. Vertical profiles of selected constituents within the ice cover of the east lobe of Lake Bonney. POC = particulate organic carbon; DIC = dissolved inorganic carbon; SRP = soluble reactive phosphorus; DOC = dissolved organic carbon; bacterial activity represents the rate of tritiated thymidine incorporation (TdR) into cellular DNA. Methodological details can be found in Priscu et al. (74) and Fritsen and Priscu (27).

DOC, presumably from extracellular photosynthate released from cyanobacteria following photosynthetic inorganic carbon fixation, can supply the heterotrophic component of the microbial assemblage with an energy and carbon source for growth which, in turn, recycles carbon dioxide to support photosynthesis by the cyanobacteria. The ammonium maximum associated with the sediment layer implies active regeneration of nitrogen, ammonium leakage following atmospheric nitrogen fixation, leaching from sediments, or a combination of these sources (in excess of sinks). The lack of a clear vertical trend in soluble reactive phosphorus (SRP) is related to adsorptive processes that occur between SRP and the inorganic sediment material. The general trends in the east lobe of Lake Bonney are evident in most lake ice covers in the McMurdo Dry Valleys. Combined data from seven lakes show increases in POC and particulate organic nitrogen (PON), primary and bacterial production, chlorophyll *a*, and bacterial biomass with sediment concentration (Figure 3.2).

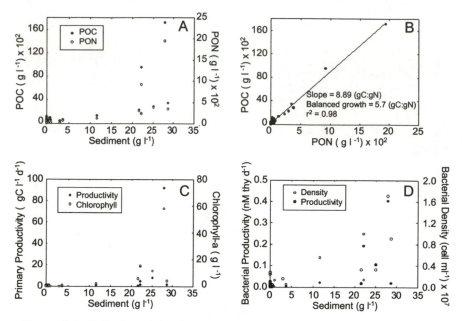

FIGURE 3.2. Collective relationships between sediment concentration and various cyanobacterial (photosynthesis; chlorophyll *a*) and bacterial (thymidine incorporation; cell density) parameters from seven lakes in the McMurdo Dry Valleys. Relationships between sediment concentration and POC also are shown. The regression of POC on PON has a slope of 8.89 (g:g), which is higher than that reported for balanced microbial growth.

The ratio of POC to PON for the combined data from all ice covers in the region averages 8.9 (g:g), which is higher than the ratio of 5.7 (g:g) that occurs during balanced growth of photoautotrophs (80).

Growth Rates and Photosynthetic Activity

Cyanobacteria dominate the biomass (e.g., POC) within the ice covers. Pinckney and Paerl (66), using chlorophyll *a* normalized pigment markers, showed that myxoxanthophyll, echinenone, zeaxanthin, and canthaxanthin, which are all markers for cyanobacteria, were the most abundant phytopigments in ice aggregates from the Lake Bonney ice cover. These pigments resemble those measured in shoreline soil samples more closely than in the underlying lake water, implying a terrestrial origin for the lake ice assemblage. Pinckney and Paerl used $^{14}CO_2$ photopigment-labeling experiments to further show that carbon-specific growth rates are less than 0.10 d^{-1} during the first 24 hours after thawing. Fritsen and Priscu (27), using $^{14}CO_2$ labeling of cellular macromolecules, showed that despite low photosynthetic rates a large proportion (41%) of the photosynthate was incorporated into protein, indicating that the cyanobacteria were undergoing efficient net cellular growth. Photosynthesis versus irradiance experiments revealed that the cyanobacterial assemblages had variable maximum biomass-specific rates of photosynthesis, ranging from 0.0043 to 0.0406 μgC (μg chlorophyll $a)^{-1} \cdot h^{-1}$ among the six ice covers studied. $^{14}CO_2$ incorporation into protein was used to compute carbon-specific growth rates ranging from 0.001 to 0.012 d^{-1}, in agreement with the pigment labeling results (66).

During the months of continuous sunlight, in situ irradiances are above that required to saturate the photosynthetic capacity of the cyanobacteria without causing inhibition (26). Hence, in situ growth rates are not likely to be light-limited when liquid water is present. The light-saturated rate of photosynthesis normalized to chlorophyll a for the cyanobacterial assemblage in Lake Fryxell increased about ten-fold when incubation temperature was increased from 2 to 20 °C corresponding to a Q_{10} value of 3.46. The optimum temperature response of light-saturated photosynthesis is near 20 °C, with about 10% of the maximum rate occurring at in situ growth temperatures (26, 27). These data support the results of Tang et al. (89) and Tang and Vincent (90), who contend that few true cyanobacterial psychrophiles are associated with polar freshwater systems. Nadeau and Castenholz (54, 55) recently isolated one of the first true polar psychrophilic cyanobacteria from a pond on the Ross Ice Shelf, Antarctica. Based on genetic analysis, Nadeau and Castenholz concluded that psychrotolerant forms are more closely related to temperate cyanobacteria, whereas the true psychrophiles probably evolved in the polar habitat. The temperature response of light-saturated photosynthesis in the permanent lake ice and stream cyanobacteria of the McMurdo Dry Valleys indicates that these assemblages are psychrotrophic or psychrotolerant whereas sea ice diatoms and the chlorophyte Chlamydomonas subcauda isolated from the liquid water column of Lake Bonney (53) are true psychrophiles (Figure 3.3). The lack of cyanobacterial psychrophiles in lake ice assemblages may be related to selection factors other than temperature (e.g., freeze-thaw tolerance, tolerance to high fluxes of solar radiation) that dictate which organisms survive and grow in icy environments (89, 92, 96).

In contrast to most of the Dry Valley lakes that have maximum concentrations of cyanobacteria and sediment located 1 to 2 m below the ice surface, the cyanobacterial aggregate layer in Lake Fryxell ice is concentrated in the upper 0.5 m. Presumably this difference is the result of elevated snow accumulation on Lake Fryxell, which lies close to the relatively moisture-rich coast of McMurdo Sound (21, 22). The snow effectively blocks incident radiation at the ice surface, hindering the downward melting of sediments and associated microorganisms. As a result of near-surface sediment accumulation, surface melt pools up to 20 cm deep and 1 m in diameter form during the summer and support dense cyanobacterial mats. These near-surface cyanobacteria exist as

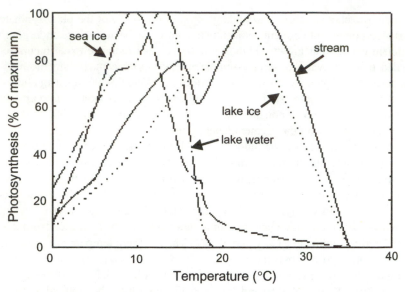

FIGURE 3.3. Photosynthetic temperature response by various microbial assemblages in the McMurdo Sound area. The samples of McMurdo Sound fast ice, lake water from Lake Bonney, stream samples from the Fryxell basin, and lake ice from Lake Bonney were dominated by diatom algae, the chlorophyte *Chlamydomonas subcaudata*, *Phormidium* plus *Nostoc*, and *Phormidium*, respectively. Experimental details can be found in references 27 and 75. Note the similarity in temperature responses between stream and lake ice cyanobacteria, both of which show psychrotolerant characteristics; sea ice and lake water algae show typical psychrophilic characteristics.

loose aggregates or flakes measuring 0.5 to 2 cm in diameter that coat 60 to 90% of the ice surface within the melt pools. Records of in situ temperatures within the shallow melt pools at the surface of Lake Fryxell show diurnal changes of 10 °C within the ice cyanobacterial aggregates (Figure 3.4 A). The absolute value of temperatures within the aggregates is dependent on the orientation of the mat relative to the solar disk throughout the day. The aggregates within the north-facing melt pool experienced temperatures about 4 °C higher than those in the south-facing pools, reaching a maximum of 6.3 °C. Since the temperature probes were placed horizontally 0.5 to 1 cm inside the mats, these temperatures are not likely to be artifacts caused by direct solar heating of the probes. Cores (2.8 cm diameter) of aggregates in the small-surface pools yielded chlorophyll *a* concentrations of 15.3 µg chl*a* · cm^{-2} (standard error = 2.4; n = 6). When extrapolated from the centimeter to meter scale, this represents about 153 mg chl*a* · m^{-2}.

Photosynthesis-irradiance relationships changed in response to freezing (Figure 3.4 B). Maximum rates of photosynthesis were approached at irradi-

ances greater than 600 μmoles of photons m^{-2} · s^{-1} in the experiments conducted. Compensation irradiances (where photosynthesis = respiration) were
202 μmol photons · m^{-2} · s^{-1} before and 308 μmol photons · m^{-2} · s^{-1} after experimental freezing, which equals a 53% increase induced by freezing. The rates
of respiration (oxygen consumption in the dark) were 33% lower following
freezing (−1.6 × · 0^{-3} mg O$_2$ · mg l^{-1} · s^{-1} after freezing versus −2.15 × 10^{-3} mg
O$_2$ · l^{-1} · s^{-1} before freezing). Despite the changes that occurred following an
experimental freeze-thaw cycle, freezing and thawing do not appear to have a
major adverse affect on the cyanobacteria in the ice over long timescales.
Rather, freezing may be the environmental parameter that allow biota and ecosystems to live within polar environments where low light (< 200 μmol pho-

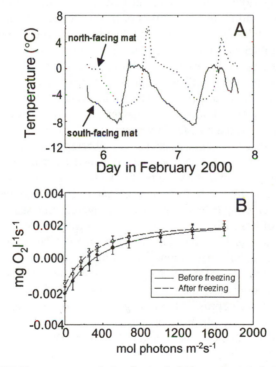

FIGURE 3.4. (A) Temperature variation for north-facing and south-facing cyanobacterial mats on the surface of Lake Fryxell during a 2.5-day period in February 2000
(day labels are centered at 0000h). (B) Rates of oxygen exchange (± one standard
deviation) measured at 0 °C in a gas-tight chamber for cyanobacterial mats collected
from the surface of Lake Fryxell before freezing and following freezing for 10 hours
at −15 °C. The curves were fitted with a modified equation for a hyperbola where
oxygen change = a−b/(1+ c · irradiance) $^{(1/d)}$; a (y-intercept), b, c, and d are parameters
fitted with the Marquadt algorithm.

tons · m^{-2} · s^{-1}) or complete darkness (spanning days to months) would otherwise be dominated by the loss of reduced carbon and ecosystem energy. Freezing presumably plays a major role in the proposed existence of a thriving microbial ecosystem within Antarctic subglacial lakes such as Lake Vostok (71, 83).

Phylogenetic Affiliation

The phylogenetic diversity of bacteria and cyanobacteria colonizing sediment particles at a depth of 2.5 m in the permanent ice cover of Lake Bonney was characterized by analyses of 16S rRNA genes amplified from environmental DNA (30, 31). An rRNA gene clone library of 198 clones was made and characterized by sequencing and oligonucleotide probe hybridization. The library was dominated by representatives of the cyanobacteria, proteobacteria, and planctomycetales, but also contained diverse clones representing the Acidobacterium-Holophaga division, the green nonsulfur division, and the actinobacteria. Of the cyanobacterial gene clusters characterized, only one was closely (> 97% similarity) affiliated with a well-characterized cyanobacterial species, *Chamaesiphon subglobosus*. The remaining cyanobacterial gene clusters were less than 93% similar to any characterized sequences in public databases although they resembled *Leptolyngbya* sp. and *Phormidium* sp. Oligonucleotide probes made from three lake ice cyanobacterial clusters were used to screen environmental 16S rDNA samples obtained from the terrestrial (soil and stream) environment in the vicinity of Lake Bonney and Lake Fryxell. The probes designed to hybridize to cyanobacterial 16S rRNA genes effectively hybridized to each sample, indicating that the cyanobacterial sequences present in the lake ice of Lake Bonney are also found in terrestrial cyanobacterial mat samples. Sequence analysis together with physiological data indicates that the cyanobacterial (and bacterial) community within the lake ice is dominated by organisms not uniquely adapted to the lake ice ecosystem. Instead, the strong katabatic winds common to the region act to disperse microorganisms in the desert environment and provide the biological seed for the lake ice microbial assemblage (31, 74). Aeolian dispersion also appears to seed the microbial assemblages in cryoconite holes located in McMurdo Dry Valley glaciers (13) and may also supply the biological seed for Antarctic glacial ice and subglacial lakes (71, 83). The close similarity between phylotypes in Taylor Valley cryoconites and lake ice, but not the Vostok system, implies that biological seeding patterns are geographically different within the Antarctic continent.

Molecular characterization (PCR amplification of the *nifH* fragments) of the *nifH* gene (encoding for the highly conserved iron-protein subunit) of nitrogenase in lake ice sediments from Lake Bonney also demonstrated the presence of a diverse diazotrophic assemblage (58). The *nifH* analysis suggested that phototrophic cyanobacteria and heterotrophic microorganisms have the potential to fix atmospheric nitrogen when liquid water is present in the ice cover. The expression of nitrogenase was confirmed by the acetylene reduction assay for nitrogenase activity (34, 58, 60).

Biogeochemistry

Nutrient bioassay experiments showed that cyanobacterial photosynthesis was stimulated by the addition of inorganic nitrogen, as either ammonium or nitrate (60). Iron in the presence of a chelator and phosphorus did not stimulate photosynthesis. These results imply that cyanobacterial photosynthesis, and presumably growth, is limited by inorganic nitrogen, a contention that is supported by the POC:PON ratios in the ice (see Figure 3.2). The ability to fix atmospheric nitrogen could offset the apparent photosynthetic inorganic nitrogen deficiency. Nitrogenase activity was stimulated by phosphorus and iron addition and showed little effect from the addition of mannitol (60). Even though molecular analysis showed that both cyanobacteria and bacteria in the ice aggregates have the potential to fix atmospheric nitrogen, the stimulatory effect of nitrogenase by light implies that cyanobacteria are responsible for a majority of the phenotypic expression of this enzyme in nature.

Another source of inorganic nutrients to the ice assemblage is the sediment. A considerable amount of ammonium and phosphate can be leached from soils surrounding the lake (Figure 3.5). These data, particularly the more complete phosphorus data set, show that most of the adsorbed nutrient is bound to small soil particles. Interestingly, a significantly higher quantity of phosphorus was leached from the soils when salty deep water (see 86) from the Lake Bonney water column was used as the solvent relative to freshwater from just beneath the ice cover. Sediments within the ice cover itself have relatively little leachable phosphorus, implying that most of the phosphorus was desorbed as the sediments melted through the ice, a process that could explain the vertical profile of inorganic phosphorus in the lake ice (see Figure 3.1). The apparent nitrogen deficiency in the ice cyanobacterial assemblage can be explained by the relative amounts of nitrogen and phosphorus that can be leached from the terrestrial soils following aeolian deposition on the ice surface by wind. Based on the available data, the average amounts of ammonium and phosphorus that can be leached from surrounding soils is 7.1 µgN · g sediment^{-1} (16.2 in potassium chloride extracts) and 4.1 µgP · g sediment^{-1}, respectively. The ratio of leachable N:P is 1.7 (3.9 if potassium chloride exchangeable ammonium is included), which is well below that required for balanced cyanobacterial growth. The low N:P ratio resulting from differential nitrogen and phosphorus leaching would provide a selective advantage for microorganisms that have the ability to fix atmospheric nitrogen.

Despite apparent cyanobacterial nitrogen deficiency in the ice, the ammonium maximum associated with the sediment layer in Figure 3.1 indicates that sources of ammonium exceed sinks. Results from ^{15}N-ammonium based isotope dilution experiments show that uptake of ammonium either equaled or exceeded microbial ammonium regeneration for the samples and size fractions analyzed (Table 3.1). Under these conditions the ammonium pool should be depleted. The absolute rates of ammonium uptake and regeneration relative to the PON pool range from 0.7% d^{-1} to 22.0% d^{-1} and 0.1% d^{-1} to 22.1% d^{-1}, respectively. The lowest activity occurred in the < 63 μm size class from the Lake Hoare ice cover. These values are similar to those measured within the

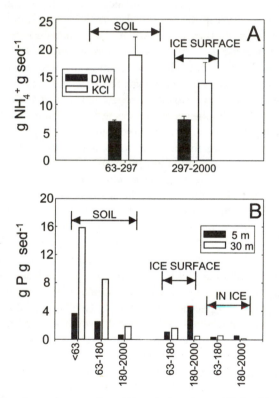

FIGURE 3.5. Experimental ammonium (A) and phosphorus (B) leaching data from terrestrial soils surrounding Lake Bonney, and soils and sediment on and within (2 m) the ice cover of Lake Bonney. Deionized water (DIW) and KCl (5 mg · l^{-1}) were used as the solvents in the ammonium experiments. Water from just beneath the ice cover (5 m) and 30 m beneath the ice cover of Lake Bonney was used for phosphorus leaching experiments. The values on the x-axes represent the size fraction of sediments (μm) analyzed in each experiment. Error bars = standard deviation.

TABLE 3.1.

Ammonium uptake (U) and regeneration (R) rates determined by ^{15}N isotope dilution experiments for lake ice assemblages in Lakes Bonney, Fryxell, Hoare, Miers, and Vida

Sample	Uptake ($\mu m\ h^{-1}$)	Uptake (h^{-1})	Regeneration ($\mu m\ h^{-1}$)	Regeneration (h^{-1})	U:R	K_s (μm)	V_{max} (h^{-1})
East Bonney < 63 μm	0.0030	0.00383	0.0031	0.00399	0.98	0.82	0.0003
East Bonney < 297 μm	0.0051	0.00384	0.0011	0.00084	4.73	—	—
Fryxell < 297 μm	0.0209	0.00224	0.0031	0.00033	6.81	—	—
Hoare < 63 μm	0.0059	0.00029	0.0011	0.00005	5.53	2.68	0.0006
Miers Total	0.0754	0.00899	0.0183	0.00219	4.11	11.09	0.0010
Vida Total	0.0403	0.00917	0.0405	0.00912	0.99	4.67	0.0003

Note: Half-saturation constants (K_s) and maximum uptake velocities (V_{max}) were determined using ^{15}N-ammonium incorporation and the Michaelis-Menten model. Isotope dilution experiments were conducted on < 63 μm, < 297 μm and unfractionated (total) samples. Michaelis-Menten parameters were determined on unfractionated samples only.

water column of Lake Vanda (76). Apparently, a combination of ammonium leaching and ammonium release following atmospheric nitrogen fixation replenishes the ammonium pool faster than it is consumed. Based on the size-fractionated data from Lake Bonney, the bacterial (< 63 μm) fraction is responsible for a majority of the ammonium regeneration. Half-saturation constants for ammonium are generally below the ammonium levels within the ice (compare Figure 3.1 and Table 3.1), corroborating the results of the ^{14}C-based bioassay experiments, which show nitrogen deficiency.

The peak in dissolved inorganic carbon (DIC) associated with the sediment layer also suggests active cycling of the organic and inorganic carbon pools. Leaching experiments revealed little DIC production from the sediments alone (Fritsen and Priscu unpublished data). Conversely, [14]CO_2 release experiments showed pronounced mineralization of organic carbon from radiolabeled cyanobacteria collected from various soil locations in the Dry Valleys (Figure 3.6). Interestingly, *Nostoc*- and *Phormidium*-dominated mats from the Lake Bonney basin were more labile than a *Phormidium*-dominated mat from the Lake Vida basin. The turnover of particulate organic carbon in these mat samples ranged from 17.5% d^{-1} in the Nostoc mat to 57.8% d^{-1} in the *Phormidium* mat collected from the Lake Vida basin, which are on average higher than the turnover rates of PON.

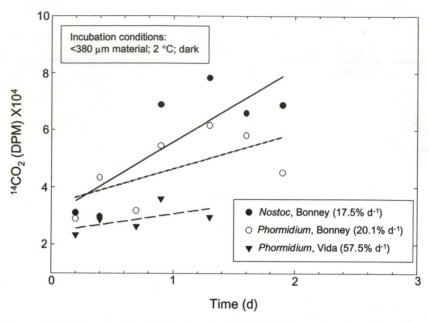

FIGURE 3.6. Mineralization ($^{14}CO_2$ release) of ^{14}C-labeled cyanobacteria collected from the Bonney or Vida lake basins. Incubations were conducted in filtered (ca. 0.7µm) surface water (5 m) from Lake Bonney. The slope of each response in concert with the specific activity of the labeled cyanobacteria was used to compute the turn-over rates for the individual species.

Microbial Consortia

A majority of the cyanobacterial and bacterial activity was associated with sediment aggregates, as opposed to individual microorganisms embedded in the ice matrix (47). A core of Lake Bonney ice collected from 0.2 to 0.25 m showed considerable differences between clean and sediment-laden layers based on epifluorescence counts of material stained with SYBR gold, a probe specific for DNA (12). Based on the SYBR gold-staining study, the average bacterial and virus-like particle (VLP) densities in clear ice were 2.29×10^3 cells · ml^{-1} and 1.23×10^4 VLPs · ml^{-1}, respectively. The sediment-laden portion of this ice core had a bacterial density of 1.15×10^4 cell · ml^{-1} and a VLP count of 2.77×10^4 ml^{-1}. The virus:bacteria ratio was 5.37 for the clear ice and 2.41 for the sediment sections of the core. These ratios are within the range of those observed in more temperate climates that include eutrophic and oligotrophic waters (106) but lower than those in the water columns of freshwater lakes of Signey Island, Antarctica (104), and Lake Hoare, Antarctica (45). Tomato mosaic tobamovirus (ToMV) RNA also has been detected but not quantified using molecular methods in 140,000-year-old Greenland ice cores (11), and viral particles have been observed but not identified in Vostok ice cores more than 300,000 years old (Priscu unpublished data). There were on average 5.02-fold more bacteria and 2.25-fold more VLPs in the Lake Bonney ice core sample that included the sediment than in the clear ice section of the core. The virus:bacteria ratio for sediment-laden ice was less than half of that observed in the clear ice, implying there are fewer viruses per bacterium in the sediment-containing section of the core. It remains unclear if the VLPs were bacteriophage or cyanobacteriophage. However, the presence of viral particles in the ice indicates that phage may play a major role in genetic transfer and overall survival of prokaryotes in the ice. Phage isolated by Castello et al. (10) from *B. subtilis* in deep ice cores from Greenland may have a similar role if cells are metabolically active within the ice.

Microautoradiographic studies reveal that both bacterial and cyanobacterial activities were tightly associated with sediment particles, corroborating experimental results (60). Microautoradiographs also indicated that virtually all of the incorporation of radiolabeled organic substrates was mediated by non-autofluorescent (non-chlorophyll containing) bacterial-size rods (0.5–1 µm length) and filaments (0.5 µm width) closely associated with aggregates whereas $^{14}CO_2$ incorporation was limited to filamentous cyanobacteria. Heterotrophic bacteria were attached to soil particles and associated with cyanobacterial colonies and aggregates. These observations are similar to those reported for temperate and tropical cyanobacteria-dominated systems (59). A majority of bacteria found in Vostok accretion ice collected from 3590 m beneath the surface were also attached to sediment particles (71). Tetrazolium salt (TTC) reduction assays further revealed that, when melting occurs, localized oxygen consumption associated with aggregates is sufficient to create reduced microzones. These microzones are associated with regions colonized by bacteria and cyanobacteria, suggesting they may be potential sites for oxygen-sensitive processes such as atmospheric nitrogen fixation (58, 60). Pinkney and Paerl (66) showed that cyanobacterial and bacterial biomass and activities in Lake Bonney ice were heterogeneously distributed among aggregates, promoting the development of oxygen and, possibly, other biogeochemical gradients. Biogeochemical zonation and diffusional oxygen and nutrient concentration gradients likely result from microscale patchiness in microbial metabolic activities (i.e., photosynthesis, respiration). These gradients, in turn, promote metabolic diversity and differential photosynthetic and heterotrophic growth rates.

Phototrophy, heterotrophy, and diazotrophy (nitrogen fixation) can occur simultaneously in ice-aggregate microbial communities. Key environmental factors controlling the rates and biogeochemical significance of these processes include: (i) the presence of radiant energy and liquid water; (ii) nitrogen, phosphorus, and trace metals sufficient for phototrophy; (iii) adequate organic matter for heterotrophic activity; and (iv) energy (light or organic carbon), phosphorus, iron, and other trace metals sufficient for diazotrophy (60).

Mineralization of POC and PON is highly dependent on organic matter availability, the main source being cyanobacterial photosynthesis. Therefore, close spatial proximity of heterotrophs to phototrophs is essential for completion of carbon, nitrogen, and phosphorus cycling. The paucity of higher trophic levels (e.g., protozoans) in the ice magnifies the importance of microbial interactions within the ice assemblage and amplifies the role played by viruses in terms of microbial survival and possibly diversity. Clearly, the spatial and temporal relationships within the ice produce a microbial consortium that is of fundamental importance for initiating, maintaining, and optimizing essential life-sustaining production and nutrient transformation processes. Close spatial and temporal coupling of metabolite exchange among producers and consumers of organic matter existing within the ice appears to be the limiting ecological factor enabling microbial processes to coexist in what appears to be an otherwise inhospitable environment. To accomplish this feat, the microbes must exist in a highly cooperative and efficient manner. We stress that without these close spatial linkages, the microbial assemblage may not be able to survive the extreme conditions posed by the ice environment.

Liquid Water Production and Detection

Virtually all of the metabolic data that exist on the Dry Valley lake ice are based on melted ice samples. The same can be said for glacial ice though isotopic signatures and modeling of deep glacial ice have suggested there is in situ activity within "solid" ice (7, 67). Clearly, no biological or biogeochemical process would occur without the presence of liquid water, a contention that has been confirmed by the lack of activity obtained in experiments on cyanobacterial photosynthesis and atmospheric nitrogen fixation in solid ice samples (60, Priscu and Fritsen unpublished data). Since the discovery of a cyanobacterial/bacterial consortium within the permanent ice covers of the Dry Valley lakes internal water production in permanent ice covers has become an important issue. In addition to life support within the ice, the permanent ice covers influence the exchange of momentum, thermal energy, and materials between the water column and the atmosphere (e.g., 28, 42, 73, 100). The first evidence for the presence of liquid water in the ice covers was reported by Henderson et al. (37), who noted that a liquid "water table" developed within the ice cover of Lake Fryxell during summer. Subsequently, Squyres et al. (87) encountered liquid water 2.1 m below the surface while constructing sampling holes in Lake Hoare during early summer, and noted that the depth of the liquid water coincided with the bottom of the in-ice sediment. Adams et al. (1) observed the presence of liquid water near a sediment inclusion 2.5 m below the ice surface in otherwise dry ice from the Lake Bonney ice cover. Within days of the observation, liquid-saturated ice was encountered approximately 1.0 m below the surface in the same vicinity. Adams et al. (2) later noted dynamic relationships in gas bubbles within the ice cover indicating differential melting associated with the sediment layer.

Typical temperature profiles in the ice cover of the east lobe of Lake Bonney are shown for the period January 1995 to January 1996 (Figure 3.7 A). The most significant feature of these data in terms of liquid water production is the almost isothermal temperatures near 0 °C during the summer. These isothermal

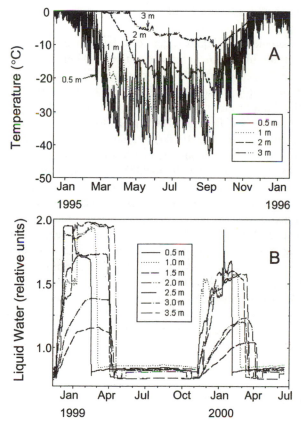

FIGURE 3.7. A: Vertical profiles of ice temperature in the east lobe of Lake Bonney for 1995. Note the near isothermal temperatures that are reached during the summer months (late November through February). (B) Relative water estimates obtained from time domain reflectometers (TDR) deployed in the ice cover of Lake Bonney during 1999 and 2000. The TDR data show the presence of liquid water when ice temperatures are isothermal near 0 °C.

temperatures persist for approximately eighty-five days from November to February. In contrast, midwinter temperature gradients are typically 6 °C · m^{-1}. Fritsen et al. (24) used ice temperature records to develop an energy budget for the period when freezing fronts were propagating through the ice in the austral autumn. According to the model output, the highest fractional volume of liquid water in Lake Bonney (ca. 25%) occurred near 2 m, coinciding with the depth of maximum sediment concentration. Lake Hoare showed a similar trend for temperatures collected between 1986 and 1988, with fractional volumes reaching 60% in the sediment layer. The differences in fractional water content presumably reflect the cooling trend that occurred in the McMurdo Dry Valleys over the past decade (19). The thermodynamic data indicate that the absorption of solar radiation by the lithogenic material is a primary process generating liquid water in the ice covers.

Direct measurements of liquid water have been made in the ice cover of Lake Bonney using time-domain reflectometers (TDR) (Figure 3.7 B), which detect changing dielectric permittivity in porous media allowing indirect measurements of volumetric water content. TDR data from late 1998 through July 2000 show that liquid water is present in the ice cover of Lake Bonney from mid-November through early May. The 2 m TDR sensor, placed directly in a sediment aggregate during the 1998–1999 austral summer, shows that this layer is one of the first to melt and the last to freeze. The extended period of melt at 2 m also is evident in the 1999–2000 austral summer.

Thermodynamic modeling and TDR data show that the liquid water produced by solar radiation–induced melting associated with lithogenic material exists for approximately 150 days during the austral summer in Lake Bonney. Based on the length of this period, the cyanobacterial doubling time has been estimated to be about nine years in Lake Bonney. Estimates of cyanobacterial doubling time, growth rates, growing days, and annual primary production for a number of lakes in the McMurdo Dry Valleys are summarized in Table 3.2. These data show that cyanobacterial growth in the ice covers is slow, with generation times ranging from 0.43 to 9 years. The long generation times are directly related to the amount of time liquid water remains in the ice covers and vary with climatic conditions.

Conclusions

The permanent ice covers of liquid water-based lakes in the McMurdo Dry Valleys are thermodynamically active and display a well-defined but transitory stratigraphy in sediment content, bubble morphology, and liquid water. The unique combination of physical features produces liquid water deep within the permanent ice covers when the air temperature is well below the freezing point.

TABLE 3.2.

Growth-related measurements for cyanobacterial assemblages associated with the sediment layer of various lakes in the McMurdo Dry Valleys

Lake	Growth Rate μ	Growing Days	Generation Time	Annual Primary Production
EL Bonney	0.001	150	9.0	86–340
WL Bonney	0.003	130	2.0	120
Hoare	0.006	130	0.96	843
Fryxell	0.008	80	0.64	5891
Joyce	—	130	—	—
Vanda	NA	NA	NA	0
Vida	2.71	130	1.92	617
Miers	12.0	100	0.43	10,447

Note: μ (d^{-1}) = growth rate based on the rate of protein synthesis; growing days = estimated number of days during a year when liquid water is available for growth; generation time (years) = *in situ* generation time based on growth rates and the number of growing days; annual primary production (mg C m^{-2} y^{-1}) = *in situ* primary production based on $^{14}CO_2$ fixation rates and the number of growing days. Growth rate and primary production were measured when cyanobacteria were exposed to liquid water at irradiances saturating to photosynthesis (ca. 200 μmol photons m^{-2} d^{-1}) and 0–2 °C. NA = not applicable; — = not measured. Modified from [26]).

About 30 cm of new ice growth takes place annually on the bottom of the ice cover with roughly an equal amount lost each year from ablation at the top (1, 15). Terrestrial sediment, including cyanobacterial and bacterial cells, is deposited on the ice surface by aeolian processes, primarily by strong katabatic winds that occur during winter (25). As the sun rises above the horizon in late September, the sediments absorb radiant energy and melt through the surface layers of the ice. The sediments reach a level in the ice where downward melting is balanced by the general upward movement of the ice due to accretion of meltwater at the bottom and ablation at the surface. The sediments generally accumulate in pockets about 2 m beneath the ice surface with "clean" ice above and below. During the summer, an "aquifer" is created within the ice, with its lower boundary marked by the sediment layer. The aquifer is connected to the lake water through discrete conduits and the lower ice remains essentially dry. The presence of liquid water and adequate solar radiation during austral summer produces a cascade of microbial processes within the ice cover. Cyanobacterial photosynthesis and atmospheric nitrogen fixation produces reduced carbon and nitrogen that in turn fuel heterotrophic processes.

The heterotrophs remineralize the organic matter, producing raw materials to support further cyanobacterial activity. The close spatial and temporal coupling of metabolites within the microbial consortium appears to be essential for the microbes to survive and replicate in what has been characterized as "the edge of life" (60). Data on microbial activity for the ice assemblage indicate that metabolic complementation among functionally diverse, but structurally simple, prokaryotic consortia along microscale biogeochemical gradients is a unique and effective strategy for meeting the requirements of life in this extreme habitat. Sediment mounds noted on the bottom of certain lakes (56, 84, 87) and coarse sediments collected in sediment traps deployed in the lake water indicate that sediments and associated organic matter are lost from the ice covers, presumably during late summer and early autumn, when the liquid water content of the ice is at a maximum. The exact mechanism for this loss remains unclear, but owing to the patchiness of the benthic sediment mounds, it seems that the loss is through discrete cracks or conduits within the ice cover rather than downward melting.

Given the dynamic processes associated with permanent Antarctic lake ice, it is conceivable that the organic matter accumulates within the ice covers by physical processes alone (i.e., aeolian sources exceed loss to the lake water) rather than via cyanobacterial production. This is an important distinction in terms of defining the viability of the ecosystem. Are the microorganisms merely "freeloaders" passing through the ice, or do they form a viable and self-sustaining ecosystem? The balance between biological and physical processes was determined by Priscu et al. (74) and showed that POC accumulated at a rate of 229 mg $C \cdot m^{-2} \cdot y^{-1}$, which is within the range of values for annual production shown in Table 3.2. The contribution from cyanobacterial primary production is about 80% of the total annual accumulation of POC in the ice, indicating that biological production exceeds POC accumulation via physical processes. Hence, the unique combination of physical and microbial processes that occur in the permanent lake ice in the McMurdo Dry Valleys produces an oasis for life in a polar desert where there is an overall paucity of liquid water and associated production of new carbon and nitrogen.

Acknowledgments

We thank B. Vaughn, L. Kester, and T. Stevens for assistance with manuscript preparation and editing. S. Konley conducted the ammonium leaching experiments. This research was supported by National Science Foundation grants OPP 0096250, OPP 9815998, OPP 9815512, OPP 9419413, OPP 0085400, and OPP/MCB 0237335.

Literature Cited

1. Adams, E.A., and J.C. Priscu. 1995. Some metamorphic processes in the lake ice in the McMurdo Dry Valleys. Ant. J. U.S. 30: 307–309.
2. Adams, E.E., J.C. Priscu, C.H. Fritsen, S.R. Smith, and S.L. Brackman. 1998. Permanent ice covers of the McMurdo Dry Valley Lakes, Antarctica: Bubble formation and metamorphism. In J.C. Priscu (ed.), Ecosystem Dynamics in a Polar Desert: The McMurdo Dry Valleys, Antarctica. Antarctic Res. Ser., 72: 281–296. American Geophysical Union.
3. Anderson, D.A., R.A. Wharton Jr., and S.W. Squyres. 1993. Terrigenous clastic sedimentation in Antarctic Dry Valley lakes. In W.J. Green and E.I. Friedmann (eds.), Physical and Biogeochemical Processes in Antarctic Lakes. Antarctic Res. Series 59: 71–82. American Geophysical Union.
4. Archibald, J.M., M.B. Rogers, M. Toop, K. Ishida, and P.J. Keeling. 2003. Lateral gene transfer and the evolution of plastid-targeted proteins in the secondary plastid-containing alga *Bigelowiella natans*. Proc. Nat. Acad. Sci. 100: 7678–7683.
5. Baltcr, M. 1999. Did life begin in hot water? Science 280: 31.
6. Brambilla, E., H. Hippe, A. Hagelstein, B.J. Tindall, and E. Stackbrandt. 2001. 16S rDNA diversity of cultured and uncultured prokaryotes of a mat sample from Lake Fryxell, McMurdo Dry Valleys, Antarctica. Extremophiles 5: 23–33.
7. Campen, R.K., T. Sowers, and R.B. Alley. 2003. Evidence of microbial consortia metabolizing within a low-latitude mountain glacier. Geology 31: 231–234.
8. Carr, M.H., et al. (twenty-one co-authors). 1998. Evidence for a subsurface ocean on Europa. Nature 391: 363–365.
9. Carpenter, E.J., S. Lin, and D.G. Capone. 2000. Bacterial activity in South Pole snow. Appl. Environ. Microbiol. 66: 4514–4517.
10. Castello, J.D., S.O. Rogers, J.E. Smith, W.T. Starmer, and Y. Zhao. 2005. Plant and bacterial viruses in the Greenland ice sheet. Chapter 13, this volume.
11. Castello, J.D., S.O. Rogers, W.T. Starmer, C.M. Catranis, L. Ma, G.D. Bachand, Y. Zhao, and J.E. Smith. 1999. Detection of tomato mosaic tobamovirus RNA in ancient glacial ice. Polar Biol. 22: 207–212.
12. Chen, F., J. Lu, B. Binder, Y. Liu, and R. Hodson. 2001. Application of digital image analysis and flow cytometry to enumerate marine viruses stained with SYBR Gold. Appl. Environ. Microbiol. 67: 539–545.
13. Christner, B.C., B.H. Kvitco II, and J.N. Reeve. 2003. Molecular identification of bacteria and eukarya inhabiting an Antarctic cryoconite hole. Extremophiles 7:177–183.
14. Christner, B.C., E. Mosley-Thompson, L.G. Thompson, and J.N. Reeve. 2001. Isolation of bacteria and 16S rDNAs from Lake Vostok accretion ice. Environ. Microbiol. 3(9): 570–577.
15. Clow, G. D., C.P. McKay, G.M. Simmons Jr., and R.A Wharton Jr. 1988. Climatological observations and predicted sublimation rates at Lake Hoare, Antarctica. J. Climate 1: 715–728.
16. Craig, H., R.A Wharton, and C.P. McKay 1992. Oxygen supersaturation in ice-covered Antarctic lakes: Biological versus physical contributions. Science 255: 218–221.

17. Doolittle, W.E. 1999. Phylogenetic classification and the universal tree. Science 284: 2124–2128.
18. Doran, P.T., C.P. McKay, G.D. Clow, G.L. Dana, A.G. Fountain, T. Nylen, and W.B. Lyons. 2002. Valley floor climate observations in the McMurdo Dry Valleys, Antarctica, 1986–2000. J. Geophys. Res.-Atmospheres 107 (D24), 4772: ACL 13–1–ACL 13–12.
19. Doran, P.T., J.C. Priscu, W.B. Lyons, J.E. Walsh, A.G. Fountain, D.M. McKnight, D.L. Moorhead, R.A. Virginia, D.H. Wall, G.D. Clow, C.H. Fritsen, C.P. McKay, and A.N. Parsons. 2002. Antarctic climate cooling and terrestrial ecosystem response. Nature 415: 517–520.
20. Eschenmoser, A. 1999. Chemical etiology of nucleic acid structure. Science 284: 2118–2124.
21. Fountain, A.G., G.L. Dana, K.J. Lewis, B.H. Vaughn, and D.M. McKnight. 1998. Glaciers of the McMurdo Dry Valleys, southern Victoria Land, Antarctica. In J.C. Priscu (ed.), Ecosystem Dynamics in a Polar Desert: The McMurdo Dry Valleys, Antarctica. Antarctic Res. Ser. 72: 65–76. American Geophysical Union.
22. Fountain, A.G., W.B. Lyons, M.B. Burkins, G.L. Dana, P.T. Doran, K.J. Lewis, D.M. McKnight, D.L. Moorhead, A.N. Parsons, J.C. Priscu, D.H. Wall, R.W. Wharton, and R.A. Virginia. 1999. Physical controls on the Taylor Valley Ecosystem, Antarctica. BioScience 49: 961–971.
23. Friedmann, E.I., L. Kappen, M.A. Meyer, and J.A. Nienow. 1993. Long-term productivity in the cryptoendolithic microbial community of the Ross Desert, Antarctica. Microbial Ecol. 25: 51–69.
24. Fritsen, C.H., E.E. Adams, C.M. McKay, and J.C. Priscu. 1998. Permanent ice covers of the McMurdo Dry Valley Lakes, Antarctica: Liquid water content. In J.C. Priscu (ed.), Ecosystem Dynamics in a Polar Desert: The McMurdo Dry Valleys, Antarctica. Antarctic Res. Ser. 72: 269–280. American Geophysical Union.
25. Fritsen, C.H., A. Grue, and J.C. Priscu. 2000. Distribution of organic carbon and nitrogen in surface soils in the McMurdo Dry Valleys, Antarctica. Polar Biol. 23: 121–128.
26. Fritsen, C.H., and J.C. Priscu. 1996. Photosynthetic characteristics of cyanobacteria in permanent ice-covers on lakes in the McMurdo Dry Valleys, Antarctica. Ant. J. U.S. 31: 216–218.
27. Fritsen, C.H., and J.C. Priscu. 1998. Cyanobacterial assemblages in permanent ice covers of Antarctic lakes: Distribution, growth rate, and temperature response of photosynthesis. J. Phycol. 34: 587–597.
28. Fritsen, C.H., and J.C. Priscu. 1999. Seasonal change in the optical properties of the permanent ice cover on Lake Bonney, Antarctica: Consequences for lake productivity and dynamics. Limnol. and Oceanogr. 44: 447–454.
29. Galtier N., N. Tourasse, and M. Gouy. 1999. A non-hyperthermophilic common ancestor to extant life forms. Science 283: 220–222.
30. Gordon, D.A., B. Lanoil, S. Giovannoni, and J.C. Priscu. 1996. Cyanobacterial communities associated with mineral particles in Antarctic lake ice. Ant. J. U.S. 31: 224–225.

31. Gordon, D.A., J.C. Priscu, and S. Giovannoni. 2000. Distribution and phylogeny of bacterial communities associated with mineral particles in Antarctic lake ice. Microbial Ecol. 39: 197–202.

32. Green, W.J. and E.I. Friedmann (eds.). 1993. Physical and biogeochemical processes in Antarctic lakes. Antarctic Res. Ser. 59: 1–216. American Geophysical Union.

33. Greenberg, R., et al. (nine co-authors) 1998. Tectonic processes on Europa: Tidal stresses, mechanical response, and visible features. Icarus 135: 64–78.

34. Grue, A.M., C.H. Fritsen, and J.C. Priscu. 1996. Nitrogen fixation within permanent ice covers on lakes in the McMurdo Dry Valleys, Antarctica. Ant. J. U.S. 32: 218–220.

35. Hawes, I. 1993. Photosynthesis in thick cyanobacterial films: A comparison of annual and perennial Antarctic mat communities. Hydrobiologia 252: 203–209.

36. Hawes, I., C. Howard-Williams, A-M. Schwarz, and M.T. Downes. 1997. Environment and microbial communities in a tidal lagoon at Bratina Island, McMurdo Ice Shelf, Antarctica. pp. 178–184. In B. Battaglia, J. Valencia, and D. Walton (eds.), Antarctic Communities: Species, Structure and Survival. Cambridge University Press, Cambridge.

37. Henderson, R.A., W.M. Prebble, R.A. Hoare, K.B. Popplewell, D.A. House, and A.T. Wilson. 1996. An ablation rate for Lake Fryxell, Victoria Land, Antarctica. J. Glaciol. 6: 129–133.

38. Hoffman, P.F., A.J. Kaufman, G.P. Halverson, and D.P. Schrag. 1998. A neo-proterozoic snowball Earth. Science 281: 1342–1346.

39. Hoffman, P.F., and D.P. Schrag. 2000. Snowball Earth. Scientific American, January, 68–75.

40. Howard-Williams, C., R.D. Pridmore, P.A. Brody, and W.F. Vincent. 1990. Environmental and biological variability in the McMurdo Ice Shelf ecosystem. pp. 23–31. In K.R. Kerry and G. Hempel (eds.), Antarctic Ecosystems: Ecological Change and Conservation. Springer-Verlag, New York.

41. Howard-Williams, C., J.C. Priscu, and W.F. Vincent. 1989. Nitrogen dynamics in two Antarctic streams. In W.F. Vincent and E. Ellis-Evans (eds.), High Latitude Limnology. Hydrobiologia, 172: 51–62. Kluwer Publications.

42. Howard-Williams, C., A. Schwarz, I. Hawes, and J.C. Priscu. 1998. Optical properties of lakes of the McMurdo Dry Valleys. In J.C. Priscu (ed.), Ecosystem Dynamics in a Polar Desert: The McMurdo Dry Valleys, Antarctica. Antarctic Res. Ser. 72: 189–204. American Geophysical Union.

43. Huber, R., H. Huber, and K.O. Stetter. 2000. Towards the ecology of hyperthermophiles: Biotopes, new isolation strategies and novel metabolic properties. FEMS Microbiol. Rev. 24: 615–623.

44. Karl, D.M., D.F. Bird, K. Bjorkman, T. Houlihan, R. Shackelford, and L. Tupas. 1999. Microorganisms in the accreted Ice of Lake Vostok, Antarctica. Science 286: 2144–2147.

45. Kepner, R.L., R.A. Wharton Jr., and C.A. Suttle. 1998. Viruses in Antarctic lakes. Limnol. and Oceanogr. 43: 1754–1761.

46. Knoll, A.H. 1994. Proterozoic and early Cambrian proteins: Evidence for accelerating evolutionary tempo. Proc. Nat. Acad. Sci. 91: 6743–6750.

47. Lisle, J.T., and J.C. Priscu. 2004. The occurrence of lysogenic bacteria and micro-bial aggregates in the lakes of the McMurdo Dry Valleys, Antarctica. Microbial Ecol. 47: 427–439.

48. Lizotte, M.P., and J.C. Priscu. 1992. Spectral irradiance and bio-optical properties in perennially ice-covered lakes of the Dry Valleys (McMurdo Sound, Antarc-tica). Antarctic Res. Ser. 57: 1–14.

49. Lyons, W.B., K.A. Welch, J.C. Priscu, J. Labourn-Parry, D. Moorhead, D.M. McKnight, P.T. Doran, and M. Tranter. 2001. The McMurdo Dry Valleys long-term ecological research program: New understanding of the biogeochemistry of the Dry Valley lakes: A review. Polar Geogr. 25: 202–217.

50. Margulis, L., and D. Sagan. 1997. Micro-cosmos: Four Billion Years of Microbial Evolution. University of California Press, Berkeley.

51. McKay, C.P., G. Clow, R.A. Wharton Jr., and S.W. Squyres. 1985. Thickness of ice on perennially frozen lakes. Nature 313: 561–562.

52. Moorhead, D.L., and J.C. Priscu. 1998. Linkages among ecosystem components within the McMurdo Dry Valleys: A synthesis. In J.C. Priscu (ed.), Ecosystem Dynamics in a Polar Desert: The McMurdo Dry Valleys. Antarctic Res. Ser.: 72: 351–364, American Geophysical Union.

53. Morgan, R.M., A.G. Ivanov, J.C. Priscu, D.P. Maxwell, and N.P.A. Huner. 1998. Structure and composition of the photochemical apparatus of the Antarctic green algae, *Chlamydomonas subcaudata*. Photosynthesis Res. 56: 303–314.

54. Nadeau, T., and R.W. Castenholz. 2000. Characterization of psychrophilic oscilla-torians (cyanobacteria) from Antarctic meltwater ponds. J. Phycol. 36: 914–923.

55. Nadeau, T., and R.W. Castenholz. 2001. Evolutionary relationships of cultivated Antarctic oscillatorians (cyanobacteria). J. Phycol. 37: 650–654.

56. Nedell, S.S., D.W. Anderson, S.W. Squyres, and F.G. Love. 1987. Sedimentation in ice-covered Lake Hoare, Antarctica. Sedimentology 34: 1093–1106.

57. Nelson, K.E. (twenty-eight co-authors). 1999. Evidence for lateral gene transfer between archaea and bacteria from genome sequence of Thermotoga maritime. Nature 399: 323–328.

58. Olson, J.B., T.F. Steppe, R.W. Litaker, and H.W. Paerl. 1998. N_2-fixing microbial consortia associated with the ice cover of Lake Bonney, Antarctica. Microbial Ecol. 36: 231–238.

59. Paerl, H.W., and J.L. Pinckney. 1996. Ice aggregates as a microbial habitat in Lake Bonney, Dry Valley lakes, Antarctica: Nutrient-rich micro-ozones in an oligotro-phic ecosystem. Ant. J. U.S. 31: 220–222.

60. Paerl, H.W., and J.C. Priscu. 1998. Microbial phototrophic, heterotrophic and di-azotrophic activities associated with aggregates in the permanent ice cover of Lake Bonney, Antarctica. Microbial Ecol. 36: 221–230.

61. Palmisano, A.C., and G.M. Simmons Jr. 1987. Spectral down-welling irradiance in an Antarctic lake. Polar Biol. 7: 145–151.

62. Parker, B.C., and R.A. Wharton Jr. 1985. Physiological ecology of blue-green algal mats (modern stromatolites) in Antarctic oasis lakes. Archiv. Hydrobiol. Suppl. 71: 331–348.

63. Pederson, K. 1997. Microbial life in deep granitic rock. FEMS Microbiol. Rev. 20: 399–414.

64. Pennisi, E. 1998. Genome data shake tree of life. Science 280: 672–674.
65. Pennisi, E. 1999. Is it time to uproot the tree of life? Science 284: 1305–1307.
66. Pinckney, J.L., and H.W. Paerl. 1996. Lake ice algal phototroph community composition and growth rates, Lake Bonney, Dry Valley Lakes, Antarctica. Ant. J. U.S. 31: 215–216.
67. Price, P.B. 2000. A habitat for psychrophiles in deep Antarctic ice. Proc. Nat. Acad. Sci. 97: 1247–1251.
68. Priscu, J.C. 1997. The biogeochemistry of nitrous oxide in permanently ice-covered lakes of the McMurdo Dry Valleys, Antarctica. Global Change Biol. 3: 301–305.
69. Priscu, J.C. 1998. (ed.) Ecosystem dynamics in a polar desert: The McMurdo Dry Valleys, Antarctica. Antarctic Res. Ser. 72: 1–369. American Geophysical Union.
70. Priscu, J.C. 1999. Life in the valley of the "dead." BioScience 49: 959.
71. Priscu, J.C., E.E. Adams, W.B. Lyons, M.A. Voytek, D.W. Mogk, R.L. Brown, C.P. McKay, C.D. Takacs, K.A. Welch, C.F. Wolf, J.D. Kirstein, and R. Avci. 1999. Geomicrobiology of sub-glacial ice above Vostok Station. Science 286: 2141–2144.
72. Priscu, J.C., and B.C. Christner. 2004. Earth's icy biosphere, pp. 130–145. In A. Bull (ed.), Microbial Diversity and Bioprospecting. ASM Press, Washington, D.C.
73. Priscu, J.C., M.T. Downes, and C.P. McKay. 1996. Extreme super-saturation of nitrous oxide in a permanently ice-covered Antarctic Lake. Limnol. and Oceanogr. 41: 1544–1551.
74. Priscu, J.C., C.H. Fritsen, E.E. Adams, S.J. Giovannoni, H.W. Paerl, C.P. McKay, P.T. Doran, D.A. Gordon, B.D. Lanoil, and J.L. Pinckney. 1998. Perennial Antarctic lake ice: An oasis for life in a polar desert. Science 280: 2095–2098.
75. Priscu, J.C., A.C. Palmisano, L.R. Priscu, and C.W. Sullivan. 1989. Temperature dependence of inorganic nitrogen uptake and assimilation in Antarctic Sea-ice microalgae. Polar Biol. 9: 443–446.
76. Priscu, J.C., W.F. Vincent, and C. Howard-Williams. 1989. Inorganic nitrogen uptake and regeneration in perennially ice-covered lakes Fryxell and Vanda, Antarctica. J. Plankton Res. 11: 335–351.
77. Priscu, J.C., C.F. Wolf, C.D. Takacs, C.H. Fritsen, J. Laybourn-Parry, E.C. Roberts, and W.B. Lyons. 1999. Carbon transformations in the water column of a perennially ice-covered Antarctic Lake. BioScience 49: 997–1008.
78. Psenner, R., B. Sattler, A. Willie, C.H. Fritsen, J.C. Priscu, M. Felip, and J. Catalan. 1999. Lake ice microbial communities in alpine and Antarctic lakes. pp. 17–31. In P. Schinner and R. Margesin (eds.), Adaptations of Organisms to Cold Environments. Springer-Verlag, New York.
79. Raymond, J., and R.E. Blankenship. 2003. Horizontal gene transfer in eukaryotic algal evolution. Proc. Nat. Acad. Sci. 100: 7419–7420.
80. Redfield, A.C. 1958. The biological control of chemical factors in the environment. American Scientist, September, 205–221.
81. Schrag, D.P., and P.F. Hoffman. 2001. Life, geology and snowball Earth. Nature 409: 306.
82. Scott, R.F. 1905. The Voyage of the Discovery. Vol II. Macmillan Publishers, London.

83. Siegert, M.J., C. Ellis-Evans, M. Tranter, C. Mayer, J.R. Petit, A. Salamatin, and J.C. Priscu. 2001. Physical, chemical and biological processes in Lake Vostok and other subglacial lakes. Nature 414: 603–609.
84. Simmons, G.W., Jr., J. R. Vestal, and R.A. Wharton Jr. 1993. Environmental regulators of microbial activity in continental Antarctic lakes. In W.J. Green and E.I. Friedmann (eds.), Physical and Biogeochemical Processes in Antarctic Lakes. Antarctic Res. Ser., 59: 197–214. American Geophysical Union.
85. Spigel, R.H., and J.C. Priscu. 1996. Evolution of temperature and salt structure of Lake Bonney, a chemically stratified Antarctic lake. Hydrobiology 321: 177–190.
86. Spigel, R.H., and J.C. Priscu. 1998. Physical limnology of the McMurdo Dry Valley lakes. In J.C. Priscu (ed.), Ecosystem Dynamics in a Polar Desert: The McMurdo Dry Valleys, Antarctica. Antarctic Res. Ser. 72: 153–188. American Geophysical Union.
87. Squyres, S.W., D.W. Andersen, S.S. Nedell, and R.A. Wharton Jr. 1991. Lake Hoare, Antarctica: Sedimentation through a thick perennial ice cover. Sedimentology 38: 363–379.
88. Studinger, M., R.E. Bell, G.D. Karner, A.A. Tikku, J.W. Holt, D.L. Morse, T.G. Righter, S.D. Kempf, M.E. Peters, D.D. Blankenship, R.E. Sweeney, and V.L. Rystrom. 2003. Ice cover, landscape setting, and geological framework of Lake Vostok, East Antarctica. Earth and Planetary Sci. Letters 205: 195–210.
89. Tang, E.P.Y., R. Tremblay, and W.F. Vincent. 1997. Cyanobacterial dominance of polar freshwater ecosystems: Are high latitude-mat-formers adapted to low temperature? J. Phycol. 33: 171–181.
90. Tang, E.P.Y., and W. Vincent. 1999. Strategies of thermal adaptation by high-latitude cyanobacteria. New Phytol. 142: 315–323.
91. Taylor, G. 1916. With Scott: The Silver Lining. Dodd, Mead and Company, New York.
92. Vincent, W.F. 2000. Cyanobacterial dominance in the polar regions. pp. 321–340. In B. A. Whitton and M. Potts (eds.), The Ecology of Cyanobacteria: Their Diversity in Time and Space. Kluwer Publishers.
93. Vincent, W.F., R.W. Castenholz, M.T. Downes, and C. Howard-Williams. 1993. Antarctic cyanobacteria: Light, nutrients, and photosynthesis in the microbial mat environment. J. Phycol. 29: 745–755.
94. Vincent W.F., J.A.E. Gibson, R. Pienitz, and V. Villeneuve. 2000. Ice shelf microbial ecosystems in the high Arctic and implications for life on snowball Earth. Naturwissenshaften 87: 137–141.
95. Vincent, W.F., and C. Howard-Williams. 2001. Life on snowball Earth. Science 287: 2421.
96. Vincent, W.F., R. Rae, I. Laurion, C. Howard-Williams, and J.C. Priscu. 1997. Transparency of Antarctic ice-covered lakes to solar UV radiation. Limnol. and Oceanogr. 43: 618–624.
97. Wall, D., and R.A. Virginia. 1998. Soil biodiversity and community structure in the McMurdo Dry Valleys, Antarctica. In J.C. Priscu (ed.), Ecosystem Processes in a Polar Desert: The McMurdo Dry Valleys, Antarctica. Antarctic Res. Ser. 72: 323–335. American Geophysical Union.
98. Wharton, R.A., Jr., R.A. Jamison, M. Crosby, C.P. McKay, and J.W. Rice Jr. 1995. Paleolakes on Mars. J. Paleolimnol. 13: 267–283.

99. Wharton, R.A., Jr., B.C. Parker, and G.M. Simmons Jr. 1983. Distribution, species composition and morphology of algal mats in Antarctic Dry Valley lakes. Phycologia 22: 355–365.

100. Wharton, R. A., Jr., C.P. McKay, G.D. Clow, and D.T. Anderson. 1993. Perennial ice covers and their influence on Antarctic lake ecosystems. In W.J. Green and E.I. Friedmann (eds.), Physical and Biogeochemical Processes in Antarctic Lakes. Antarctic Res. Ser. 59: 53–70. American Geophysical Union.

101. Wharton, R.A., Jr., C.P. McKay, G.M. Simmons, Jr., and B.C. Parker. 1986. Oxygen budget of a perennially ice-covered Antarctic lake. Limnol. and Oceanogr. 31: 437–443.

102. Williams, D.M., J.F. Kasting, and L.A. Frakes. 1998. Low-latitude glaciation and rapid changes in the Earth's obliquity explained by obliquity-oblateness feedback. Nature 396: 453–455.

103. Wilson, A.T. 1965. Escape of algae from frozen lakes and ponds. Ecology 46: 376.

104. Wilson, W.H., D. Lane, D.A. Pearce, and J.S. Ellis-Evans. 2000. Transmission electron microscope analysis of virus-like particles in freshwater lakes of Signy Island, Antarctica. Polar Biol. 23: 657–660.

105. Wing, K.T., and J.C. Priscu. 1993. Microbial communities in the permanent ice cap of Lake Bonney, Antarctica: Relationships among chlorophyll a, gravel and nutrients. Ant. J. U.S. 28: 246–249.

106. Wommack, E., and R. Colwell. 2000. Virioplankton: Viruses in aquatic ecosystems. Microbiol. and Mol. Biol. Rev. 64: 69–114.

4

The Growth of Prokaryotes in Antarctic Sea Ice:
Implications for Ancient Ice Communities

David S. Nichols

THE DEVELOPMENT of permanently cold environments on Earth is a relatively recent event in terms of aerobic prokaryotic evolution. Diversification of early amphiaerobes is believed to have occurred 1.8 to 2.4 billion years ago (56), while the formation of Earth's first permanently cold environment, the cold deep ocean (less than 10 °C), is thought to have occurred only 38 million years ago (30). Presently over 80% of the Earth's biosphere is permanently cold, that is, less than 5 °C (53). While the evolution of bacteria capable of surviving and requiring permanently cold temperatures for growth may be relatively recent, these bacteria now have an enormous proportion of Earth's biosphere available in which to proliferate and influence by microbially mediated processes. Psychrophilic bacteria should, therefore, be recognized as potential key components in global ecological and biogeochemical cycles.

Ice environments formed from the freezing of freshwater (glaciers, permafrost) differ in many ways from those formed from the freezing of the world's oceans (sea ice). However, due to the fundamental physico-chemical processes involved, sea ice and glacial ice environments do share significant similarities that influence the survival and/or growth of microorganisms. This chapter will highlight the fact that sea ice surrounding Antarctica represents an unusual case in bacterial colonization and evolution. Antarctica has not always been permanently cold, and did not begin to cool until the formation of the Southern Ocean circulation 90 million years ago. Currently, less than 2% of the continent is ice-free. Microbial life is restricted to areas where free water is available and temperature fluctuations are less severe (19). One such refuge, where microbial life not only survives but dominates the ecosystem, is the sea ice that surrounds the Antarctic coast.

This chapter will compare and contrast the bacterial communities from Antarctic sea ice to those described from glacial ice and permafrost elsewhere in this volume. The differences in habitat structure influencing community development will be described, and hypotheses regarding the selective factors responsible for the differences or similarities in sea ice versus glacial and permafrost communities will be explored.

Temperature and Ice Bacteria

An obvious environmental factor influencing ice communities is temperature, which not only determines ice structure, but also the biological potential for bacterial growth. As a generalization, all three ice habitats (glacial ice, sea ice, and permafrost) can be considered to experience a similar temperature magnitude, although differences are apparent in temperature variations. While permafrost is very stable in temperature profile, overall temperature can vary between −1 °C and −12 °C, depending on geographical location. In contrast, both glacial ice and sea ice exhibit a temperature profile. For example, Antarctic glacial ice temperature ranges from the mean annual air temperature at the ice surface to warmer temperatures toward the bedrock due to geothermal heating. Typical values may be −22 °C to −6 °C over 1.5 km of ice thickness from Law Dome, Eastern Antarctica, and −50 °C to −3 °C (melting point) over 3.6 km of ice at Vostok, central Antarctica. Likewise the temperature of sea ice varies from mean air temperature at the surface to −2 °C (water temperature) at the ice-water interface. However, in sea ice this gradient occurs over only ca. 2 m of ice thickness. These differences are of significance in defining the regions within ice habitats where microorganisms have the potential for growth and/or survival (see chapter 7). Whether bacteria can or do grow in such regions may depend on their ability to cope with low temperatures. In this respect, not all bacteria are equal.

The term *psychrophile* (*psychro-*, cold; *phile-*, loving) was first coined by Schmidt-Nielson in 1902 to describe bacteria capable of growth at or below 0 °C (52). For bacteria, the temperature range of growth usually spans 35 to 40 °C, irrespective of the preferred region. However, it soon became apparent that a more stringent definition was required to separate those organisms requiring cold temperatures for growth (true psychrophiles) from others that were capable of growth at or near 0 °C, but possessed quite high optimal growth temperatures (around 20–30 °C). Such organisms became known as psychrotolerant (or psychrotrophic). The term *psychrophile* is therefore an operational definition, to describe the temperature/growth relationship of microorganisms. In practice, there is a continuum of organisms between those that can be considered as "extreme" psychrophiles, with optimal growth temperatures below 7 °C, to moderate psychrophiles, to psychrotolerant species (17).

The definition of psychrophiles and psychrotolerant species relies upon the temperature limits for growth, or cardinal temperatures. This tells us nothing regarding the optimal growth temperature or potential maximal growth rate at that temperature; both are important factors when considering the potential importance of bacterial functionality in an environment. Despite the fact that organisms have adapted to different temperature ranges for growth, bacteria do not exhibit complete growth rate compensation for that preferred temperature

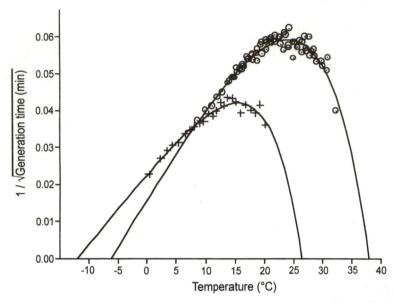

FIGURE 4.1. Plot of the inverse of the square root of the generation time (in minutes) versus temperature for the Antarctic sea ice bacteria *Glaciecola punicea* (+) (a psychrophile) and *Gelidibacter* sp. IC158 (☉) (a psychrotolerant bacterium). Points represent experimentally determined data, while the line represents the predicted value for the entire biokinetic temperature range as determined from fitting the Ratkowsky (square root) growth rate model to the experimental data. Adapted from Nichols et al. (42).

range. A typical example is given by the comparison of the growth characteristics of two sea ice bacteria, *Glaciecola punicea* (a psychrophile) and *Gelidibacter* sp. IC158 (a psychrotolerant bacterium). *Gelidibacter* sp. IC158 has an optimal growth temperature (T_{opt}) of 24 °C, where a maximal generation time of 277 minutes is achieved. In contrast, the maximal generation time of *G. punicea* reaches only 566 minutes at its T_{opt} of 15 °C. However, comparison of the growth response over the entire biokinetic range (Figure 4.1) reveals that below 8 °C *G. punicea* maintains a higher rate than the psychrotolerant species. Hence, the comparative ecological importance of psychrophilic versus psychrotolerant populations should not be limited to an assessment of which possesses a higher maximum growth rate, but which bacterium may grow faster (and presumably be more metabolically active) at environmental temperature.

For example, the apparent growth rate advantage of psychrophiles at extremely low temperatures (see Figure 4.1) does not explain the dominance of psychrophilic isolates from Antarctic sea ice as ice temperature, particularly in the lower core region where 80% of the sea ice microbial community (SIMCO) biomass exists, is comparable to that of the surrounding seawater where psychrophilic bacteria are a minority (20, 21).

Salinity, Ice Channels, and Bacteria

In addition to restrictions of temperature, microbial growth may only continue where liquid water is available. At the temperatures discussed above for ice environments, this usually necessitates the presence of solute molecules (usually salts) to lower the freezing point of water below the environmental temperature. The process of ice formation results in the expulsion of dissolved ionic species (including salts) from the ice crystal matrix into the surrounding inter-crystal solution. Hence a "ready-made" habitat of concentrated liquid is formed between ice crystals. However, the size and extent of this network differs greatly between ice environments. In sea ice, due to the relatively slow freezing process and the interface with seawater, large (ca. 50 mm diameter) networks of brine channels are formed between ice crystals, which may change in volume and concentration with ice temperature over time (37). In glacial ice, the rapid freezing of deposited material results in the formation of small, closely packed ice crystals. Instead of concentrating salts, the freezing of these crystals deposits droplets of acidic species. A network of small (ca. 7 mm diameter) liquid veins along the triple boundaries of ice crystals is formed, which may serve as a habitat for microbial growth. Within mature permafrost, the great majority of the soil matrix is frozen (92–98%). However, nanometer-thick unfrozen water films provide a coating around permafrost soil particles where, similar to sea ice, salts and ionic species are concentrated (see chapter 7).

Hence, while all three ice environments are subject to similar temperature regimes, significant differences are apparent in the availability of free water, acidity, and salinity, which may influence the composition, growth, and/or survival of microbial communities.

Comparative Biodiversity

While metabolic activity has been demonstrated from permafrost communities in the laboratory (see chapter 7), sea ice remains the most active example of bacterial ice communities. First, then, let us consider what is known concerning the biodiversity of Antarctic sea ice bacteria, and how this community differs from the underlying seawater from which it was formed.

Bacteria isolated from Antarctic seawater and sea ice were associated with four phylogenetic groups, the α and γ subdivisions of the proteobacteria, the

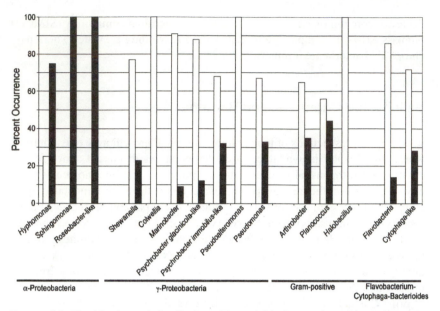

FIGURE 4 2. The identity and distribution of bacterial isolates cultured from Antarctic sea ice (□) and underlying seawater (■). Adapted from Bowman et al. (9).

Gram-positive branch, and the *Cytophaga-Flexibacter-Bacteroides* (CFB) group (5, 9). However, seawater and sea ice isolates differed in two major respects. Sea ice samples yielded significantly higher proportions of psychrophiles (45%), while under-ice seawater samples yielded predominantly psychrotolerant isolates (9, 28). Secondly, the psychrophilic taxa isolated from under-ice water were in general phylogenetically distinct from those in sea ice. The under-ice psychrophilic populations consisted of a gas vacuolate strain closely related to *Roseobacter algicola* and a *Planococcus* phenotype (9, 23). In contrast, psychrophilic isolates from the phylogenetic groupings *Colwellia*, *Shewanella*, and *Psychrobacter* were isolated from sea ice (9) (Figure 4.2).

Psychrotolerant isolates were taxonomically similar from each location, consisting of species of the following genera: *Shewanella*, *Psychrobacter*, *Pseudomonas*, *Planococcus*, and *Arthrobacter* (9). However, some psychrotolerant genera were associated with specific environments; *Halomonas*, *Hyphomonas*, and *Sphingomonas* were from under-ice water while *Pseudoalteromonas* and *Marinobacter* were isolated almost exclusively from sea ice. Of further interest was the division between taxonomic groupings in general. Only members of the α-proteobacteria were dominant in under-ice water while sea ice harbored comparatively high proportions of genera from the γ-proteobacteria, the Gram-positive branch, and the CFB phylum (Figure 4.2).

Abyzov et al. (see chapter 16) have detailed the biodiversity of culturable isolates from the profile of the Vostok ice core. They report the dominance of Gram-positive cocci, small rods, cyanobacteria, and actinobacteria from all horizons of the core profile. In addition, species of the CFB group were associated with specific regions of ice containing high levels of particulates. The dominance of Gram-positive spore-forming species (around 50% of isolates) was supported by the molecular survey of Christner et al. (see chapter 15), while isolates of *Arthrobacter* and *Sphingomonas* also were detected, similar to those found from Antarctic seawater and occasionally sea ice.

The culture studies of Antarctic sea ice have been supported by one molecular survey (12). While heterogeneity at the bacterial genus/species-level between ice samples was highlighted, it is of interest to note that the majority of phylotypes identified from sea ice represented known heterotrophic species, as was the case from glacial ice (see chapter 15). This is in contrast to molecular studies of seawater environments, which usually yield a high proportion of phylotypes with no cultivated relatives (41).

In conclusion, glacial ice appears dominated by Gram-positive, spore-forming bacteria, which occur to a lesser extent in sea ice and seawater. Members of the CFB group also occur in both ice types, yet examples of the α-proteobacteria appear limited to glacial ice and seawater only. In contrast, γ-proteobacteria dominate sea ice. It is also salient to note that the majority of isolates from glacial ice can be classed as psychrotolerant bacteria (as is the case for seawater) while sea ice is dominated by psychrophilic populations. It may be relevant to consider these differences in biodiversity in terms of how bacteria enter each ice environment, and how such bacteria respond to the developing habitat over time.

Psychrophiles versus Psychrotolerants in Ice

Community Establishment

Glacial ice represents an environment in which bacteria are "seeded" into the ice matrix from the deposition of airborne particles onto the ice surface, a "top-down" process in which particles and cells are then incorporated into the ice matrix. Sea ice is the opposite, where bacteria and seawater particulates are incorporated into forming sea ice from the ice-water interface at the bottom of the ice profile.

While it may be facile to assume that ice will select a unique microbial community from the microorganisms that are incorporated into it, the selective pressures that result in this fact are not readily apparent. The establishment of the sea ice bacterial "inocula" may be a largely unselective physical process, although factors such as cell size or bacterial association with algae may be involved. The results of Helmke and Weyland (28) may be indicative of community development. This study reported the dominance of psychrotolerant isolates from seawater samples, in which ice formation had occurred. Dominance of psychrotolerant isolates also was apparent in young ice, supporting the proposition that bacterial inclusion in forming sea ice is unselective at least in terms of community temperature response. However, isolates from older ice were predominantly psychrophilic. This shift in population distribution was more evident in consolidated sea ice cores, where greater than 90% of the isolates were psychrophiles. A similarly high percentage of psychrophiles (ca. 87%) was isolated from consolidated fast ice (5, 9). While much fewer psychrophiles (21%) were reported from a combination of water, sediment, and sea ice samples (33), the majority of evidence points to a selection of psychrophilic bacteria in established sea ice. Therefore, a bacterial community distinct from that of the underlying water column develops within the sea ice.

Where the initial bacteria in sea ice are derived from physical concentration processes, bacterial growth then is responsible for the proliferation of bacterial numbers and biomass. The bacterial community appears to be metabolically active from an early stage of ice development. This may be a key difference in the development of sea ice and glacial ice communities. A laboratory study of ice formation led to the conclusion that algae and particularly bacteria experienced metabolic inhibition during the ice freezing and formation process. However, after two weeks of incubation, bacterial and algal activity was reestablished in the artificial sea ice (26). Analogous studies of glacial ice systems have yet to be undertaken. Active metabolism of sea ice bacteria under simulated in situ conditions also was demonstrated by Sullivan et al. (60). Numerous studies have indicated that established sea ice represents a concentrated source of actively growing bacteria, compared with the underlying water column. Within pack ice from the Weddell Sea, bacterial concentrations were two to nineteen times higher than in underlying waters, with bacterial growth rates from autumn pack ice similar to that from underlying seawater (35). Within the bottom section of fast ice in McMurdo Sound during summer, bacterial production was equivalent to that in several meters of the underlying seawater, with growth rates similar to or exceeding that of the bacterioplankton beneath the ice (34). A central difference in driving the formation of such a metabolically active community may be the primary production provided by sea ice diatom populations. Primary photosynthetic producers have been found in both glacial ice and permafrost communities (see chapters 10 and 16). However, it appears that photosynthetic metabolism is not apparent under in situ conditions. The continued provision of photosynthetically derived nutrients may significantly influence the comparative development of sea ice versus permafrost or glacial ice communities. This factor may be further compounded by the differences in the size and extent of solution channels in glacial ice limiting the diffusion and replenishment of which nutrients are present.

Thus, sea ice supports a metabolically active bacterial community of high cell density that is dominated by culturable psychrophiles. The unanswered question is, what factor(s) of the forming sea ice environment lead to the favored development of psychrophiles in sea ice, when they represent only a minority of the culturable isolates in the underlying seawater?

Nutrient or Physical Selection?

There is an implied linkage between the provision of nutrients by microalgae and the biodiversity of sea ice bacterial communities (5, 9, 12, 28, 46). This association derived from the pioneering work of Sullivan and others in discerning the food web interaction between microalgal primary production and bacterial secondary production in Antarctic sea ice over the summer bloom period. Total bacterial cell numbers and activity closely followed trends in chlorophyll *a* content in sea ice, while many bacterial cells appeared intimately associated with microalgae (25, 28, 34, 59). This finding also is true of Antarctic lake ice and permafrost ecosystems (see chapters 3 and 7). Many workers then extrapolated the link between microalgal primary production and total bacterial numbers to imply an association between high nutrient levels and the composition of the bacterial community within sea ice.

Such a hypothesis bears some analysis. There are three salient questions: (i) does sea ice contain high levels of available organic matter? (ii) what is the temporal and spatial distribution of organic nutrient, and is it available to influence sea ice microbial community development? and (iii) what is the evidence for a selective effect on microbial community development in the presence of high organic levels?

Sea ice is characterized by high organic nutrient levels for significant periods of the annual cycle due to algal primary production. Microalgal communities are extremely active, particularly during spring, when major blooms occur due to increasing levels of sunlight (36). Estimates of total dissolved organic matter (DOM) released by sea ice microalgal communities range from 4 to 32% of primary production (45), creating an environment where dissolved organic carbon (DOC) and dissolved organic nitrogen (DON) levels in sea ice can be up to thirty and eight times higher than surface waters, respectively (49, 63).

What is the evidence for the selection of psychrophiles in environments with high organic nutrient levels? Antarctic marine bacteria also are exposed to periods of high nutrient levels in association with phytoplankton blooms. In particular, blooms of *Phaeocystis* spp. are common in Antarctic coastal waters, and are responsible for massive releases of DOC, in excess of 100 mg C · L^{-1} compared to average nonbloom concentrations of 1.5 mg C · L^{-1} (15, 57). Blooms of *Phaeocystis* spp. also have shown a close coupling with pelagic and epiphytic bacterial populations on several occasions (2, 47). Attached bacteria from seawater occur in a similar proportion to those found in sea ice, with a seasonal change in the ratio of free-living and particle-associated cells in seawater that parallel total bacterial abundance (16). Particle-associated bacteria are closely linked to the CFB phylum in nonpolar marine areas (14, 18), where they are actively involved in the degradation of complex organic material and a poor utilizer of simple DOC (1, 13, 51). Interestingly, representatives of the CFB phylum are commonly isolated from particulates frozen within the accretion ice of Lake Vostok, where they are thought to have been transferred from glacial bedrock (see chaptesr 16 and 17).

From observations of springtime algal blooms in the Arctic, it was hypothesized that psychrophilic and oligotrophic bacteria with high affinity uptake mechanisms are the first group to respond to increased DOM. However, as DOM levels increased, this population was soon out-competed by psychrotolerant copitrophs that possess uptake mechanisms with higher maximum rates (63). Hence, evidence exists to suggest that high organic nutrient levels may in fact encourage the dominance of psychrotolerant populations. But this does not explain the dominance of psychrophiles in the sea ice ecosystem. If nutrients are not a central selective factor, alternatives may include physico-chemical factors, such as variation in salinity during ice formation (43).

Physical Factors

As discussed earlier, the process of ice formation includes the expulsion of dissolved salts/ionic species from the ice crystal matrix into the surrounding intercrystal solution. For glacial ice this depends on the composition of deposited (snowfall) material and for permafrost the nature of the soil type. In sea ice, the amount of salts initially entrapped by the enlarging ice matrix depends heavily on the interaction between water salinity and the ice growth rate. While few studies have been able to assess the *in situ* salinity of ice brines, the salinity is broadly dependent on the ice temperature and age. The salinity directly reflects the dependence on ice temperature for "closed" brine networks. For "open" networks there may be significant gravity drainage of brines from other areas of the ice profile and into the water column as the ice ages. During spring/summer the process is reversed, with meltwater forming distinct freshwater layers under the ice (37) (Figure 4.3). Hence while the magnitude of brine salinity is largely determined during ice formation, the resultant brines may remain relatively stable in response to stable ice temperatures.

In the under-ice water, where brine drainage and meltwater impact, variations in the salinity regime is important to the survival and viability of sea ice microalgae (50) and marine bacteria, particularly psychrophiles (39). Within the Kiel Bight, water bodies of different salinity (seawater and terrestrial fresh-

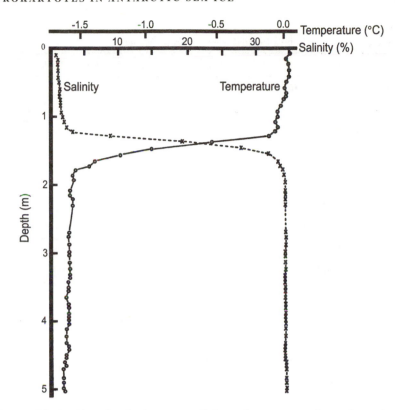

FIGURE 4.3. Profiles of under-ice water salinity and temperature observed at a site near Syowa Station during summer. Adapted from Satoh et al. (55).

water runoff) meet and mix. Water bodies of different salinity contained distinct bacterial populations with different salinity optima, which were termed "marine" (salinity optima 30–4⁰/oo) and "brackish water" (optima 10–25⁰/oo). Similar effects were noted for the Baltic and North Seas (48). The magnitude of salinity changes described in the above examples, for both estuarine and brackish environments, also occurs in Antarctic coastal waters associated with the annual ice cycle (Figure 4.3). A similar, but more dynamic, system may exist in the relationship between glacial ice and subglacial lakes. For example, glacial ice is known to melt into Lake Vostok along the northern ice-water interface and accretion ice freezes onto the glacial surface in central and southern regions of the lake (see chapter 17). The effects of melting and/or freezing may further influence the community composition of glacial accretion ice and subglacial lakes (see chapters 6, 15, 16 and 17).

The effects of salinity variation on bacteria in sea ice or under-ice water also may be significant. The majority of novel taxa isolated from sea ice are

TABLE 4.1.
Characteristics of novel taxa isolated from Antarctic sea ice

	Psychrophile	Halophile*	PUFA†	Reference
Cellulophaga algicola	+	+	–	4
Colwellia demingiae	+	+	+	6
Colwellia hornerae	+	+	+	6
Colwellia rossensis	+	+	+	6
Flavobacterium gillisiae	–	–	–	48
Gelidibacter algens	+	+	–	7
Gelidibacter sp. IC158	–	–	–	9
Glaciecola punicea	+	+	–	8
Glaciecola pallidula	+	+	–	8
Octadecabacter antarcticus	+	+	–	27
Planococcus mcmeekinii	–	–	–	35
Polaribacter franzmannii	+	+	–	28
Psychrobacter antarcticus	+	+	?	51
Psychroflexus torquis	+	+	+	10
Psychroflexus gondwanense	–	–	–	10
Psychroserpes burtonensis	+	+	–	7
Shewanella gelidimarina	+	+	+	11
Shewanella frigidimarina	–	+	+	11

* Requires sodium chloride for growth.
† Produces polyunsaturated fatty acids.

not only psychrophilic but also require sodium chloride for growth (Table 4.1). Psychrophiles from both sea ice and under-ice water also are characterized by a narrow salinity range for growth, while psychrotolerant isolates exhibit a much broader range and no requirement for sodium chloride (5, 9). This is demonstrated by the comparison of the growth characteristics of the two sea ice bacteria, *Glaciecola punicea* (a psychrophile) and *Gelidibacter* sp. IC158 (a psychrotolerant bacterium) (Figure 4.4). The water activity range for growth of *Gl. punicea* was significantly extended when grown at suboptimal compared to optimal growth temperature. No such extension was observed for the psychrotolerant *Gelidibacter* sp. IC158 grown at analogous temperatures. This may indicate an adaptive preference of the psychrophile to a broad salinity range at suboptimal temperature. The implication is that the salinity regime within sea ice may be an important parameter controlling bacterial community development. Likewise, the concentration of ionic substances within the grain boundaries of glacial ice and the liquid water layer of permafrost may significantly influence the growth and/or survival of bacteria.

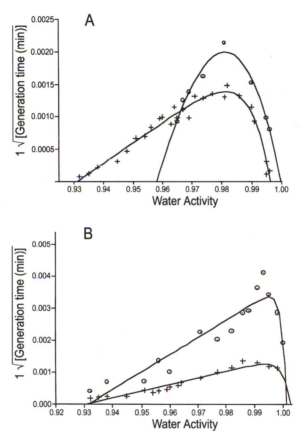

FIGURE 4.4. Square root plot of 1/growth rate versus water activity for (A) *Glaciecola punicea* ACAM 611[T] grown at: suboptimal [8.0 °C (+)] and optimal [15.0 °C ⊙] growth temperatures and (B) *Gelidibacter* sp. IC158 grown at suboptimal [8.0 °C (+)] and optimal [25.0 °C ⊙] growth temperatures. Points represent experimentally determined data while the line represents the predicted value for the entire biokinetic temperature range as determined from fitting the Ratkowsky (square root) growth rate model to the experimental data. Adapted from Nichols et al. (42).

Adaptation of Ice Psychrophiles

Cell membrane function requires an effective ("fluid") bilayer under environmental conditions (58). The functionality of bacterial membranes is modulated through the manipulation of membrane phospholipid composition. Many adaptive strategies are employed by bacteria to achieve this end, most involving the modification of phospholipid fatty acid composition (52). A common strategy utilized by sea ice psychrophiles is the production of polyunsaturated fatty acids (PUFA) (Table 4.1), which possess a low melting point that allows the membrane to remain more "fluid" at lower temperatures (53).

The proportion of PUFA in bacterial membrane lipids changes with growth temperature, and may be maximal at the lowest growth temperatures (44). However, using model lipid systems, the presence of more than two double bonds in a membrane lipid has little additional effect on lowering the melting temperature and even monounsaturated phospholipid species in bacterial membranes remain fluid at a few degrees below zero (52). Multiple double bonds restrict acyl-chain rotation and create a structural element that influences the packing order of the acyl chains (31, 32). PUFA may be more relevant to the maintenance of the correct lipid phase (i.e., a homeophasic adaptive response), as many common bacterial phospholipid classes (including unsaturated phosphatidylethanolamine, a major lipid in most Gram-negative species) favor the formation of nonbilayer phases (e.g., hexagonal-II phase) that would disrupt membrane packing and permeability. The multiple double bonds in PUFA may allow sufficient molecular motion of the acyl chains to lower the melting temperature, thereby keeping the membrane fluid, but concomitantly provide a higher degree of packing order to prevent nonbilayer phases (54). The important factor of cell membrane regulation may lay not in a single parameter (namely, "fluidity") but in the adjustment of membrane lipid composition to balance factors driving the melting transition with those favoring the bilayer to nonbilayer phase transition (27).

The association of PUFA-producing isolates with Antarctic sea ice may not be with the environment per se, but instead reflect the association of specific bacterial populations (*Colwellia*, *Shewanella*, and *Cytophaga-Flavobacterium-Bacteroides* [CFB] species) with a common environmental niche of constant low temperature, and possibly other factors (54). The description of the novel sea ice taxa *Shewanella gelidimarina* and *S. frigidimarina* (11) highlighted an interesting correlation between environment and physiology. Comparison of the 16S rDNA phylogenetic tree for the genus *Shewanella* with the known ability of different species to produce PUFA implies that the progenitor of the genus possessed this ability (Figure 4.5). Although the ability to synthesize PUFA has been retained by all the psychrophilic and several of the psychrotrophic species, this ability (or the expression of PUFA synthesis) has disap-

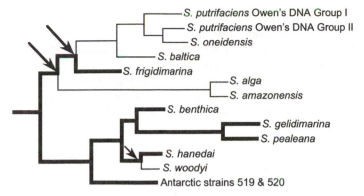

FIGURE 4.5. Schematic representation of the unrooted 16S rDNA phylogenetic tree
for the genus *Shewanella*. Species known to produce polyunsaturated fatty acids
(PUFA) and their lines of descent are in bold. Species and lines of descent known not
to produce PUFA are thin. Adapted from Russell and Nichols (54) and Bowman (3).

peared on two distinct occasions during the divergence of psychrotolerant and/
or mesophilic species. Another apparent coincidence is that the divergent mesophilic species also have lost their halophilic character. The apparent conservation of PUFA genes in this genus may also point to the potential utilization
of these "functional genes" as phylogenetic markers in culture-independent
ecological studies (54). Hence, the strategy of PUFA production appears phylogenetically linked to a group of marine-related bacteria, which often come to
dominate sea ice microbial communities. Chapter 15 deals with apparent survival strategies for common nonpsychrophilic glacial bacteria, including spore
formation and thicker cell walls. It would be interesting to investigate the
species of CFB bacteria isolated from the accretion ice of Lake Vostok (see
chapter 16) with many known psychrophilic, PUFA-producing species. Similar
strategies may be utilized by this group of bacteria.

Conclusions

In conclusion, studies of Antarctic sea ice have revealed that a unique microbial environment, which forms and dissipates in an annual cycle. During formation a microbial community distinct from that of the underlying water column develops that is dominated by bacterial populations of psychrophilic
character. The selective forces responsible for this domination of the bacterial
community by psychrophilic species is yet to be fully resolved. However, the
evidence presented argues for the significant role of physico-chemical parameters in the selection of psychrophilic bacterial populations. Physiological in-

vestigations of these bacteria have revealed that a common adaptation to the cell membrane of psychrophiles is the inclusion of polyunsaturated fatty acids.

While sea ice represents a transient ice environment influenced by large physico-chemical changes (nutrient, salinity) over comparatively short time-scales, it provides us with a microcosm in which to observe how changes in an ice environment may influence microbial communities and may serve as a model system for environments with a dynamic ice-water interface.

Acknowledgments

The National Science Foundation and Bowling Green State University are gratefully acknowledged for financial support to attend the workshop.

Literature Cited

1. Amon, R.M.W., H.-P. Fitznar, and R. Benner. 2000. Linkages among the bioreactivity, chemical composition, and diagenetic state of marine dissolved organic matter. Limnol. and Oceanogr. 46: 287–297.
2. Billen, G., and A. Fontigny. 1987. Dynamics of a Phaeocystis-dominated spring bloom in Belgian coastal waters. II. Bacterioplankton dynamics. Mar. Ecol. Prog. Ser. 37: 249–257.
3. Bowman, J.P. 2002. Genus *Shewanella*. In N.R. Krieg and J.G. Holt (eds.), Bergey's Manual of Systematic Bacteriology. Williams & Wilkins, Baltimore.
4. Bowman, J.P. 2000. Description of *Cellulophaga algicola* sp. nov., isolated from the surfaces of Antarctic algae, and reclassification of *Cytophaga uliginosa* (ZoBell and Upham 1944) Reichenbach 1989 as *Cellulophaga uliginosa* comb. nov. Int. J. Syst. Evol. Microbiol. 50: 1861–1868.
5. Bowman, J.P., M.V. Brown, and D.S. Nichols. 1997. Biodiversity and ecophysiology of bacteria associated with sea ice. Ant. Sci. 9: 134–142.
6. Bowman, J.P., J.J. Gosink, S.A. McCammon, T.E. Lewis, D.S. Nichols, P.D. Nichols, J.H. Skerratt, J.T. Staley, and T.A. McMeekin. 1998. *Colwellia demingae* sp. nov., *Colwellia hornerae* sp. nov., *Colwellia rossensis* and *Colwellia psychrotropica* sp. nov.: psychrophilic, marine bacteria with the ability to synthesize docosahexaenoic acid (22:6w3). Int. J. Syst. Bacteriol. 48: 1171–1180.
7. Bowman, J.P., S.A. McCammon, J.A. Brown, P.D. Nichols, and T.A. McMeekin. 1997. *Psychroserpens burtonensis* gen. nov., sp. nov., and *Gelidibacter algens* gen. nov., sp. nov., psychrophilic bacteria isolated from Antarctic lacustrine and sea ice habitats. Int. J. Syst. Bacteriol. 47: 670–677.
8. Bowman, J.P., S.A. McCammon, J.L. Brown, and T.A. McMeekin. 1998. *Glaciecola punicea* gen. nov., sp. nov., and *Glaciecola pallidula* gen. nov., sp. nov.: Psychrophilic bacteria from Antarctic sea-ice habitats. Int. J. Syst. Bacteriol. 48: 1213–1222.

9. Bowman, J.P., S.A. McCammon, M.V. Brown, D.S. Nichols, and T.A. McMeekin. 1997. Diversity and association of psychrophilic bacteria in Antarctic sea ice. Appl. Environ. Microbiol. 63: 3068–3078.

10. Bowman, J.P., S.A. McCammon, T.E. Lewis, J.H. Skerratt, J.L. Brown, D.S. Nichols, and T.A. McMeekin. 1998. *Psychroflexus torquis* gen. nov., a psychrophilic species from Antarctic sea ice, and reclassification of *Flavobacterium gondwanense* (Dobson et al. 1993) as *Psychroflexus gondwanense* gen. nov., comb. nov. Microbiology 144: 1601–1609.

11. Bowman, J.P., S.A. McCammon, D.S. Nichols, J.S. Skerratt, S.M. Rea, P.D. Nichols, and T.A. McMeekin. 1997. *Shewanella gelidimarina* sp. nov. and *Shewanella frigidimarina* sp. nov. novel species with the ability to produce eicosapentaenoic acid (20:5w3) and grow anaerobically by dissimilatory Fe(III) reduction. Int. J. Syst. Bacteriol. 47: 1040–1047.

12. Brown, M.V., and J.P. Bowman. 2001. A molecular phylogenetic survey of sea ice microbial communities (SIMCO). FEMS Microbiol. Ecol. 1226: 1–9.

13. Cottrell, M.T., and D.L. Kirchman. 2000. Natural assemblages of marine proteobacteria and members of the Cytophaga-Flavobacter cluster consuming low- and high-molecular-weight dissolved organic matter. Appl. Environ. Microbiol. 66: 1692–1697.

14. Crump, B.C., E.A. Armbrust, and J.A. Baross. 1999. Phylogenetic analysis of particle-attached and free-living bacterial communities in the Columbia River, its estuary, and the adjacent coastal ocean. Appl. Environ. Microbiol. 65: 3192–3204.

15. Davidson, A.T., and H.J. Marchant. 1992. Protist abundance and carbon concentration during a Phaeocystis-associated bloom at an Antarctic coastal site. Polar Biol. 12: 387–395.

16. Delille, D. 1993. Seasonal changes in the abundance and composition of marine heterotrophic bacterial communities in an Antarctic coastal area. Polar Biol. 13: 463–470.

17. Delille, D., and E. Perret. 1989. Influence of temperature on the growth potential of southern polar marine bacteria. Microbial Ecol. 18: 117–123.

18. DeLong, E.F., D.G. Franks, and A.L. Alldredge. 1993. Phylogenetic diversity of aggregate-attached vs. free-living marine bacterial assemblages. Limnol. and Oceanogr. 38: 924–934.

19. Franzmann, P.D., and S.J. Dobson. 1993. The phylogeny of bacteria from a modern Antarctic refuge. Ant. Sci. 5: 267–270.

20. Garrison, D.L., and K.R. Buck. 1991. Surface-layer sea ice assemblages in Antarctic pack ice during the austral spring: Environmental conditions, primary production and community structure. Mar. Ecol. Prog. Ser. 75: 161–172.

21. Garrison, D.L,. and A.R. Close. 1993. Winter ecology of the biota in Weddell Sea pack ice. Mar. Ecol. Prog. Ser. 96: 17–31.

22. Gosink, J.J., R.P. Herwig, and J.T. Staley. 1997. *Octadecabacter arcticus* gen. nov., sp. nov., and *O. antarcticus*, sp. nov., nonpigmented, psychrophilic gas vacuolate bacteria from polar sea ice and water. Syst. Appl. Microbiol. 20: 356–365.

23. Gosink, J.J., and J.T. Staley. 1995. Biodiversity of gas vacuolate bacteria from Antarctic sea ice and water. Appl. Environ. Microbiol. 61: 3486–3489.

24. Gosink, J.J., C.R. Woese, and J.T. Staley. 1998. *Polaribacter* gen. nov., with three new species, *P. irgensii* sp. nov., *P. franzmannii* sp. nov. and *P. filamentus* sp.

nov., gas vacuolate polar marine bacteria of the Cytophaga-Flavobacterium-Bacterioides group and reclassification of "*Flectobacillus glomeratus*" as *Polaribacter glomeratus* comb. nov. Int. J. Syst. Bacteriol. 48: 223–235.

25. Grossi, S.McG., S.T. Kottmeier, and C.W. Sullivan. 1984. Sea ice microbial communities III. Seasonal abundance of microalgae and associated bacteria, McMurdo Sound, Antarctica. Microbial Ecol. 10: 231–242.

26. Grossmann, S., and M. Gleitz. 1993. Microbial responses to experimental sea ice formation: Implications for the establishment of Antarctic sea ice communities. J. Exp. Mar. Biol. Ecol. 173: 273–289.

27. Hazel, J.R. 1995. Thermal adaptation in biological membranes: Is homeoviscous adaptation the explanation? Ann. Rev. Physiol. 57: 19–42.

28. Helmke, E., and H. Weyland. 1995. Bacteria in sea ice and underlying water of the eastern Weddell Sea in midwinter. Mar. Ecol. Prog. Ser. 117: 269–287.

29. Junge, K., J.J. Gosink, H.-G. Hoppe, and J.T. Staley. 1998. *Arthrobacter, Brachybacterium* and *Planococcus* isolates identified from Antarctic sea ice brine. Description of *Planococcus mcmeekinii* sp. nov. Syst. Appl. Microbiol. 21: 306–314.

30. Kennet, J.P., and N.C. Shackleton. 1976. Oxygen isotope evidence for the development of the psychrosphere 38 Myr ago. Nature 260: 513–515.

31. Keough, K.M.W., B. Giffin, and N. Kariel. 1987. The influence of unsaturation on the phase transition temperatures of a series of heteroacid phosphatidylcholines containing twenty-carbon chains. Biochim. Biophys. Acta 902: 1–10.

32. Keough, K.M.W., and N. Kariel. 1987. Differential scanning calorimetric studies of aqueous dispersions of phosphatidylcholines containing two polyenoic chains. Biochim. Biophys. Acta 902: 11–18.

33. Kobori, H., C.W. Sullivan, and H. Shizuya. 1984. Bacterial plasmids in Antarctic natural microbial assemblages. Appl. Environ. Microbiol. 48: 515–518.

34. Kottmeier, S.T., S.M. Grossi, and C.W. Sullivan. 1987. Sea ice microbial communities. VIII. Bacterial production in annual sea ice of McMurdo Sound, Antarctica. Mar. Ecol. Prog. Ser. 35: 175–186.

35. Kottmeier, S.T., and C.W. Sullivan. 1990. Bacterial biomass and production in pack ice of Antarctic marginal ice edge zones. Deep-Sea Res. 37: 1131–1330.

36. Krist, G.O., and C. Wienke. 1995. Ecophysiology of polar algae. J. Phycol. 31: 181–199.

37. Maykut, G.A. 1985. The ice environment. pp. 21–82. In R.A. Horner (ed.), Sea Ice Biota. CRC Press, Boca Raton, Florida.

38. McCammon, S.A., and J.P. Bowman. 2000. Taxonomy of Antarctic *Flavobacterium* species: Description of *Flavobacterium gillisiae* sp. nov., *Flavobacterium tegetincola* sp. nov., and *Flavobacterium xanthum* sp. nov., nom. rev. and reclassification of *Flavobacterium salegens* as *Salegentibacter salegens* gen. nov., comb. nov. Int. J. Syst. Evol. Microbiol. 50: 1055–1063.

39. Morita, R.Y., L.P. Jones, R.P. Griffiths, and T.E. Staley. 1973. Salinity and temperature interactions and their relationship to the microbiology of the estuarine environment. pp. 221–232. In L.H. Stevenson and R.R. Colwell (eds.) Estuarine Microbial Ecology. University of South Carolina Press, Columbia.

40. Mountfort, D.O., F.A. Rainey, J. Burghardt, J.H. Kaspar, and E. Stackebrandt. 1998. *Psychromonas antarcticus* gen. nov., sp. nov., a new aerotolerant anaero-

bic, halophilic psychrophile isolated from pond sediment of the McMurdo Ice Shelf, Antarctica. Arch. Microbiol. 169: 231–238.

41. Mullins, T.D., T.B. Britischgi, R.L. Krest, and S.J. Giovannoni. 1995. Genetic comparisons reveal the same unknown bacterial lineages in Atlantic and Pacific bacterioplankton communities. Limnol. and Oceanogr. 40: 148–158.

42. Nichols, D.S., A.R. Greenhill, C.T., Shadbolt, T. Ross, and T.A. McMeekin. 1999. Physico-chemical parameters for growth of the sea ice bacteria Glacincola punicea ACAM 611[T] and Gelidibacter sp. IC158. Appl. Environ. Microbiol. 65: 3757–3760.

43. Nichols, D.S., P.D. Nichols, and T.A. McMeekin. 1995. Ecology and physiology of psychrophilic bacteria from Antarctic saline lakes and sea ice. Sci. Prog. 78: 311–347.

44. Nichols, D.S., J. Olley, H. Garda, R.R. Brenner, and T.A. McMeekin. 2000. Effect of temperature and salinity stress on the growth and lipid composition of *Shewanella gelidimarina*. Appl. Environ. Microbiol. 66: 2422–2429.

45. Palmisano, A.C., and D.L. Garrison. 1993. Microorganisms in Antarctic sea ice. pp. 167–218. In E.I. Friedmann (ed.), Antarctic Microbiology. Wiley, New York.

46. Poremba, K., C.-D. Drselen, and T. Stoeck. 1999. Succession of bacterial abundance, activity and temperature adaptation during winter 1996 in parts of the German Wadden Sea and adjacent coastal waters. J. Sea Res. 42: 1–10.

47. Putt, M., G. Miceli, and D.K. Stoecker. 1994. Association of bacteria with *Phaeocystis* sp. in McMurdo Sound, Antarctica. Mar. Ecol. Prog. Ser. 105: 179–189.

48. Rheinheimer, G. 1977. Regional and seasonal distribution of saprophytic and coliform bacteria. pp. 121–137. In G. Rheinheimer (ed.), Microbial Ecology of a Brackish Water Environment. Springer-Verlag, Berlin.

49. Riaux-Gobin, C., P. Tréguer, M. Poulin, and G. Vétion. 2000. Nutrients, algal biomass and communities in land-fast ice and seawater off Adélie Land (Antarctica). Ant. Sci. 12: 160–171.

50. Robineau, B., L. Legendre, M. Kishino, and S. Kudoh. 1997. Horizontal heterogeneity of microalgal biomass in the first-year sea ice of Saroma-ko Lagoon (Hokkaido, Japan). J. Mar. Syst. 11: 81–91.

51. Rossello-Mora, R., Thamdrup, H. Schøfer, R. Weller, and R. Amann. 1999. The response of the microbial community of marine sediments to organic carbon input under anaerobic conditions. Syst. Appl. Microbiol. 22: 237–248.

52. Russell, N.J. 1989. Functions of lipids: Structural roles and membrane functions. pp. 279–365. In C. Ratledge and S.G. Wilkinson (eds.), Microbial Lipids, vol. 2. Academic Press, London.

53. Russell, N.J. 1992. Physiology and molecular biology of psychrophilic micro-organisms. pp. 203–224. In R.A. Herbert and R.J. Sharpe (eds.), Molecular Biology and Biotechnology of Extremophiles. Blackie, Glasgow.

54. Russell, N.J., and D.S. Nichols. 1999. Polyunsaturated fatty acids in marine bacteria—a dogma rewritten. Microbiology 145: 767–779.

55. Satoh, H., K. Fukami, K. Watanabe, and E. Takshashi. 1989. Seasonal changes in heterotrophic bacteria under fast ice near Syowa Station, Antarctica. Can. J. Microbiol. 35: 329–333.

56. Schopf, J.W., J.M. Hayes, and M.R. Walter. 1983. Evolution of Earth's earliest ecosystems: Recent progress and unsolved problems. pp. 361–384. In J.W.

Schopf (ed.), Earth's Earliest Biosphere, its Origin and Evolution. Princeton University Press, Princeton, New Jersey.

57. Scott, F.J., A.T. Davidson, and H.J. Marchant. 2000. Seasonal variation in plankton, submicrometre particles and size-fractionated dissolved organic carbon in Antarctic coastal waters. Polar Biol. 23: 635–643.

58. Singer, S.J., and G.L. Nicolson. 1972. The fluid mosaic model of the structure of cell membranes. Science 175: 720–731.

59. Sullivan, C.W., and C.A. Palmisano. 1984. Sea ice microbial communities: Distribution, abundance and diversity of ice bacteria in McMurdo Sound, Antarctica in 1980. Appl. Environ. Microbiol. 47: 788–795.

60. Sullivan, C.W., A.C. Palmisano, S. Kottmeier, S. McGrath-Grossi, and R. Moe. 1985. The influence of light on growth and development of the sea-ice microbial community of McMurdo Sound. pp. 78–83. In W.R. Siegfried, P.R. Condy, and R.M. Laws (eds.), Antarctic Nutrient Cycles and Food Webs. Springer-Verlag, Berlin.

61. Thomas, D.N., H. Kennedy, C. Haas, and G.S. Dieckmann. 2001. Dissolved carbohydrates in Antarctic sea ice. Ant. Sci. 13: 119–125.

62. Uphoff, H.U., A. Felske, W. Fehr, and I. Wagner-Döbler. 2001. The microbial diversity in picoplankton enrichment cultures: A molecular screening of marine isolates. FEMS Microbiol. Ecol. 35: 249–258.

63. Yager, P.L., T.L. Connelly, B. Mortazavi, K.E. Wommack, N. Bano, J.E. Bauer, S. Opsahl, and J.T. Hollibaugh. 2001. Dynamic bacterial and viral responses to an algal bloom at subzero temperatures. Limnol. and Oceanogr. 46: 790–801.

5

Frozen in Time: The Diatom Record in Ice Cores from Remote Drilling Sites on the Antarctic Ice Sheets

Davida E. Kellogg and Thomas B. Kellogg

DIATOMS are single-celled algae belonging to the Bacillariophyta. Their two-piece silica frustules, resembling Petri dishes in basic structure and distinguished by species-specific markings, pores, and spines, make them objects of beauty. The resistance of those silica frustules to dissolution in cold waters, the long geologic history and widespread distribution of the class, and the varying ecologic tolerances and consequent geographic provinciality and limited stratigraphic ranges of many species render them tailor-made for paleoclimate research in the polar regions. Diatoms live and prosper in both marine and the entire range of nonmarine environments, from super-saline through brackish to fresh waters, and at temperatures ranging from those prevailing in hot springs at Yellowstone Park to those in the interstices of pack ice in polar seas.

The oldest known diatoms are marine species from the Cretaceous (57). Freshwater species do not seem to have appeared until the Miocene, although the range of one species, *Aulacosira granulata*, may extend from the Recent all the way back into the Oligocene. Most marine species are relatively short-lived, making their stratigraphic ranges a standard tool of stratigraphers. Fossil diatomaceous deposits occur on all continents, including Antarctica. The same species-specific ecological tolerances and provenances that allow forensic pathologists to reconstruct the scene of a drowning allow paleoclimatologists to reconstruct climate change (8, 55). The discovery of diatoms in ice cores from the polar ice sheets is a much more recent development, but one that may turn out to be of equal importance for paleoclimatology. Because diatom samples were chemically processed and fixed, we have no way to determine if any of them were alive at the time of collection.

In this chapter, we first review the history of diatom finds in polar ice cores and then discuss their paleoclimatological significance and uses for both current and future research.

Geographic and Temporal Distribution of Diatoms in Polar Ice Cores

The study of diatoms in polar ice cores is in its infancy. This situation is partly due to the recent commencement of drilling operations on the polar ice sheets. So far only a few long ice cores have been made available for micropaleonto-logical studies, mostly as an afterthought to chemical, isotopic, and electrocon-ductivity analyses. There were few indications when drilling was first proposed that polar ice cores would contain diatoms. When serendipity struck in 1995 in the form of an offer of the remains of the 227 PICO (Polar Ice Coring Office) ice core drilled at the South Pole in 1980–1982, reports of diatoms in polar ice could be numbered on the fingers of one hand. Gayley and Ram (21, 22) reported small concentrations of unidentified diatom frustules in a 200-year-old ice core from Crête, Greenland. Harwood (29, 30), Ram et al. (59), and Gayley et al. (23) all reported finding diatoms at the bottom of Greenland ice cores, though Kellogg (unpublished) found none in samples from GISP 2 (Greenland Ice Sheet Project 2).

Vostok and Dome C

Gayley and Ram (21) mentioned that C. Lorius, T.G. Thompson, and L.H. Burckle, had each told them of diatom finds in Antarctic ice cores. Burckle et al. (10, 11) reported small numbers of diatoms in ice cores from Dome C and Vostok. Though always low, concentrations of these species were about twenty times greater in ice dated to the Last Glacial Maximum (LGM) than in Holo-cene ice, raising the possibility that changes in diatom assemblages through time down polar ice cores may indicate large-scale climate modification. Counter to expectations that any diatoms found at remote locations on the polar ice sheets would have been derived from subaerially exposed fossil ma-rine deposits around the coasts of Antarctica and/or the Southern Ocean, 98% of the diatoms from the Dome C and Vostok cores were nonmarine species, some of them (e.g., *Pinnularia shackletonii* and *Navicula deltaica*) endemic to Antarctica, and others (e.g., *Luticola cohnii* and *Cyclotella stelligera*) more wide-ranging. There are no local sources of diatoms near either of these sites because it is too cold for even ephemeral melt pools to develop. Therefore, the authors reasoned, the diatoms in these cores point to both a distant source or sources, possibly as far away as southern South America, and an aeolian mode of transport, namely, entrainment by storm winds spiraling southward from the Polar Front to these remote sites on the Antarctic ice sheets (Figure 5.1). Abyzov et al. (1) found an unidentified nonmarine diatom from the Vostok ice core. Most recently, Sambrotto and Burckle found rare fragmented specimens of *Fragilariopsis kerguelensis*, *Thalassionema nitzschioides*, and *Thalassio-*

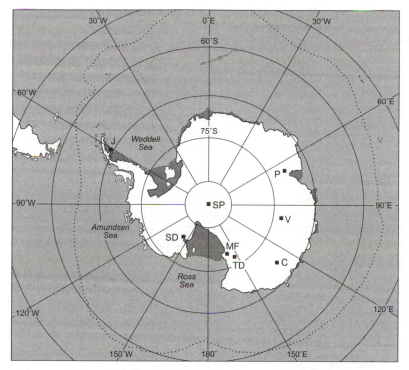

FIGURE 5.1. Map of Antarctica showing locations of ice cores (C = Dome C, SD = Siple Dome, SP = South Pole, TD = Taylor Dome, V = Vostok) and other high-altitude deposits where nonmarine diatoms have been reported (J = James Ross Island; MF = Mt. Feather, Table Mountain and Mt. Crean; P = Pagodroma Formation and Fisher Massif).

sira lentiginosa (Neogene Southern Ocean marine forms), along with nonmarine diatoms in the Vostok ice core (Burckle, pers. comm.).

The available database on which to test these preliminary hypotheses currently consists of only two detailed studies of long ice cores, from the South Pole (40, 42) and Taylor Dome (41).

The South Pole

Diatom concentrations varied between 0 and > 450 specimens/l of melted ice in a patchy distribution down the entire length of the South Pole ice core spanning the past 2000 years, with highest concentrations occurring in levels dated to the Little Ice Age (ca. A.D. 1400–1750) (40). Among the over forty species identified from the South Pole core were marine and nonmarine species known from sediments, lakes, ponds, and seasonal melt pools around Antarctica.

At the South Pole, as at Vostok and Dome C, there are no nearby sources for diatoms, which must be transported to these remote locations by the episodic storms that occur several times annually (6, 62). Independent evidence for aerial transport of diatom frustules to the interior of Antarctica was provided by Burckle and Potter (13), who reported many of the same Plio-Pleistocene marine species in cracks in Antarctic sedimentary and igneous rocks of Devonian to Cretaceous age. Diatoms are mostly under 100 µm in diameter or length, and most of the frustules retrieved from Antarctic ice cores are fragments measuring less than 10 µm. The high degree of fragmentation and dissolution of the majority of South Pole core specimens suggests that they were reworked from subaerially exposed sediments around Antarctica. A few cosmopolitan nonmarine species could have been transported from extra-Antarctic locations. Because individual species could have been transported from more than one source area, we cannot pinpoint the direction from which the winds that deposited them were coming other than to say that they were blowing from the coasts onto the Antarctic ice sheets, in a direction opposite to the predominant katabatic winds that blow down off the polar plateau.

The South Pole Water Well

The South Pole Water Well (SPWW) is a 24-m-diameter subsurface meltwater pool, originally excavated to provide potable water for Amundsen-Scott Station (Figure 5.2). Atmospheric particles falling on the snow surface at the Pole are incorporated into horizontal ice layers, and later liberated as the ice melts downward in the well. The sediment left on the bottom of this pool is a lag deposit of atmospheric particles containing the most concentrated unbiased source of micrometeorites known on Earth (74), but diatoms also are concentrated. Moreover, these deposits are datable based on age-depth relationships of ice strata in the South Pole ice core (46). Ice layers now melting out fell on the snow surface between A.D. 400 and 1500, and the residue from each year's ice melt represents ca. 1000 years of atmospheric deposition.

Meltwater from the SPWW collected in 1995 (from ice dated at A.D. 1100–1500) contains concentrations of nonmarine and marine diatoms similar to that in an entire liter of melted ice core (from a 1-m-long section of a quarter-diameter split of ice core) representing ca. 1000 years' accumulation (40, 41). All of the nonmarine diatoms recovered from this sample are known from Antarctic locations (39, 45). They include *P. shackletonii*, *N. deltaica*, and possibly *Nitzschia westii* (all Antarctic endemics with modern distributions that indicate that the air masses that carried them to the South Pole crossed the coastline in the McMurdo/ Dry Valleys to Northern Victoria Land sector) and *A. granulata* (known from the Antarctic Peninsula and Schirmacher Oasis, but also from circum-Antarctic locations). Marine taxa include several species of

Fragilariopsis, including *F. cylindrus*, fragments of at least two species of *Thalassiosira*, and possible *Coscinodiscus* species.

Taylor Dome and Siple Dome

We were curious as to whether wind-blown diatoms are routinely incorporated into ice at other remote Antarctic locations. Specifically, how widespread are diatoms, how do diatom assemblages and abundances differ between the East (EAIS) and West (WAIS) Antarctic ice sheets, and how do diatom fluxes vary down long ice cores, especially across glacial-interglacial transitions? Tropospheric and stratospheric air masses tend to subside and flow outward at the surface of the EAIS (60), bringing with them any entrained particulates. Snowfall here may be treated as direct samples of Antarctic atmospheric particulates at the time of deposition, and polar ice cores may be viewed as continuous archives of thousands of individual samples of the paleoatmosphere (56). Like the inorganic particulates, diatoms from the Antarctic ice sheets should provide information about the path and strength of the winds that transported them.

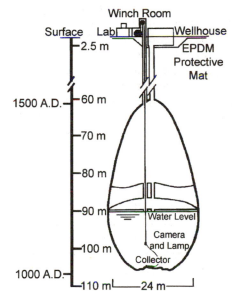

FIGURE 5.2. Diagram of the SPWW (South Pole Water Well) (from S. Taylor et al. [74]). Excess water withdrawn for use at Pole Station is heated and returned to the well to continue melting. Melting occurs at the bottom, at a rate of ca. 10 m/yr equivalent to 100 years of deposition, releasing progressively older diatoms and inorganic particles, which accumulate on the bottom.

This information could provide both independent corroboration for atmospheric circulation models and an independent criterion for choosing between competing hypotheses about climatic history and ice sheet dynamics. The availability of material from snow pits dug preliminary to drilling the long Siple Dome core, and from preliminary snow pits and the 554 m PICO ice core at Taylor Dome (77°47.7′ S, 158°43.1′ E, elevation 2400–2420 m) during 1993–1994 (25) permitted us to begin to pursue these avenues of investigation by extending the search for diatoms to additional locations on the WAIS and EAIS, respectively.

Taylor Dome Ice Core

As at the South Pole, diatoms were distributed throughout the Taylor Dome core, which spans the last 155,000 years back into isotope stage 7 (66). However, because all samples for diatom analysis were taken from below the firn-to-ice transition at 70 to 80 m, the time period sampled does not overlap the interval sampled at the South Pole. Lowest diatom abundances in the Taylor Dome core occurred during glacial intervals (isotope stages 5a-2), and highest abundances were during interstadials (isotope stages 5b, 5d, and 7–6) (41). In all, 6 marine and 45 nonmarine species were recorded. Total abundances ranged from nil to over 660 specimens per sample. Of the 78 samples examined, 26 (33.3%) contained more than 75% marine specimens, 16 (20.5%) are more than 75% nonmarine, and 21 (26.9%) were barren of diatoms. Marine species dominated the Taylor Dome diatom assemblages during cold intervals (e.g., isotope stage 2, at the 5a–4, 5c–5b, and 5d–5c transitions, and stage 6), while nonmarine taxa dominated in the Holocene, at sporadic points in stage 3, and in stages 5b-5a, 5d, and 7. Most species encountered were present in the South Pole core (41), and also have been reported from a variety of other Antarctic sites.

Siple Dome Snow Pits

Several 2-m snow pits were dug at the corners of a local array of markers near the summit of Siple Dome, and at a centrally located camp, in preparation for drilling at Siple Dome. Bulk snow samples from these pits, yielding over 2 l of meltwater/sample, were provided by Pieter Grootes. These snow pit samples represent deposition within the past decade, allowing us to compare patterns of recent windborne diatom deposition at sites on the EAIS (Taylor Dome) and WAIS (Siple Dome). Diatom assemblages in all five snow pit samples are dominated by *Thalassiothrix/Thalassionema* fragments (most probably *Thalassiothrix longissima*), and fragments of a variety of marine centric diatoms. The southwestern, northwestern, and central pits also contained a distinctive mixture of the common Antarctic nonmarine species *Luticola muticopsis* and marine species. The northwestern pits, in which diatoms were most abundant, contained fresh-looking specimens of the Ross Sea Pliocene-Recent indicator species *Fragilariopsis curta*, as well as very fresh specimens of *F. cylindra* still articulated in short chains, the way they grow attached to the underside of pack ice. Both these species are known dominants in the Southern Ocean. Also observed were the nonmarine species *Tabellaria fenestrata*, *Nitzschia gracilis*, and the possibly extra-Antarctic species *Brachysira microcephala*. Most of the species from the snow pits also were found in the South Pole and Taylor Dome ice cores.

Diatom Provenance and Transport

Ninety-one taxa have been reported, most identified to the species level, from the Antarctic ice cores and snow pits (40–42). Many of these taxa are present in one or just a few samples. Of the ninety-one taxa, twenty-six are marine, and sixty-five are nonmarine. The marine species are all known from waters adjacent to Antarctica, and many also have been reported from uplifted marine sediments exposed in ice-free areas of the continent. All but five of the nonmarine taxa have been reported from locations along the Antarctic coastline, from offshore circum-Antarctic islands, or from other southern land masses (e.g., Kerguelen, South America, etc.). The physical condition of individual frustules ranges from poor (broken, abraded, partly dissolved) to excellent (whole, fresh-looking, still connected in chains).

Marine Species

The closest sources for marine diatoms recorded from Antarctic ice cores and snow pits are the oceans adjacent to Antarctica and uplifted diatom-bearing marine deposits of the Late Wisconsin Ross Sea Drift from the Dry Valleys (19, 71), and similar deposits elsewhere on the continent or off-lying Sub-Antarctic islands. Species recorded from the South Pole and Siple and Taylor Domes have been common constituents of the Antarctic marine assemblage for the past million years or more. The one exception is *Denticulopsis hustedtii*, a Miocene species known from sediments in the Dry Valleys. A list of the marine taxa in the Taylor Dome and Siple Dome snow pits and the Taylor Dome and South Pole ice cores was given by D. Kellogg and T. Kellogg (40). We list marine species reported in the literature from other high-altitude Antarctic deposits in Table 5.1. Many of these species, for example, *F. curta* and *F. kerguelensis*, are endemic to Southern Ocean waters. Others, for example, *T. nitzschioides* and *Paralia sulcata*, are well known from waters outside the Antarctic. Even *Pseudoeunotia doliolus*, a subtropical Pleistocene species, was

TABLE 5.1.
Marine diatoms in high-altitude Antarctic deposits

Taxon	1	2	3	4	5	6	7	8	9	10	11	12	13	14	15	16	17	18	19
Actinocyclus actinochilus		x	x				x								x				
Actinocyclus ingens	x		x				x				t?				x				
Actinoptychus spp.									x	x									
Actinoptychus senarius															x				
Arachnoidiscus spp.															x				
Auliscus spp.									x	x									
Biddulphia spp.									x										
Biddulphia weissflogii															x				
Cestodiscus spp.	x																		
Chaetoceros spp.							x								x				
Chaetoceros spp. spores and setae		x													x	x			
Cocconeis spp.													c	f			c		
Coscinodiscus spp.							x				f				x	x			x
Coscinodiscus elliptopora		x																	
Coscinodiscus furcatus								x											
Coscinodiscus kolbei		x																	
Coscinodiscus lineatus											f?		x						
											t?		?						
Coscinodiscus marginatus							x				f?		x		x		x		
Coscinodiscus oculus-iridis															x				
Coscinodiscus oligocenicus						x													
Coscinodiscus perforatus											f		x						
Coscinodiscus radiatus											c?								
Coscinodiscus sublineatus											t?								
Coscinodiscus superbus		x																	
Coscinodiscus vulnificus	x	x													x				
Cosmiodiscus insignis		x																	
Cymatosira biharensis							x												
Dactyliosolen antarcticus															x				
Denticula spp.											t								
Denticulopsis dimorpha		x															x		
Denticulopsis hustedtii					x		x												
Denticulopsis simonsenii															x				
Diploneis spp.															x				
Eucampia spp.															x				
Eucampia antarctica		x			x						t				x	x			

TABLE 5.1. (*cont'd*)
Marine diatoms in high-altitude Antarctic deposits

Taxon	1	2	3	4	5	6	7	8	9	10	11	12	13	14	15	16	17	18	19
Fragilariopsis curta														t					
Fragilariopsis cylindra																		v	
Fragilariopsis kerguelensis					x						t					x		c	
Fragilariopsis obliquecostata														t					
Fragilariopsis ritscheri											t								
Fragilariopsis separanda											t		x						
Fragilariopsis sublinearis														t					
Hemiaulus spp.		x																	
Hemiaulus caracteristicus		x																	
Hemiaulus incisus		x																	
Hemiaulus pacificus		x																	
Hemiaulus polycistinorum		x																	
Hemidiscus karsteni				x															
Hyalodiscus obsoletus																x			
Isthmia spp.															x				
Kisseleviella carina		x																	
Melosira charcotii									x	x	f?								
Naviculopsis biapiculata		x																	
Nitzschia spp.							x		x	x	ft		x			x		v	
Nitzschia angulata		x																	
Nitzschia grunowii											t?								
											c?								
Nitzschia scabra																x			
Nitzschia turgiduloides														t					
Paralia sulcata							x									x			
Pinnularia quadratarea																x			x
Podosira spp.									x		t?								
Pseudopyxilla dubia		x																	
Pseudotriceratium radiosoreticulatum		x																	
Pyxilla spp.		x																	
Pyxilla johnsoniana		x																	
Pyxilla prolungata		x																	
Pyxilla reticulata		x																	
Rhabdonema spp.																x			
Rhaphoneis spp.																x			
Rhizosolenia spp.							x					f							

TABLE 5.1. (*cont'd*)
Marine diatoms in high-altitude Antarctic deposits

Taxon	1	2	3	4	5	6	7	8	9	10	11	12	13	14	15	16	17	18	19
Rhizosolenia styliformis							x												
Rocella gelida						x													
Rocella praenitida		x																	
Rocella vigilans						x													
Rossiella symmetrica						x													
Rouxia spp.							x												
Rouxia peragalli v. antarctica					x														
Sceptroneis spp.		x																	
Simonseniella barboi																x			
Stellarima microtrias							x								x	x			
Stephanopyxis spp.														f	x				
Stephanopyxis eocenica		x																	
Stephanopyxis grunowii		x					x				t?								
Stephanopyxis hyalomarginata		x																	
Stephanopyxis superba		x																	
Stephanopyxis turris							x												
Synedra bradyi							x												
Synedra jouseana		x				x													
Thalassionema spp.		x													x				
Thalassionema nitzschioides							x		x	x	ftc	x	x				x		
Thalassionema/ Thalassiothrix spp															x				
Thalassiosira spp.							x						x		x				
Thalassiosira frenguelliopsis											f?	x							
Thalassiosira gracilis																	x		
Thalassiosira inura															x				
Thalassiosira kolbeii															x				
Thalassiosira lentiginosa	x	x		x	x						t?		x		x		x		
Thalassiosira lineata																	x		
Thalassiosira oestrupii			x	x							t?								
Thalassiosira oliverana											t				x				
Thalassiosira torokina	x	x									t				x				

TABLE 5.1. (*cont'd*)
Marine diatoms in high-altitude Antarctic deposits

Taxon	1	2	3	4	5	6	7	8	9	10	11	12	13	14	15	16	17	18	19
Thalassiosira trifulta									f?		x								
Thalassiosira tumida											x								
Thalassiothrix spp.																x			
Thalassiothrix antarctica									tc	x									
Thalassiothrix longissima					x														
Triceratium unguiculatum	x																		
Trichotoxon reinboldii							x	x	ft		x								
Trinacria excavata	x				x														
Vulcanella hannae	x																		
Xanthiopyxis ovalis					x														

Note: c = common; f = few; t = trace; x = present; ? = questionable identification.
Locations and References:
1—Horlick Mts. Sirius (36).
2—Tillite Spur Sirius (28, 31–33).
3—Quartz Hills Sirius (28, 31–33).
4—Plunket Point Dominion Range Sirius (36).
5—Mt. Feather Sirius (36, 70).
6—Mt. Sirius (36, 42).
7—Elephant Moraine Northern Victorialand (27).
8—Mt. Feather Sirius (48).
9—Sirius Till Mt. Feather (4).
10—Sirius Group Mt. Feather (2).
11—Regolith Mt. Feather (f), Table Mt. (t), or Mt. Crean (c)(4).
12—Mt. Feather Permian sandstone (2).
13—Mt. Feather ablation till (2).
14—Snow, Mt. Feather (f) and/or Table Mountain (t) and/or Mt. Crean (c)(2, 4).
15—Mt. Fleming Sirius.
16—Fisher Massif Prince Charles Mountains (60).
17—Meteorites, Allen Hills/Queen Alexandra Range (12).
18—Vostok Ice Core (not identified by authors) (1) (v) ; Dome C Ice Core (17, 18) (c).
19—James Ross Island igneous rocks (16).
*—This occurrence may not represent just wind-transported diatoms but rather probably represents original deposition on the ice sheet from winds, followed by flow within the ice sheet following the mechanism proposed by (20, 57).

reported in Antarctic plankton by van Heurck (77). Only *Chaetoceros diadema*, represented by a single specimen in one of the Taylor Dome snow pits, has not yet been reported from the Antarctic. Because most marine specimens in the ice-core samples are damaged, we suspect that the majority of them came from uplifted marine deposits, such as outcroppings of Ross Sea Drift in the Dry Valleys (71). Ross Sea Drift contains abundant diatoms and diatom fragments (Kellogg unpublished), along with siliceous and calcareous marine fossils (27, 44). Sponge spicule fragments are commonly found in the ice core and snow pit samples along with diatoms. We attribute the fragmentation, corrosion, and partial dissolution of marine diatoms and sponge spicules to reworking, possibly during multiple episodes of transport by grounded ice, during former ice advances prior to deposition in sites susceptible to wind erosion. Abrasion during atmospheric transport could also partly explain these features. The fresh-looking marine specimens in some of the snow pits might have been picked up by winds directly from the ocean.

Nonmarine Species

Non-marine taxa are more diverse than marine taxa in the ice core samples. We found seventy-four taxa in the ice cores and other high-altitude sites which we suspect represent wind-blown diatoms (e.g., Mt. Feather, Mt. Crean, and Table Mountain in the Transantarctic Mountains, Pagodroma Formation in the Fisher Massif; and James Ross Island). Many of these nonmarine taxa are wide-ranging species known from other continents and/or the circum-Antarctic islands.

Although seventy-one of the seventy-four taxa we report here have also been reported from one or more Antarctic and/or circum-Antarctic locations, three taxa, *Cyclotella bodanica* v. *lemanica*, *Fragilaria berolinensis*, and *Navicula incertata*, have been reported only from high-elevation sites at Mt. Feather and Table Mountain. in Southern Victorialand, and the Pagodroma Formation in the Amery Ice Shelf area of East Antarctica, respectively. These three taxa may be misidentifications, as they have not been reported anywhere else in the Antarctic or circum-Antarctic or, less likely, they may have been transported by winds from a location far to the north.

Transport by Birds

Schlichting et al. (61) showed that birds collected at Palmer Station and in the Weddell Sea carried nine species of algae, nine of protozoa, and twelve fungi. They also noted that all species of blue-green algae sampled from Arctic terns in Antarctica are known from pools on Ellesmere Island, less than 800 km from the North Pole (16), as well as from Antarctic soils (15), suggesting that birds may carry blue-green algae between the polar regions. Schlichting et al. (61) did not report diatoms from their bird samples, but this may reflect the small number of birds sampled. Birds probably disperse diatoms around the periphery of Antarctica, and between South America and the Antarctic Peninsula. Transport to locations on the polar plateau is less likely because birds rarely visit this area. Although polar skuas have been reported at the South Pole, no doubt drawn there by the dump, this food source did not exist before

Pole Station was built. Thus winds remain the probable transport mechanism for diatoms in the ice cores and even in the South Pole Water Well samples (discussed below), all of which predate the construction of Pole Station.

Antarctic Meteorology

Antarctic meteorology is poorly known because observation stations are few and far between. Long-term observations are available from only a few coastal research stations (McMurdo, Palmer, etc.) and even fewer interior stations (South Pole, Vostok). These stations are supplemented by automatic weather stations (AWS) installed in remote areas (64). AWS data, collected by the ARGOS Data Collection System on NOAA polar-orbiting satellites, have been used to construct synoptic climate reconstructions and to develop atmosphere-snow transfer functions for interpreting indications of climate change in long ice cores.

Because of the high topographic gradient near the coast, most storm systems in the Southern Ocean tend to move clockwise around the continent, and storm precipitation is usually confined to coastal regions. Occasional strong storms cross the coastline and head poleward. The accumulation rate over Antarctica can be determined indirectly from atmospheric numerical analyses produced by the European Center for Medium-Range Weather Forecasts (7, 17, 76). On average, about 40% of the water vapor falling as snow on Antarctica enters the continent as storms pass through West Antarctica (7). This sector is subject to the largest interannual variability in Antarctica, particularly in conjunction with the El Niño-Southern Oscillations (ENSO) weather phenomenon. The derived accumulation rate over part of West Antarctica (180°-120°W, 75°-90°S facing the South Pacific Ocean) varied in phase with the Southern Oscillation Index (a tropical Pacific pressure index related to ENSO) from 1982 to 1990 (17), and then became anticorrelated after 1990 during the prolonged series of El Niño events of the early 1990s (75). Further, ENSO events observed in the South Pacific (78) appear to affect both the sea ice cover in this sector (24) and pressure and temperature on the continent (63).

As already noted, almost half of the moisture precipitating on Antarctica moves poleward across the coast of Antarctica between the Ross Ice Shelf and the Antarctic Peninsula. This poleward advection of moist, warm air has a profound impact on the surface temperature regime over West Antarctica, as well as on boundary layer dynamics near the Siple Coast (5). The potential temperature pattern derived from an AWS array deployed upslope from the Siple Coast shows that warm air is most noticeable near the ice divide upslope from Siple Coast, and that its effects steadily decrease downslope toward the Ross Ice Shelf (35, 58). This warm air signature is the inland expression of the large poleward flux of warm, moist air across the coast, and is thought to arise because of the combined radiative effects of clouds and moist air on the surface energy balance (5, 49). In the long-term average, this weather feature follows the ridge line into East Antarctica (34). But, because of the strong ENSO variability in the area discussed above, there is likely to have been strong interannual variability in the location and intensity of this warm air advection pattern.

Except for poleward transport along the above-mentioned ridge line, storms rarely penetrate to the interior of East Antarctica, where precipitation rates are much lower than in West Antarctica and moisture is transported poleward in the stratosphere by winds that converge on the high East Antarctic domes to replace cold air formed at the surface (60) that flows outward and downward off the polar plateau during episodic katabatic events (4). What we would like to determine is how well the diatom content of recent snowfall across Antarctica functions as a proxy indicator of current atmospheric circulation patterns. Then changes in the pattern of diatom distribution through time down long ice cores could be used to infer past atmospheric fluctuations.

Nonbiogenic Proxies of Atmospheric Transport

The principal atmospheric transport proxies studied in ice cores that are indicative of atmospheric circulation strength and direction are dust particles (18), glaciochemical signatures (sulfate, chlorine, and other salts derived from the ocean [53]), and changes in accumulation rate (calculated from dated horizons, or based on measurements of ^{10}Be [65], cosmogenic ^{14}C [48], ECM or electrical conductivity [a measure of the acidity of the ice, suited for discerning the presence of volcanic material [73], or extrapolated from surface measurements [50]). Sea salt and dust are incorporated in the atmosphere in regions of strong cyclogenesis that form along marine and atmospheric thermal gradients. For Antarctica, the gradual drop of LGM (Last Glacial Maximum) calcium values in dust from ca. 15,600 to 14,600 calendar years ago at Taylor Dome is thought to result from a gradual decrease in influx of continental material transported from arid regions of the southern continents (51), corresponding to the termination of near-glacial atmospheric conditions. While Southern Hemisphere atmospheric circulation was not strong enough to incorporate significant amounts of dust between 14,600 and ca. 10,000 years ago, the Taylor Dome chloride series data indicate that it was still vigorous enough to increase the transport of sea salt to Antarctica relative to modern values (51).

High-resolution glacio-chemical records from the GISP 2 (Greenland Ice Sheet Project 2) and Siple Dome ice cores show synchronous changes in atmospheric circulation during the most recent rapid climate change, the Little Ice Age (ca. A.D. 1400–1900). But changes in atmospheric circulation are out of synchrony with similar records from the Taylor Dome core. The decrease seen in sea salt concentration in Taylor Dome ice at that time was attributed to a shift in the position and strength of the Amundsen Sea Low Pressure System (52), which shifts local pathways to Taylor Dome from the Ross Sea (when the ASL is weak) to the East Antarctic Plateau (when the ASL is strong during periods of intensified atmospheric circulation). Mayewski et al. (52) reported that the annual sea salt record in the long ice core from Law Dome on the EAIS is correlated with variability in the ASL and sea level pressure in the adjoining regions of the Southern Ocean from September to November. Over the past 500 years, the Davis Sea Low (DSL) has behaved similarly to the ASL, deepening and migrating eastward when temperatures at Law Dome decline.

Biogenic Proxies

Shifts in strength and direction of winds delivering sea salts from the coast to interior locations should also be reflected in the record of marine diatoms from polar ice cores. Ice samples in which nonmarine diatom species known from coastal Antarctic and extra-Antarctic locations co-occur with elevated concentrations of certain nonbiogenic particulates are consistent with transport by the relatively infrequent strong winds from the coast that bring the "warmest, wettest, most particle-laden air" to the South Pole (36). Other samples, in which the highest concentrations of total particulates contain few or no diatoms, are consistent with surface air arriving from the Transantarctic Mountains. The Taylor Dome diatom record, for example, is consistent with the chloride series record on a broad scale. Unfortunately, because of the low diatom abundances in polar ice, and the small diameter of the cores, most of the ice samples studied represent about 1000 years of deposition. The time-averaged nature of these samples has frustrated attempts to distinguish rapid changes in deposition by coastal storms versus katabatic winds using diatoms present in (or absent from) ice cores. However, analysis of five large-volume snow samples, each yielding ca. 100 cm³ of meltwater, from horizons only 2 to 3 cm thick in Siple Dome snow pit D, revealed rapid fluctuations in diatom content on a timescale of a few years.

The highest diatom fluxes in the Taylor Dome core occur in the Holocene and isotope stages 5b, 5c, and 7. Comparisons of downcore variations in diatom concentration with the data of Mosely-Thompson (56) and Taylor et al. (74) for nonbiogenic particulates can be used to test the hypothesis that these two types of particulates co-vary throughout the LGM at sites on the EAIS. Ram et al. (59) found a twenty-fold increase in concentrations of both nonmarine diatoms and nonbiogenic particulates in two samples of LGM ice compared to Holocene ice from Dome C. Burckle (pers. comm.) hypothesized that greater concentrations of both types of particulates co-occurring during the LGM were due mainly to increased wind strength from the same source areas.

Now that we have demonstrated different source areas for many of the non-marine species incorporated in Antarctic ice cores (43), we anticipate that they will have considerable use as independent means of support and confirmation for other data bearing on paleo-wind strength and direction, such as Sr/Nd ratios and size sorting of particulates, and for testing hypotheses about major changes in wind regimes across glacial/interglacial transitions. Comparison of changes in glacial/interglacial diatom and particulate abundance patterns in the Taylor Dome core on the EAIS with the Siple Dome core from a lower altitude on the WAIS should allow us to determine if airborne particulates transported to these ice sheets followed different pathways or changed significantly in strength during this critical climate transition. Of particular interest for paleoclimate reconstruction may be those times when the conclusions reached from independent dust particle and diatom data do not agree.

Time Averaging of Samples

The low concentrations of diatoms in ice cores, and the consequent necessity for time averaging of samples, has frustrated expectations of relating rapid down-core changes in diatom abundance and species content to decadal, or even faster, changes in atmospheric circulation patterns observed in Taylor and Siple Dome cores by ECM (72) and $\delta^{18}O$ analyses (53). Because such a small cross-section of ice is available from ice cores, long core segments must be used to get sufficient diatoms for statistical treatment, and time resolution at better than century scale is unlikely. What is needed is large-volume samples from discrete levels. The residue from each year's ice melt in the SPWW represents ca. 100 years of deposition from the atmosphere. Samples taken at one month intervals could resolve variations in atmospheric particle flux on a decadal timescale. The problem is that it will take many years of sampling to get a long time series. Closely-spaced large-volume snow pit samples have the potential to provide diatom time series on annual or even seasonal timescales, but the length of these time series is limited to a few decades.

New Data on Diatom Provenances

If atmospheric transport is responsible for carrying diatoms to remote locations in the Antarctic interior, then the diatoms found at these locations have the potential to provide important information on transport paths through time if (i) the diatom assemblage at a given location on the ice sheet varies through time, and (ii) the source locations from which the diatoms were transported can be established. Our analyses of ice cores at the South Pole and Taylor

Dome confirm that the species composition of the diatom assemblage varies through time (40, 41).

Establishing provenances for Antarctic nonmarine diatoms requires first that the modern distributions of these taxa be determined, and the data analyzed to determine if any of the species are restricted to specific areas. Provenance data derived primarily for use in interpretation of ice core diatoms should also be useful for determining provenances for other diatom-bearing deposits (Vostok [1] and Dome C [10, 11] ice cores, Table Mountain, Mt. Feather and Mt. Crean in the Transantarctic Mountains [2, 3, 38], Fisher Massif in the Prince Charles Mountains [47], and cracks in igneous rocks on James Ross Island [19]). We regard all of these diatom assemblages as resulting from atmospheric transport.

Of the mixed marine/nonmarine diatom assemblage in Antarctic ice cores, only nonmarine species are likely to produce useful provenance information because it is impossible to trace the marine taxa back to specific locations in the Southern Ocean. The nonmarine species, on the other hand, may be traced to specific locations if modern species distributions are known. The first part of this process was completed recently, and required an exhaustive examination of the literature on Antarctic nonmarine diatom distributions. In addition to noting the Antarctic location(s) from which each taxon has been reported, we also compiled data for the Antarctic and circum-Antarctic islands, and for southern South America, New Zealand, Tasmania, and Australia because analyses of dust in ice cores suggest extracontinental transport, specifically from South America (26, 53). The names of many of these species have been changed since diatoms were first reported from Antarctica, often because taxonomists adopted electron microscopy for studying specimens starting in the 1960s. An extensive review of the literature on diatom taxonomy was required to combine forms that have been reported under different names at different times and by different diatomists and to determine the distribution of each taxon. Some of our species have been reported under as many as five or six different names. The resulting list of over 1575 species, reported by workers over the last 120 years at about 350 locations constitutes the database used by T. Kellogg and D. Kellogg (43) to establish nonmarine diatom sources.

Because a complete discussion of our provenance results is beyond the scope of this chapter, we summarize here the results of this provenance analysis. A total of 293 taxa were found to be restricted to Antarctica and the islands south of the Antarctic Convergence (e.g., South Shetlands, South Orkneys, South Georgia): 62 taxa are restricted to locations in East Antarctica; 30 taxa are restricted to locations in the Ross Embayment; 162 taxa have been reported only from the Antarctic Peninsula area and the adjacent South Shetland and South Orkney island groups and South Georgia; 37 taxa have wider distributions that make them less useful as provenance indicators; an additional 17 taxa have been reported at locations in all three sectors. Many of these species are known from locations elsewhere in the world, but they have not been reported from the circum-Antarctic areas included in our database.

We define the circum-Antarctic region as the continental areas and islands south of about 30°S to the Antarctic Convergence. Seven hundred forty species occurred only at circum-Antarctic sites. Species with restricted distributions and hence probable utility as provenance indicators occurred at New Zealand (136 spp.), Tasmania (35 spp.), South America (205 spp.), and the circum-Antarctic Islands (132 spp.). The latter include Campbell Island (5 spp.), Falkland Islands (8 spp.), Gough Island (8 spp.), Kerguelen (59 spp.), Kerguelen + Crozet (40 spp.), and Macquarie Island (6 spp.). Four-hundred-thirty-four species occur at two or more circum-Antarctic locations and are of limited use as provenance indicators.

Finally, 542 species have been reported in both the Antarctic and circum-Antarctic areas, and 173 species have been reported from sites in more than one Antarctic sector, as well as more than one circum-Antarctic location. Included in this last group are some of the most common diatoms found in Antarctica: *Achnanthes brevipes* (21 Antarctic locations), *A. delicatula* (24 locations), *A. lanceolata* (37 locations), *Cyclotella stelligera* (15 locations), *Encyonema minutum* (19 locations), *Fragilaria construens* v. *venter* (15 locations), *Hantzschia amphioxys* (102 locations), *Luticola cohnii* (22 locations), *L. murrayi* (47 locations), *L. mutica* (45 locations), *L. muticopsis* (168 locations), *Muelleria gibbula* (25 locations), *Pinnularia borealis* (74 locations), and *Sellaphora seminulum* (41 locations). These 14 species and 22 additional taxa from this group have been reported from the South Pole (10 reports), Siple Dome (2), Taylor Dome (18), Table Mountain (6), Mountain Feather (12), Mt. Crean (2), Dome C (3), Fisher Massif, and igneous rocks on James Ross Island (3). None of these taxa is likely to yield useful provenance information.

Our data are provisional because the database of sites lacks sufficient geographic coverage to establish provenances with certainty, and because of uncertainties in species identifications published in the literature. Overcoming these problems will require careful work over many years to establish the modern distributions of the nonmarine diatoms. Our data allow us to determine provenances for nonmarine diatoms in ice from specific sites on the Antarctic ice sheets, and begin to realize the potential of diatoms in ice cores for resolving ongoing questions about icesheet history. Of the 293 nonmarine diatom species that live in Antarctic coastal locations, 239 are sufficiently limited geographically to make them useful as provenance indicators. Additionally, 306 taxa restricted to specific circum-Antarctic locations also should be important for identifying transport from South America, New Zealand, Tasmania, and the circum-Antarctic Islands. Despite the potential for tracing air masses presented by these results, some problems remain unresolved. (i) Most of the strictly Antarctic species are relatively rare. Thus, their transport to high-altitude inland sites is less likely than for the more abundant species that commonly live in both Antarctic and circum-Antarctic locations. Unfortunately, use of the more abundant species produces results that are tainted by the uncertainty introduced by their wide range of "possible" sources. (ii) None of these results helps to resolve the problem of low diatom abundance in most ice core samples. Finding a sufficient number of specimens for analysis usually requires large-volume ice samples, which can only be obtained from ice cores by combining material from a thick core segment, representing a long time interval. High time-resolution work will probably be impossible using ice from a single ice core, but may be possible if ice can be obtained from the same stratigraphic levels in two or more cores from the same location. The best opportunity for high-resolution work is from snow pit samples, which can provide a large volume of melt water from discrete levels if sampling is done carefully. While this means that high-resolution, even interannual, results are most likely for only the last few decades, this information can still provide important results showing shifts in moisture source related to ENSO phenomena and resultant shifts in the location of the Amundsen Sea Low.

Contamination

Because diatom counts in individual ice samples are low, we are concerned about possible contamination that might be introduced during drilling, examination, or packing of cores in the field, during processing at the National Ice Core Laboratory (NICL) in Denver where ice cores are stored and sampled, or in our laboratory. If diatoms were carried to sampling sites by scientists, or introduced during processing at NICL, we should see a significant extra-Antarctic component in the diatom flora. We did not find any such component in our ice core samples. While contamination could occur in our laboratory because of the hundreds of samples stored there, we doubt that contamination has occurred for two reasons. First, our ice core sample-processing method avoids many of the steps that might lead to contamination. Diatoms are filtered from melted ice (at NICL or at field sites). The filters are shipped to Orono in individual sealed sterile containers, where they are placed on sterile glass slides, made transparent with a drop of acetic acid, dried in a closed desiccator, and then covered with mounting medium and a sterile cover slip. None of the sieving steps or multiple centrifugations required for processing diatoms from water or sediment samples are used. Instead, individual sample filters are exposed to ambient laboratory atmospheric conditions for no more than a minute or two before the cover slip is mounted permanently. Second, for the four years prior to starting work with Antarctic ice core samples, we worked exclusively with North Atlantic sediment cores. We have never found Antarctic taxa (e.g., the distinctive nonmarine assemblage contaminating our North Atlantic material. Finally, our Antarctic ice core samples contain distinct diatom assemblages (39).

We do see evidence of contamination by wood fragments, opal phytoliths, pollen, sponge spicules, and radiolaria in many ice core samples, sometimes in abundance. This is especially true of the wood and plant fibers which probably represent contamination from packing materials and other debris at the coring sites; it is present in most samples from the ice cores, whether handled by us or other workers (Burckle pers. comm.). Pollen, sponge spicules, and radiolaria occur only sporadically, and probably represent wind-blown material transported along with the diatoms.

Conclusions

The current state of the study of diatoms in polar ice is a good news–bad news story. The good news is that diatoms, though generally rare, may be found in ice and snow across the Antarctic ice sheets and down the length of long ice cores. The very presence of diatoms so far from any possible source area is itself the key to their aeolian mode of emplacement, and the full extent and nature of their spatial and temporal distributions in polar ice should have much to contribute to paleoclimatology and paleoclimate modeling, when they are eventually delineated. The bad news is that high-resolution climatic studies on a glacial/interglacial timescale will be difficult using diatoms from ice cores because large samples are needed to obtain enough diatoms for statistical treatment. Large-volume samples from snow pits and the SPWW, however, may yield sufficient diatoms for such studies within the Late Holocene.

Acknowledgments

This work was partially funded by a University of Maine Faculty Summer Research Grant and NSF grant OPP 931–6306 to DEK. We thank friends and colleagues who provided samples, data, and encouragement, especially Joan Fitzpatrick, Pieter Grootes, Beverly and Terence Hughes, Carl Kreutz, Paul Mayewski, Charles Stearns, Eric Steig, Kendrick Taylor, Susan Taylor, and the staff at NICL.

Literature Cited

1. Abyzov, S.S., I.N. Mitskevich, M.N. Poglazova, N.I. Barkov, V.Y. Lipenkov, N.E. Bobin, B.B. Koudryashov, and V.M. Pashkevich. 1998. Long-term conservation of viable microorganisms in ice sheet of central Antarctica. pp. 75–84. In SPIE Conference on Instruments, Methods, and Missions for Astrobiology, San Diego, California, July 1998, SPIE vol. 3441.

2. Barrett, P.J., N.L. Bleakley, W.W. Dickinson, M.J. Hannah, and M.A. Harper. 1997. Distribution of siliceous microfossils on Mount Feather, Antarctica, and the age of the Sirius Group. pp. 763–770. In C.A. Ricci (ed.), The Antarctic Region: Geological Evolution and Processes. Terra Antarctica Publication, Siena, Italy.

3. Bleakley, N.L. 1996. Geology of the Sirius Group at Mount Feather and Table Mountain, South Victoria Land, Antarctica, Ph.D. thesis, Victoria University of Wellington, Wellington, New Zealand.

4. Bromwich, D.H. 1989. Satellite analyses of Antarctic katabatic wind behavior. Bull. Am. Meteorol. Soc. 70: 738–749.

5. Bromwich, D.H., and Z. Liu. 1996. An observational study of the katabatic wind confluence zone near Siple Coast, West Antarctica. Monthly Weather Rev. 124: 462–477.

6. Bromwich, D.H., and F.M. Robasky. 1993. Recent precipitation trends over the polar ice sheets. Meteorol. Atmospheric Phys. 51: 259–274.

7. Bromwich, D.H., F.M. Robasky, R.I. Cullather, and M.L. van Woert. 1995. The atmospheric hydrologic cycle over the Southern Ocean and Antarctica from operational numerical analyses. Monthly Weather Rev. 123: 3518–3538.

8. Burckle, L.H. 1984. Diatom distribution and paleoceanographic reconstruction in the Southern Ocean—present and last glacial maximum. Marine Micropaleontol. 9: 241–261.

9. Burckle, L.H., and J.S. Delaney. 1999. Terrestrial microfossil in Antarctic ordinary chondrites. Meteoritics & Planetary Sci. 34: 475–478.

10. Burckle, L.H., R.I. Gayley, M. Ram, and J.-R. Petit. 1988. Diatoms in Antarctic ice cores: Some implications for the glacial history of Antarctica. Geology 16: 326–329.

11. Burckle, L.H., R.I. Gayley, M. Ram, and J.-R. Petit. 1988. Biogenic particles in Antarctic ice cores and the source of Antarctic dust. Ant. J. U.S. 23: 71–72.

12. Burckle, L.H., D.E. Kellogg, T.B. Kellogg, and J. Fastook. 1997. A new mechanism for emplacement and concentration of diatoms in subglacial deposits. Boreas 26: 55–60.

13. Burckle, L.H., and N. Potter Jr. 1996. Pliocene-Pleistocene diatoms in Paleozoic and Mesozoic sedimentary and igneous rocks from Antarctica: A Sirius problem resolved. Geology 24: 235–238.

14. Burckle, L.H., and A. Wasell. 1995. Marine and freshwater diatoms in sediment pockets in igneous rocks from James Ross Island, Antarctica. Ant. J. U.S. 30: 8–9.

15. Cameron, R.E. 1971. Antarctic soil microbiology and ecological investigations. pp. 137–189. In L.O. Quam (ed.), Research in the Antarctic. American Academy of Arts and Sciences, Washington, D.C.

16. Croasdale, H. 1973. Freshwater algae of Ellesmere Island, N.W.T., vol. 3. National Museums of Canada, Ottawa.

17. Cullather, R.I., D.H. Bromwich, and M.L. van Woert. 1996. Interannual variations in Antarctic precipitation related to El Niño-Southern Oscillation. J. Geophys. Res. 101: 19109–19118.

18. De Angelis, M., N.I. Barkov, and V.N. Petrov. 1992. Sources of continental dust over Antarctica during the last glacial cycle. J. Atmospheric Chem. 14: 233–244.

19. Denton, G.H., J.G. Bockheim, S.C. Wilson, and M. Stuiver. 1989. Late Wisconsin and early Holocene glacial history, inner Ross embayment, Antarctica. Quatern. Res. 31: 151–182.

20. Faure, G., and D.M. Harwood. 1990. Marine microfossils in till clasts of the Elephant Moraine on the East Antarctic ice sheet. Ant. J. U.S. 25: 23–25.

21. Gayley, R.I., and M. Ram. 1984. Observation of diatoms in Greenland ice. Arctic 37: 172–173.

22. Gayley, R.I., and M. Ram. 1985. Atmospheric dust in polar ice and the background aerosol. J. Geophys. Res. 90: 12921–12925.

23. Gayley, R.I., M. Ram, and E.F. Stoermer. 1989. Seasonal variations in diatom abundance and provenance in Greenland ice. J. Glaciol. 35: 290–292.

24. Gloersen, P. 1995. Modulation of hemispheric sea-ice cover by ENSO events. Nature 373: 503–506.

25. Grootes, P.M., E.J. Steig, and M. Stuiver. 1994. Taylor Ice Dome study 1993–1994: An ice core to bedrock. Ant. J. U.S. 29: 79–81.

26. Grousset, F.E., P.E. Biscaye, M. Revel, J.-R. Petit, K. Pye, S. Joussaume, and J. Jouzel. 1992. Antarctic ice-core dusts at 18 k.y. B.P.: Isotopic constraints on origins and atmospheric circulation. Earth and Planetary Sci. Letters 111: 175–182.

27. Hall, B. 1997. Geological Assessment of Antarctic Ice Sheet Stability, p. 338. University of Maine, Orono.

28. Harwood, D.M. 1983. Diatoms from the Sirius Formation. Ant. J. U.S. 18: 98–100.

29. Harwood, D.M. 1986. Do diatoms from beneath the Greenland ice sheet indicate interglacials warmer than present? Arctic 39: 304–308.

30. Harwood, D.M. 1986. The search for microfossils beneath the Greenland and West Antarctic ice sheets. Ant. J. U.S. 21: 105–106.

31. Harwood, D.M. 1986. Recycled siliceous microfossils from the Sirius Formation. Ant. J. U.S. 21: 101–103.

32. Harwood, D.M. 1986. Diatom Biostratigraphy and Paleoecology and a Cenozoic History of Antarctic Ice Sheets, Ph.D. thesis, Ohio State University.

33. Harwood, D.M. 1991. Cenozoic diatom biogeography in the southern high latitudes: Inferred biogeographic barriers and progressive endemism. pp. 667–673. In M.R.A. Thomson, J.A. Crame, and J.W. Thomson (eds.), Geological Evolution of Antarctica. Cambridge University Press, Cambridge.

34. Hogan, A.W. 1997. Synthesis of warm air advection to the South Polar Plateau. J. Geophys. Res. 102: 14009–14020.

35. Hogan, A.W., S.C. Barnard, and J.A. Samson. 1979. Meteorological transport of particulate material to south polar plateau. Ant. J. U.S. 14: 192–193.

36. Hogan, A., S. Barnard, J. Samson, and W. Winters. 1982. The transport of heat, water vapor and particulate material to the south polar plateau. J. Geophys. Res. 87: 4287–4292.

37. Kellogg, T.B., L.H. Burckle, D.E. Kellogg, and J.L. Fastook. 1997. A new mechanism for diatom emplacement and concentration in glacigenic deposits. Ant. J. U.S. 32: 29–30.

38. Kellogg, D.E., and T.B. Kellogg. 1984. Non-marine diatoms in the Sirius Formation. Ant. J. U.S. 19: 44–45.

39. Kellogg, D.E., and T.B. Kellogg. 1987. Diatoms of the McMurdo Ice Shelf, Antarctica: Implications for sediment and biotic reworking. Palaeogeogr., Palaeoclimatol., Palaeoecol. 60: 77–96.

40. Kellogg, D.E., and T.B. Kellogg. 1996. Diatoms in South Pole ice: Implications for eolian contamination of Sirius Group deposits. Geology 24: 115–118.

41. Kellogg, D.E., and T.B. Kellogg. 1996. Glacial/interglacial variations in the flux of atmospherically transported diatoms in Taylor Dome ice core. Ant. J. U.S. 31: 68–70.

42. Kellogg, D.E., and T.B. Kellogg. 1997. Diatoms in a South Pole ice core: Serious implications for the age of the Sirius Group. Ant. J. U.S. 32: 213–218.

43. Kellogg, T.B., and D.E. Kellogg. 2002. Non-Marine Diatoms from Antarctic and Subantarctic Regions: Distribution and Updated Taxonomy. Gantner Verlag, Ruggell (Liechtenstein).

44. Kellogg, T.B., D.E. Kellogg, and M. Stuiver. 1990. Late Quaternary history of the southwestern Ross Sea: Evidence from debris bands on the McMurdo Ice Shelf, Antarctica. pp. 25–56. In C.R. Bentley (ed.), Antarctic Res. Ser., 50. American Geophysical Union, Washington, D.C.

45. Kellogg, D.E., M. Stuiver, T.B. Kellogg, and G.H. Denton. 1980. Non-marine diatoms from Late Wisconsin perched deltas in Taylor Valley, Antarctica. Palaeogeogr., Palaeoclimatol., Palaeoecol. 30: 157–189.

46. Kuivinen, K.C., B.R. Koci, G.W. Holdsworth, and A.J. Gow. 1982. South Pole ice core drilling, 1981–1982. Ant. J. U.S. 17: 89–91.

47. Laiba, A.A., and Z.V. Pushina. 1997. Cenozoic glacial-marine sediments from the Fisher Massif (Prince Charles Mountains). pp. 977–984. In A.C. Ricci (ed.), The Antarctic Region: Geological Evolution and Processes. Terra Antarctica Publication, Siena, Italy.

48. Lal, D., A.J.T. Jull, G.S. Burr, and D.J. Donahue. 1997. Measurements of in situ C-14 concentrations in Greenland Ice Sheet Project 2 ice covering a 17-kyr time span: Implications to ice flow dynamics. J. Geophys. Res. 102: 26505–26510.

49. Liu, Z., and D.H. Bromwich. 1994. Satellite investigation of spring cloud cover in West Antarctica. Ant. J. U.S. 29: 293–295.

50. Marshall, H.-P., E. Waddington, and D. Morse. 1997. Taylor Dome accumulation rate project. p. 15. Collected Abstracts for a Workshop on Taylor Dome on 26–27 April 1997. University of Washington, Seattle.

51. Mayewski, P.A., K. Kreutz, M.S. Twickler, and S.I. Whitlow. 1997. Global implications of Taylor Dome glaciochemistry. pp. 19–20. Collected Abstracts for a Workshop on Taylor Dome on 26–27 April 1997. University of Washington, Seattle.

52. Mayewski, P.A., E. Meyerson, J. Souney, K. Kreutz, V. Morgan, T. van Ommen, and I. Goodwin. 2001. Antarctic multi-decadal scale climate variability over the last 500 years. p. 14. Abstracts, 9th Annual Agassiz Symposium, University of Maine, 4 May 2001.

53. Mayewski, P.A., M.S. Twickler, S.I. Whitlow, L.D. Meeker, Q. Yang, J. Thomas, K. Kreutz, P.M. Grootes, D.L. Morse, E.J. Steig, E.D. Waddington, E.S. Saltzman, P.-Y. Whung, and K.C. Taylor. 1996. Climate change during the last deglaciation in Antarctica. Science 272: 1636–1638.

54. McKelvey, B.C., J.H. Mercer, D.M. Harwood, and L.D. Stott. 1984. The Sirius Formation: Further considerations. Ant. J. U.S. 19: 42–43.

55. Moore, T.C., Jr., L.H. Burckle, K. Geitzenauer, B. Luz, B. Molnia-Cruz, J.H. Robertson, H.M. Sachs, C. Sancetta, J. Thiede, P. Thompson, and C. Wenkam. 1980. The reconstruction of sea surface temperatures in the Pacific Ocean of 18,000 B.P. Marine Micropaleontol. 5: 215–247.

56. Mosley-Thompson, E., and L.G. Thompson. 1982. Nine centuries of microparticle deposition at the South Pole. Quatern. Res. 17: 1–13.

57. Patrick, R., and C.W. Reimer. 1966. Diatoms of the United States Exclusive of Alaska. Vol. 1. Academy of Natural Sciences of Philadelphia, Philadelphia.

58. Radok, U. 1973. On the energetics of surface winds over the Antarctic ice cap. pp. 69–100. In Energy Fluxes Over Polar Surfaces, vol. 129. World Meteorological Organization, Geneva.

59. Ram, M., R.I. Gayley, and J.-R. Petit. 1988. Insoluble particles in Antarctic ice: Background aerosol size distribution and diatom concentration. J. Geophys. Res. 93: 8378–8382.

60. Reiter, E.R. 1971. Atmospheric Transport Processes, Part 2: Chemical Tracers. U.S. Atomic Energy Commission, Oak Ridge, Tennessee.

61. Schlichting, H.E., B.J. Speziale, and R.M. Zink. 1978. Dispersal of algae and protozoa by antarctic flying birds. Ant. J. U.S. 13: 147–149.

62. Sinkula, B. no date. Antarctic Composite IR Images, 1 Nov. 1992 to 1 Nov. 1993, Three Hourly Intervals (videotape). University of Wisconsin at Madison.

63. Smith, S.R., and C.R. Stearns. 1993. Antarctic climate anomalies surrounding the minimum in the Southern Oscillation index. pp. 149–174. In D.H. Bromwich and C.R. Stearns (eds.), Antarctic Res. Ser. 61. American Geophysical Union, Washington, D.C.

64. Stearns, C.R., and G.A. Weidner. 1990. Antarctic automatic weather stations: Austral summer 1989–1990. Ant. J. U.S. 25: 254–258.

65. Steig, E.J. 1997. Uses of ^{10}berylium in the Taylor Dome ice core. p. 24. Collected Abstracts for a Workshop on Taylor Dome on 26–27 April 1997. University of Washington, Seattle.

66. Steig, E.J., D.L. Morse, E.D. Waddington, M. Stuiver, P.M. Grootes, P.M. Mayewski, S.I. Whitlow, and M.S. Twickler. 1999. Wisconsinan and Holocene climate history from an ice core at Taylor Dome, western Ross Embayment, Antarctica. Geografiska Annaler 82A: 213–235.

67. Stroeven, A.P., L.H. Burckle, J. Kleman, and M.L. Prentice. (1998). Atmospheric transport of diatoms in the Antarctic Sirius Group: Pliocene deep freeze. GSA Today 8: 1, 4–5.

68. Stroeven, A.P., and M.L. Prentice. 1994. Do marine diatoms in Sirius Group tills indicate ice sheet disintegration? Geological Society of America, Abstracts with Programs 26: A143.

69. Stroeven, A.P., and M.L. Prentice. 1997. A case for Sirius Group alpine glaciation at Mount Fleming, South Victoria Land, Antarctica: A case against Pliocene East Antarctic ice sheet reduction. Geol. Soc. Am. Bull. 109: 825–840.

70. Stroeven, A.P., M.L. Prentice, and J. Kleman. 1996. On marine microfossil transport and pathways in Antarctica during the late Neogene: Evidence from the Sirius Group at Mount Fleming. Geology 24: 727–730.

71. Stuiver, M., G.H. Denton, T.J. Hughes, and J.L. Fastook. 1981. History of the marine ice sheet in West Antarctica during the last glaciation: A working hypothesis. pp. 319–436. In G.H. Denton and T.J. Hughes (eds.), The Last Great Ice Sheets. Wiley-Interscience, New York.

72. Taylor, K.C., G.W. Lamorey, G.A. Doyle, R.B. Alley, P.M. Grootes, P.A. Mayewski, J.W.C. White, and L.K. Barlow. 1993. The 'flickering switch' of late Pleistocene climate change. Nature 361: 432–436.

73. Taylor, K.C., R.B. Alley, G.W. Lamorey, and P. Mayewski. 1997. Electrical measurements on the Greenland Ice Sheet Project 2 core. J. Geophys. Res. 102: 26511–26517.

74. Taylor, S., J.H. Lever, R.P. Harvey, and J. Govoni. 1997. Collecting micrometeorites from the South Pole water well CRREL Report 97–1. Cold Regions Research and Engineering Laboratory, Hanover, New Hampshire.

75. Trenberth, K.E., and T.J. Hoar. 1996. The 1990–1995 El Niño-Southern Oscillation event: The longest on record. Geophys. Res. Letters 23: 57–60.

76. Turner, J., D.H. Bromwich, et al. 1996. The Antarctic first regional observing study of the troposphere (FROST). Bull. Am. Meteorol. Soc. 77 (9): 2007–2032.

77. Van Heurck, H. 1909. Diatomées. pp. 1–129, 13 pls., Résultats du Voyage du S.Y. Belgica en 1897–1898–1899, Rapports Scientifiques. Botanique, vol. 6. Imprimerie J.-E. Buschmann, Antwerpen.

78. White, W.B., and R.G. Peterson. 1996. An Antarctic circumpolar wave in surface pressure, wind, temperature and sea-ice extent. Nature 380: 699–702.

6

The Nature and Likely Sources of Biogenic Particles Found in Ancient Ice from Antarctica

Raymond Sambrotto and Lloyd Burckle

ANCIENT ICE has provided a wealth of information on environmental changes occurring over the past few hundred thousand years. Where the chronological age of the ice can be accurately determined, much new information has been gleaned from careful chemical and isotopic measurements extending back through the oldest ice layers (22). These time series have shown intriguing relationships between changes in atmospheric greenhouse gases and glacial cycles. However, the specific mechanisms that relate these climate signals are not clear and their exact nature has attracted extensive scientific interest. The ability to extract additional information from biogenic particles in ancient ice would complement and expand existing analyses of environmental change.

Here, we report on a preliminary analysis of biogenic material from Antarctic ice in the larger context of the established time series. Although many particles found in ancient ice are nonbiogenic (1), there are a significant number that are biological in origin. The biogenic component may hold much information relevant to the structure and function of both current and past biological systems. It may also provide information relevant to paleoclimate analyses. Much of the biogenic material in ice likely has been transported by the wind, and the use of wind-dispersed biological materials in environmental analyses is well established. For example, airborne freshwater and marine diatoms have been used to determine sediment source and depositional environments in the source area, as in the transport of dust from the Sahelian region of Africa to the equatorial Atlantic (see ref. 9 and chapter 5, this volume). This observation over the equatorial Atlantic was subsequently repeated (10, among others), while freshwater diatoms of North African origin were collected in air samples recovered from Barbados (3).

Thus, the wind is an effective, long-distance transport mechanism for particles less than 20 μm, a size class that includes a variety of biogenic components that relay environmentally important information. From a purely biological perspective, biogenic particles reflect organisms characteristic of ancient environments and perhaps extinct forms. Some biogenic particles also represent viable organisms (14, 24). Another important aspect of biogenic material captured in ice is that its analysis would augment the chemical information available from ice samples and help to clarify some of the underlying processes in environmental change. For example, a leading hypothesis for the coherence between carbon dioxide and climate invokes the dependence of the ocean's biological productivity on the aeolian transport of iron (18). The ocean's plankton productivity is the first step in the biological pumping of organic matter to deeper water, where it is isolated from the atmosphere from decades to millennia. The waning of this pumping rate could account for some of the estimated change in atmospheric carbon dioxide during the Holocene. A detailed test of this hypothesis would involve far more than we currently understand about large-scale variations in the ocean-atmosphere–continent climate system. However, Antarctic ice core records have clearly shown the correlation between estimated tropospheric temperature (from $\delta^{18}O$) and total dust (20, 22). If such large-scale ecological and climate changes are indeed linked to aeolian transport, the Antarctic continent should hold a valuable record of these variations in its ice.

We have examined two samples of ancient ice from Antarctica that contained biological particles. We were able to make a positive identification of some particles and preliminary identifications of others. Still other particles, although clearly biogenic, could not be identified. We also report on isolation experiments that suggest the viability of bacteria and fungi in the ice. Such microbial culture work can provide enough material for further genetic analyses. We conclude that Antarctic ice is an efficient collector of windborne biogenic material, and advocate a broader effort to integrate such biological information into an analysis of aeolian transport and Southern Hemisphere environmental change.

Approach

Ice samples were obtained from 3587 m depth in the Vostok 5G ice core, as well as from ice found beneath a shallow soil layer in the Beacon Valley near Taylor Dome (Figure 6.1). Various depths of the Vostok ice core have been studied (e.g., chapters 15 and 16), and the segment examined here has been dated at approximately 430,000 years. The ice remained frozen continually after collection and during its shipment from the National Science Foundation (NSF)-supported National Ice Core Laboratory (NICL), then located at the University of New Hampshire. The volume of the ice sample was approximately 1 l. Some of the ice from 3539 m to the surface of Lake Vostok (ca. 3750 m at the core site) is accreted from the lake itself (see refs. 12, 13 and

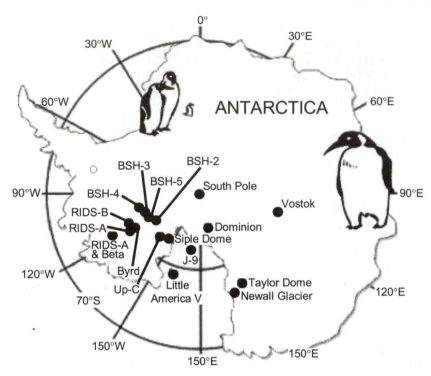

FIGURE 6.1. Antarctic ice core drill sites.

chapter 17, this volume). The approximately 5000-year turnover time of the lake water may be attributed to the latent heat exchange during the melt—freeze cycle that occurs along the sloped water-ice boundary (29). The lower ice also may contain material picked up from the region surrounding the lake (chapter 17). The exact source of the lower ice and the extent of its interaction with the continental crust are important aspects for understanding its biological content. In any event, the sample that we examined was located in a region with fewer visible inclusions compared to the layer 20 to 40 m above it (Figure 6.2). It also was located within 20 m of two other samples that have been analyzed biologically (14, 24). Both of these other samples contained microbes. The sample from 3590 m (24) contained organisms genetically similar to proteobacteria and actinobacteria, although no metabolic activity was detected. In the 3603 m core segment, active metabolism of simple carbon substrates was detected (14). The overall distribution of bacteria in the core is high in the basal layer of the ice sheet (3539 m) and decreases to the depth of the current sample before increasing again between 3590 and 3610 m (chapter 16). Thus, our sample was neither from a region of maximum visible particles, nor of bacteria.

The second ice sample that we examined was from a layer of ice buried in the Beacon Valley below Taylor Dome, Antarctica (30). It was covered by a layer of volcanic ash dated at slightly over 8 million years old. Thus, the sample may represent a Miocene ice sheet that survived under fairly stable Antarctic conditions since then. Both ice samples were analyzed for the oxygen isotopic ratios that reflect the temperature at which meteoric precipitation occurred (Table 6.1). The Vostok sample reflects an average temperature similar to that at the South Pole today, while the Miocene ice sample (found at both a lower latitude and elevation) suggests a warmer temperature. The Vostok samples not destined for incubation were thawed at 4 °C, and centrifuged to collect the particulate material in a pellet. The pellet was suspended in a small amount of the supernatant to prepare wet mounts for microscopic analysis. Some of the samples for microscopic analysis were stained with 1% eosin Y to identify proteinaceous material. For our incubation experiments, all equipment was autoclaved before use, and all transfers and manipulations were done in a positive displacement transfer hood. The surface of the ice was swabbed with 70% ethanol before thawing. Approximately 320 ml of ice was used for the incubations.

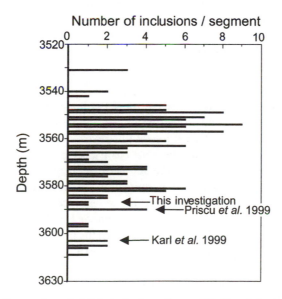

FIGURE 6.2. The number of visible inclusions observed in Vostok ice core sections at different depths was recorded by Joan Fitzpatrick (USGS, Denver, Colorado). Original data are available at www.nicl.usgs.gov/vostok/. Arrows denote the depths in the Vostok core at which samples were collected by us and others, and examined for inclusions and for biological analyses.

TABLE 6.1.
The $\delta^{18}O$ composition of the two ice samples examined here as well as that of a deeper Vostok sample

Ice Sample	$\delta^{18} O \ (0/00)$	Approximate Precipitation Temperature $(°C)^a$
Miocene glacial ice	−32	−22
Vostok 3587 m[b]	−54	−37
Vostok 3590 m[c]	−56.8	−38

[a]Data from reference 7.
[b]Data from this study.
[c]Data from reference 24.

    The meltwater was incubated in three different media: a bacterial broth consisting of 0.1% yeast and vitamins (broth A); an identical broth that also contained 0.01% dextrose (broth B); and f/2 phytoplankton medium made with freshwater (11, 25). In addition to the liquid media, several Petri plates with a 3% agar solution were made up using broth A or B. Samples were incubated at 10 °C and 1 °C. The samples for bacterial growth were placed in the dark, and the samples in f/2 medium were kept in a 16:8 hour light:dark cycle.

Results and Discussion

Particle Characterization

We have examined about two-thirds of the 1 l volume of meltwater from the Vostok ice and about 200 ml of the Beacon ice sample. Microscopic examination of the pellet from the Vostok ice sample revealed an assortment of particles, many of them biological. The most easily recognizable source is wind-blown material from elsewhere in the Southern Hemisphere. For example, the bisaccate pollen grains (likely *Podocarpus*) found in the Vostok sample (Figure 6.3 A) were almost certainly carried to the ice surface by aeolian processes. *Podocarpus* is widely distributed in the Southern Hemisphere. Pollen from these plants is typically well preserved, although the processes on the grains we found were smaller than extant forms, likely due to abrasion during transport to and within the ice sheet.

    Diatoms also were found in the Vostok core as well as in the Beacon ice (Figures 6.3 B–C). The species *Fragilariopsis kerguelensis*, *Thalassionema nitzschioides*, and *Thalassiosira lentiginosa* were found in our Vostok sample, and are extant marine species (chapter 5). The combined presence of *F. kerguelensis* and *T. lentiginosa* suggests an age of less than 2.5 million years. Their marine origin indicates that they did not originate from Lake Vostok. We

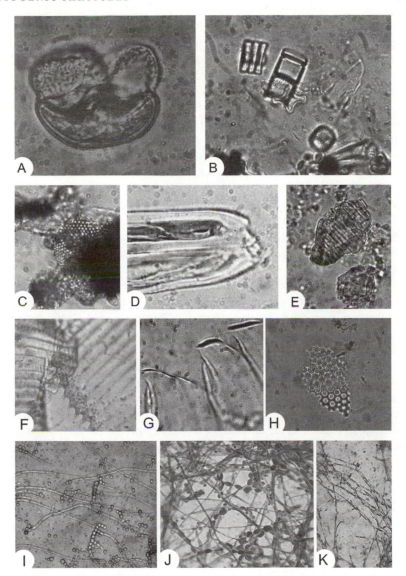

FIGURE 6.3. (A) bisaccate pollen grain (*Podocarpus*?) observed in the Vostok ice sample. (B) Diatom frustules from the Vostok ice core. (C) Diatom fragments and filamentous material from the Beacon ice sample. (D) Closeup of object resembling the pharynx of a nematode. (E) Particles resembling opal-phytoliths observed in ice collected from the Beacon Valley. (F) Ribbon-like structure with serrated edge (Vostok core). (G) Closeup of edge of similar particle in ice from Beacon Valley. (H) Unidentified diatom fragment from Vostok ice. (I, J, K) Fungal mycelium and spores observed in Vostok core. Differences among spore size and shape suggest different fungal species are present in the ice.

also found clasts of freshwater diatoms. The planktonic diatom *Aulocoseira* spp. dominates the assemblage, suggesting a deep, freshwater source. However, an ice-free Lake Vostok is not necessary to explain the presence of either the marine or freshwater diatoms. Both could have been transported to the Antarctic ice by aeolian transport and made their way to the bottom of the glacier following flow lines as suggested by Burckle et al. (6). The marine diatoms in particular require a transport mechanism because the subice topography around Lake Vostok is above the present sea level and thus, even during ice-free conditions would not be exposed to ocean water. Also, the ocean waters surrounding Antarctica support rich diatom growth (28), and thus a ready supply of such particles. Although this same argument does not eliminate freshwater diatom growth in Lake Vostok during some period in which the East Antarctic ice sheet (EAIS) receded, abundant *Aulocoseira* spp. were recovered from the surface of a granite erratic from Dronning Maud Land (4), which was clearly a case of aeolian transport. Thus, we do not view Lake Vostok as the source of the freshwater clast. The source of the diatoms recovered from Beacon Valley is less clear. These diatoms were few in number, fragmented, and unidentifiable, although certainly of marine origin. Given the fact that our Beacon ice sample is almost 20% lithics and of glacial transport origin, we suggest that the diatoms were wind-blown onto a nonglaciated surface and picked up by an advancing glacier.

We also found wormlike objects in the Vostok core (Figure 6.3 D), which resemble an extant predaceous nematode, *Eudorylaimus antarcticus*, found in the Dry Valleys. Nematodes can withstand desiccation for long periods of time, and may have originated on the seasonally ice-free periphery of the Antarctic continent. In the Beacon Valley ice, numerous particles tentatively identified as opal-phytoliths were observed (Figure 6.3 E). Phytoliths provide strength and support in grasses and other plants. They can be dispersed easily by the wind once the organic material around them has degraded. However, the precise origin of this material is uncertain. Although most of the putative phytoliths we found were in relatively poor condition, in some cases it may be possible to identify the type of plant from which they originated. The remaining putative biogenic particles we observed (Figure 6.3 F–H) have not yet been identified. However, their repeating morphologies strongly suggest biological structures. We hope to clarify the nature of at least some of these particles with further microscopic and chemical analyses, as well as in consultation with others.

A more extensive down-core analysis would indicate how the quality and quantity of biogenic particles changed through time. Such a record would be complex to interpret on an environmental basis, because it reflects the changes in both the source of the particles as well as the efficiency of their transport by air and ice. However, some success at linking biogenic particles to variations in environmental conditions has been achieved. For example, Burckle et al. (5) reported freshwater diatoms at the level of the Last Glacial Maximum (LGM) in Dome C on the central plateau of East Antarctica. Also, in their analysis of ice from Dome C, Ram et al. (26) noted that the concentration of diatoms in two samples of LGM ice was much greater than that in a sample of Holocene ice (0.01/g in Holocene ice versus 0.20/g in LGM ice). There also was a similar temporal decrease in the concentration of insoluble particles, suggesting that the mechanisms that enhanced transport of crustal material to Antarctica during the LGM were the same as those that led to the higher diatom concentrations. Thus, the increase in diatoms appears to reflect their greater dispersion by winds during the LGM.

The use of biogenic particles from ice as climate proxies could be particularly useful in Antarctic studies. For example, the ability of ice-trapped diatom abundance to record wind intensity is relevant to current biogeochemical hypotheses regarding the role of the Southern Ocean in driving atmospheric carbon dioxide changes. A leading hypothesis for the coherence between carbon dioxide and climate invokes the dependence of the ocean's biological productivity on fertilization by the aeolian transport of iron (18). In this model, dry conditions associated with glacial periods increase dust levels that nourish ocean plankton and, in turn, draw down atmospheric carbon dioxide and cool the Earth. Mineralogical analyses already have shown the major regions where dust originates en route to Antarctica (3). Our results suggest that changes in the levels and character of diatoms in ice over time would convey important additional information in that, unlike the lithogenic particles, they vary with ocean productivity. Although wind remains an important factor, any divergence between diatom and lithogenic abundance would suggest a change in Southern Ocean productivity. Thus, the ecological information in ice has the potential to yield a much more integrated view of the climate-ocean-continent linkages in the Southern Hemisphere than any single chemical or isotopic parameter. Such information would significantly improve analyses of the temporal interactions between climate and biology.

Viability Experiments

Our incubation experiments were performed with ca. 200 ml of Vostok 5G ice. Bacterial and fungal growth were observed ten days postinoculation (Table 6.2). Fungal growth was observed in f/2 medium incubated in the light or dark, but only when incubated at 10 °C (Table 6.2 and Figure 6.3 I). The f/2 medium supplied a rich mixture of inorganic nutrients and metals, which can support fungal growth. Fungal growth also was significantly greater in the light, perhaps due to the activation of fungal spores by UV radiation. The size of the hyphae and the distance between separations suggested that the fungus was a

TABLE 6.2.
Isolations from the Vostok ice sample using different media and incubation
temperatures

Culture Medium	10 °C	1 °C
f/2 medium	Many conidia and hyphae	Some bacteria
(Light)	Some bacteria	
f/2 medium	Some conidia and hyphae	Some bacteria
(Dark)	Some bacteria	
Bacterial broth A	Moderate bacteria	
Bacterial broth B	Few bacteria	Few bacteria

species of *Fusarium* (Figure 6.3 J and ref. 19). Also, differences among cell size and shape suggested that there may have been more than one species (Figure 6.3 K). We observed fungal masses in the frozen ice as well as the nematodes mentioned previously, that, together with the observed fungal growth, suggest a soil source. Ma et al. (see ref. 16 and chapter 11, this volume) isolated viable fungi from glacial ice in Greenland and Antarctica. Taken together, these results suggest that fungi can remain viable for more than 400,000 years in Antarctic ice.

Subsamples of each culture were treated with a nucleic acid stain and examined with epifluorescence microscopy. Stain-acquiring particles in the range of 1 micron were interpreted as bacteria. Bacteria grew in all media and under all culture conditions except in broth A incubated at 1 °C (Table 6.2). Most cells were rod-shaped. The growth of bacteria reproduced the findings of Karl et al. (14) and Abyzov et al. (see ref. 1 and chapter 16, this volume), who isolated bacteria from deep sections of the Vostok 5G core. The isolation of organisms from ancient ice improves our ability to characterize them genetically because large amounts of nucleic acid can be obtained from culture material. This approach has been used to clarify the taxonomic affinities of ancient organisms, and to verify that the material was ancient and free of modern contaminants (17). The presence of organisms able to maintain viability for over 400,000 years in ancient ice is an exciting biological discovery with implications for gene flow throughout the environment. It also raises further questions about how such viability is maintained, and how we can better observe the distribution of organisms in ice (23). Initial work has already demonstrated a relationship between microbial numbers and climate change over the last 250,000 years (1), and the discovery of extended viability in ice suggests that there is much more to be learned.

Acknowledgments

This work was supported by a grant from the Lamont Climate Center to L. Burckle and R. Sambrotto and NSF OPP# 0105104 to R. Sambrotto. We thank the meeting organizers and the Office of Polar Programs for inviting us to participate in this workshop. This is Lamont-Doherty Earth Observatory Contribution No. 6321.

Literature Cited

1. Abyzov, S.S., N.I. Barkov, N.E. Bobin, B.B. Kudryashov, V.Y. Lipenkov, I.N. Mitskevich, V.M. Pashkevich, and M.N. Poglazova. 1998. The ice sheet of central Antarctica as an object of study of past ecological events on the earth. Izvextiya Akademii Nauk Seriya Biologicheskaya 5: 610–616.

2. Abyzov, S.S., I.N. Mitskevich, M.N. Poglazova, N.I. Barkov, V.Y. Lipenkov, N.E. Bobin, B.B. Kudryashov, and V.M. Pashkevich. 1999. Antarctic ice sheet as an object for methodological problems of exobiology. Adv. Space Res. 23: 371–376.

3. Basile, I., F.E. Grousset, M. Revel, J-R. Petit, P.E. Biscaye, and N.I. Barkov. 1997. Patagonian origin of glacial dust deposited in East Antarctica (Vostok and Dome C) during glacial stages 2, 4 and 6. Earth and Planetary Sci. Letters 146: 573–589.

4. Burckle, L.H. 1995. Diatoms in igneous and metamorphic rocks from Dronning Maud Land, East Antarctica; improbable places, improbable surfaces. Ant. J. U.S. 30: 67–68.

5. Burckle, L.H., R.I. Gayley, M. Ram and J-R. Petit. 1988. Diatoms in Antarctic ice cores: Some implications for the glacial history of Antarctica. Geology 16: 78–82.

6. Burckle, L.H., D.E. Kellogg, T.B. Kellogg, and J.L. Fastook. 1997. A mechanism for emplacement and concentration of diatoms in glacigenic depoits. Boreas 26: 55–60.

7. Criss, R.E. 1999. Principles of Stable Isotope Distribution. Oxford University Press, Oxford.

8. Delaney, A.C., D.W. Parkin, J.J. Griffin, E.D. Goldberg, and B.E.F. Reimann. 1967. Air borne dust collected at Barbados. Geochem. Cosmochem. Acta. 31: 885–909.

9. Ehrenberg, C.G. 1845. Einige vorlaufige resultate der untersuchungen der von der Sudpolrreise des Capitain Ross, so wie in den Herrn Schayer und Darwin zugekommenen materialien uber das verhalten des kleinsten lebens in den oceanen und den grossten bisher zuganglichen tiefen des weltmeers. Bericht uber die zur Bekanntmachung geeigneten Verhandlungen der konigl. preuss. Akademie der Wissenschaften zu Berlin 844: 182–207.

10. Folger, D., L.H. Burckle, and L.H. Heezen. 1967. Opal phytoliths in a north Atlantic dust fall. Science 155: 1234–1244.

11. Guillard, R.R.L., and J.H. Ryther. 1962. Studies of marine plankton diatoms, 1. Cyclotella nana (Hustedt) and Detonula confervacea. Can. J. Microbiol. 8(2): 229–239.

12. Jean-Babtiste, P., J-R Petit, V. Lipenkov, D. Raynaud, and N.I. Barkov. 2001. Constraints on hydrothermal processes and water exchange in Lake Vostok from helium isotopes. Nature 411: 460–462.

13. Jouzel, J., J-R Petit, R. Souchez, N.I. Barkov, V. Lipenkov, D. Raynaud, M. Stievenard, N.I. Vassiliev, V. Verbeke, and F. Vimeux. 1999. More than 200 m of lake ice above subglacial Lake Vostok, Antarctica. Science 286: 2138–2141.

14. Karl, D.M., D.F. Bird, K. Björkman, T. Houlihan, R. Shackelford, and L. Tupas. 1999. Microorganisms in the accreted ice of Lake Vostok, Antarctica. Science 286: 2144.

15. Lloyd, J., and G.D. Farquhar. 1996. The CO_2 dependence of photosynthesis, plant growth responses to elevated atmospheric CO_2 concentrations and their interaction with soil nutrient status. 1. General principles and forest ecosystems. Functional Ecol. 10: 4–32.

16. Ma, L.J., C.M. Catranis, W.T. Starmer, and S.O. Rogers. 1999. Revival and characterization of fungi from ancient polar ice. Mycologist 13: 70–73.

17. Ma, L.J., S.O. Rogers, C.M. Catranis, and W.T. Starmer. 2000. Detection and characterization of ancient fungi entrapped in glacial ice. Mycologia 92: 286–295.

18. Martin, J.H. 1990. Glacial to interglacial CO_2 change: The iron hypothesis. Paleoceanography 5: 1–13.

19. Nelson, P.E., T.A. Toussoun, and W.F.O. Marassas. 1983. *Fusarium* Species, An Illustrated Manual for Identification. Pennsylvania State University, University Park.

20. Petit, J-R., I. Basile, A. Leruyuet, D. Raynaud, C. Lorius, J. Jouzel, M. Stievenard, V.Y. Lipenkov, N.I. Barkov, B.B. Kudryashov, M. Davis, E. Saltzman, and V. Kotlyakov. 1997. Four climate cycles in Vostok ice core. Nature 387: 359–360.

21. Petit, J-R., J. Jouzel, D. Raynaud, N.I. Barkov, J.M. Barnola, I. Basile, M. Bender, J. Chappellaz, M. Davis, G. Delaygue, M. Delmotte, V.M. Kotlyakov, M. Legrand, V.Y. Lipenkov, C. Lorius, L. Pepin, C. Ritz, E. Saltzman, and M. Stievenard. 1999. Climate and atmospheric history of the past 420,000 years from the Vostok ice core, Antarctica. Nature 399: 429–436.

22. Petit, J-R., L. Mounier, J. Jouzel, Y.S. Korotkevich, V.I. Kotlyakov, and C. Lorius. 1990. Paleoclimatological and chronological implications of the Vostok core dust record. Nature 343: 56–58.

23. Price, P.B. 2000. A habitat for psychrophiles in deep Antarctic ice. Proc. Nat. Acad. Sci. 97: 1247–1251.

24. Priscu, J.C., E.E. Adams, W.B. Lyons, M.A. Voytek, D.W. Mogt, R.L. Brown, C.P. McKay, C.D. Takacs, K.A. Welch, C.F. Wolf, J.D. Kirshtein, and R. Avci. 1999. Geomicrobiology of subglacial ice above Lake Vostok, Antarctica. Science 286: 2141–2144.

25. Provasoli, L., J.J.A. McLaughlin, and M.R. Droop. 1957. The development of artificial media for marine algae. Arch. Microbiol. 25: 392–428.

26. Ram, M., R.I. Gayley, and J-R. Petit. 1988. Insoluble particles in Antarctic ice: Background aerosol size distribution and diatom concentration. J. Geophys. Res. 93: 8378–8382.

27. Rogers, S.O., L.-J. Ma, C. Catranis, S. Zhou, and W.T. Starmer. 2000. Ancient Microorganisms in Arctic and Antarctic Ice. Mycological Society of America Annual Meeting.

28. Sambrotto, R.N., and B. Mace. 2000. Coupling of biological and physical regimes across the Antarctic polar front as reflected by nitrogen production and recycling. Deep-Sea Res. Part II. 47: 3339–3368.

29. Souchez, R., J-R Petit, J.-L. Tison, J. Jouzel, and V. Verbeke. 2000. Ice formation in subglacial Lake Vostok, Central Antarctica. Earth and Planetary Sci. Letters 181: 529–538.

30. Sugden, D.E., D.R. Marchant, N. Potter, R.A. Souchez, G.H. Denton, C.C. Swisher, and J.-L. Tison. 1995. Preservation of Miocene glacier ice in East Antarctica. Nature 376: 4412–4414.

31. Thompson, L.G. 1977. Variations in microparticle concentration, size distribution and elemental composition in Camp Century, Greenland, and Byrd Station, Antarctica, deep ice cores. pp. 351–364. In International Symposium on Isotopes and Impurities in Snow and Ice, Publ. 118. International Association of Hydrological Sciences, Paris.

7

Microbial Life below the Freezing Point within Permafrost

Elizaveta Rivkina, Kayastas Laurinavichyus, and
David A. Gilichinsky

VIABLE PROKARYOTES, green algae, cyanobacteria, actinobacteria, filamentous fungi, and yeasts have been isolated from glacial ice and permafrost (1, 10). Their ages correspond to those of the frozen horizons from which they were isolated. The most ancient microorganisms isolated from ice are 400,000 years old (1), and from permafrost, several million years old (10). Viable microorganisms following long-term conservation were isolated from frozen cores with permanent ground temperatures that ranged from −1 to −2 °C near the southern permafrost border in West Siberia, at temperatures from −3 to −4 °C in the interior of Alaska, −5 to −6 °C in the Mackenzie Delta, −9 to −12 °C in the Eurasian Northeast, and in Antarctic Dry Valleys, where ground temperatures are as low as −18 to −25 °C (10). In the upper 500 m of the Antarctic ice sheet, viable cells were detected at temperatures of −50 °C (1).

Microorganisms have adapted to cold environments, and now populate all ecological niches. Microbial communities were the only viable forms detected in permafrost dated to the Late Cenozoic (10). A fundamental question of importance to earth scientists, biologists, cryobiologists, astrobiologists, and paleobiologists is: How long can life remain viable? This question cannot be answered by calculations, experiments, or modeling because the time factor cannot be reproduced. Therefore, the cryosphere can provide the answer because it is a unique, natural storehouse of ancient viable systems, in which it is possible to observe the results of cryoconservation on geological timescales.

Microbial communities from permafrost have been described as "a community of survivors" (7), This "starvation-survival lifestyle" is the normal physiological state of microbial communities within permafrost, which results from the combination of cold temperature, desiccation, and starvation (17). Microbial communities in the same state of starvation-survival also have been isolated from the Antarctic ice sheet (1). Nevertheless, such terms are not clearly defined, and do not adequately describe the state of the isolated microbes in situ below the freezing point. Thawing of glacial ice and permafrost places these ancient biological systems into modern biogeochemical cycles. But the key question regarding the metabolic state of the microorganisms detected in permafrost and glacial ice (i.e., inhibited metabolism or anabiosis) still remains open to debate.

In this chapter, our objective is to provide a summary of the direct and indirect evidence of microbial activity in permafrost.

Review of Indirect Evidence

The Ability of Cells to Grow after Long-Term Exposure To Ground Radiation

The age of the cells corresponds to the longevity of the permanently frozen state of the sediments. The oldest viable microorganisms on the Kolyma lowland tundra in northeastern Siberia date back to 3 million years. Estimation of ground radiation has been made there together with Chris McKay and colleagues, using both elemental analysis of the radioactive elements in sandy and loam samples, and direct measurements in the boreholes. The radiation dose received by the permafrost bacteria is about 2 mGy per year. Taking into account the Late Pliocene age of the oldest bacteria (3 million years), the total dose received by these cells would therefore reach 0.6 kGy. The plate count of microorganisms listed in Table 7.1 demonstrates the ability of cells to produce colonies after being exposed to these doses.

The Presence of Biogenic Methane in situ

Methane was detected in specimens extracted from drill holes on the Kolyma lowland and degassed by warming. The methane concentrations were measured in the field and laboratory by headspace equilibration-gas chromatography with a flame ionization detector (2). The sensitivity limit of the methane detection was 10 ppm, and the results were reproducible to within 15%. In a generalized Late Cenozoic geological cross-section, in which all the geological layers exist in chronological order, methane-containing layers (up to 40.0 ml/ kg) were sandwiched between layers free of methane (Figure 7.1) (21). The methane distribution does not appear to be correlated to any textural or chemical characteristics of the sediments, age, or the depth of burial. The methane content does not exhibit any systematic trend with depth that would suggest a deep subsurface source, or penetration downward from the surface. The higher concentrations of methane were typical of peaty and organic rich epigenetically frozen sequences. Methanogenic archaea also were detected in the samples containing methane (19).

TABLE 7.1.
The number of viable cells isolated from Kolyma lowland Late Pliocene permafrost

Depth (m)	Incubation Temperature (°C)			
	+30	+20	+4	0
	hole 3/92			
5.7	5.0×10^4	1.0×10^4	1.8×10^4	0
27.9	2.8×10^1	1.0×10^5	0.8×10^2	2.8×10^1
55.2	4.7×10^3	2.3×10^4	3.2×10^2	7.7×10^1
	hole 1/85			
36.0	N/D	1.0×10^4	N/D	N/D
50.0		2.0×10^3		
53.0		2.3×10^3		
57.0		1.1×10^4		
63.0		2.5×10^5		
	hole 77/1			
30.0	N/D	9.0×10^3	N/D	N/D
61.0		1.2×10^4		
63.0		5.0×10^3		

N/D = not determined.

The same results were obtained by analyzing permafrost samples from the Fairbanks, Alaska area; the Mackenzie Delta; and West Siberia. Even in Antarctica, permanently frozen Late Pleistocene lacustrine sands contain methane (360 µl/kg), and this methane is of biological origin ($\delta^{13}C = -54.8^0/oo$) (27).

The Presence of Metastable Ferrous Sulfides and Nitrites

The same samples extracted from drill holes on the Kolyma lowland were tested for the presence of ammonium, nitrite, and nitrate (14). The nitrogen compounds were leached in distilled water, and 1M KCl solution was added directly in the field and fixed at high concentrations of KH_2PO_4 and K_2HPO_4. The analyses were made in the laboratory using the HPLS. In most samples, all compounds occurred together (nitrite at concentrations up to 17 ppm). Nitrite distribution in a generalized Late Cenozoic permafrost cross-section is shown in Figure 7.1. Nitrifying and denitrifying bacteria also were detected simultaneously in these samples containing nitrite (19, 25).

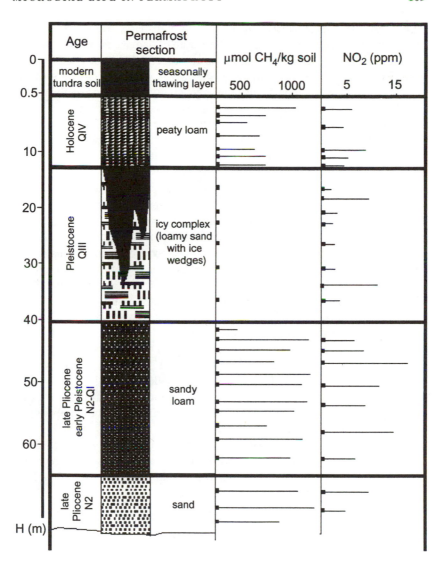

FIGURE 7.1. Methane (21) and nitrite (14) content in a cross-section of Late Cenozoic permafrost from the Eurasian northeast tundra.

Permafrost contains not only these nonstable chemical compounds, but also the metastable ferrous sulfide minerals, greigite and mackinawite (28).

We can summarize all these data, which together suggest that permafrost microorganisms are not, as previously thought, in a frozen resting state. Viable cells after exposure to ground radiation for millions of years means that these cells must retain their capacity to repair DNA damage, even in the frozen environment, at a rate of repair equal to the rate of damage accumulation. The presence of unstable ferrous sulfides or nitrite together with the nitrifying and denitrifying bacteria in permafrost, as well as the simultaneous presence of methane and ancient methanogens and their ability to produce methane (19) also indicates microbial activity. This suggestion is supported by studies that show that if nutrient medium is available, the subzero temperatures themselves are not the factor that prohibits microbial growth (12). The experiments were carried out with 10% glycerol added as a cryoprotector to starch-ammonium or R2A nutrient medium. In the presence of such protectors, the growth of ancient cells was observed at temperatures at least as low as −5 °C after 6 months and −10 °C after 1 to 1.5 years in the freezer (13). These results correlate well with the lower temperature limit for microbial growth (23). But the conditions of such experiments are totally unlike permafrost conditions *in situ*.

Other reports within the past decade or so support our conclusions. Metabolic activity has been reported in the Antarctic cryptoendolithic microbial community at temperatures between −5 and −10 °C (8, 26). In addition, the temperature limit of bacterial growth in frozen food is considered to be −8 °C (9). In Arctic and Antarctic lichens, photosynthetic activity has been observed in a similar temperature range (15), and even at −17 °C (24). Lastly, Carpenter et al. (4) determined that low rates of bacterial DNA and protein synthesis occur in South Pole snow at temperatures of −12 to −17 °C. However, no quantitative measurements of the dynamics of metabolic activity or of growth have been described. Therefore, the authors of some review articles have concluded that the reports of bacterial metabolic activity at subzero temperatures below −12 °C are unsubstantiated (6, 9, 22).

There are few data describing the metabolic activity of microorganisms below the freezing point in natural environments, and only for recent microbial communities. For ancient microorganisms, we can refer only to results of our own published experiments.

Direct Evidence

Results of Our Own Experiments

1. Experiments were conducted using free water brine lenses (cryopegs) found in Arctic permafrost derived from ancient marine sediment layers of the Polar Ocean. The cryopegs were formed and isolated from sediments about 100 to 120 kyr ago. They remain liquid at the *in situ* temperature of −10 °C as a result of their high salt content (170–300 g/l). The lenses of brine within permafrost comprise the only free water habitat on Earth that is characterized by permanently subzero temperatures, high salinity, and isolation from the influences of external factors during geological time.

A sensitive method to study microbial activity at subzero temperatures is to measure their metabolic activity with ^{14}C-labeled substrates. Therefore, the activity of the cryopeg community also was determined by measuring the uptake of ^{14}C-labeled D-glucose (sp. act, 250 mCi/mol) into the bacterial biomass. Five-ml aliquots of brines were supplemented with 2 μCi glucose, and incubated at temperatures from 20 to −15 °C for 24 hours. With decreasing incubation temperature, incorporation of ^{14}C from glucose by the cryopeg microbial biomass decreased from 200,000 cpm at room temperature, but remained significant even at −15 °C (8500 cpm), indicating that microbial metabolism at low temperatures occurs in this habitat.

2. Constructive (anabolic) metabolism, resulting in the formation of bacterial lipids at subzero temperatures, was detected in Arctic permafrost samples up to 3 million years old. Metabolic activity was measured in the laboratory of E.I. Friedmann (then located at Florida State University) at temperatures between 5 and −20 °C on the basis of incorporation of ^{14}C-labeled acetate into lipids by samples of a natural population of bacteria from permanently frozen soil. Incorporation followed a sigmoidal pattern similar to growth curves. At all temperatures, the log phase was followed, within 200 to 350 days, by a stationary phase, which was monitored until day 550 of activity. The minimum doubling times ranged from one day (@5 °C) to 20 days (@−10 °C) to ca. 160 days (@−20 °C). The curves reached the stationary phase at different levels, depending on the incubation temperature (18).

3. The radioisotope approach is a sensitive method to measure energetic metabolic activity at subzero temperatures. We now report the results of an investigation of methane formation in frozen sediments. For determination of hydrogenotrophic and acetoclastic methanogenesis, the samples were aseptically crumbled with a knife; 5 g of sample were placed in a 20 ml sterile vial and purged with N_2. During this preparation, the temperature of the sample was increased to 0 °C. 100 μl of $NaH^{14}CO_3$ (18 °Ci/mol) or $Na^{14}CH_3CO_2$ (40 Ci/mol) solution containing 10 μCi was injected into the vial. The vial was intensively mixed by shaking, and placed in a refrigerated liquid bath (VMR Scientific Products) for incubation. The temperatures of incubation were 5, −1.8, −5, −10, and −16.5 °C. The period of incubation was 3 weeks. After incubation, the samples were fixed by injection with 10 ml of solution: 200 g/l NaCl, 1N KOH, and 0.1 ml antifoam reagent (antifoam A concentrate, Sigma). Dead controls were created by injection of KOH (pH = 12) directly after the labeled substrate was added. Counts per minute in dead controls did not exceed 250. Newly formed radioactive CH_4 was removed from the experimental vials by air flow (50 ml/min). It was passed through a drexel bottle with a solution of 200 g/l NaCl, 1 N KOH, and 100 μl antifoam reagent, and combusted to $^{14}CO_2$ at 500 to 700 °C with cobalt oxide as a catalyst (16). At the final stage $^{14}CH_4$, oxidized to $^{14}CO_2$, was absorbed in the vial with a mixture of 2 ml β-phenylethylamine and 10 ml of Universal LSC cocktail (Sigma). The full re-

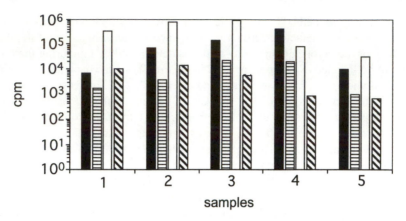

FIGURE 7.2. Incorporation of ^{14}C from different substrates at different temperatures in methane: 1 (black bars) from carbonate at 20 °C, 2 (bars with horizontal lines) from carbonate at −5 °C, 3 (open bars) from acetate at 20 °C, and 4 (bars with diagonal lines) from acetate at −5 °C.

moval time did not exceed one hour. Vials were counted on a liquid scintillation counter (LS 5000 TD, Beckman Company).

During the experiment, ^{14}CH formation was observed in all $^{14}CH_4$-containing samples, but there was no $^{14}CH_4$ formation in samples that did not originally contain CH_4. The amount of methane formed at −5 °C in the samples of Holocene age is only one to two orders of magnitude lower in comparison with the amount of $^{14}CH_4$ formed in the same samples at 20 °C (Figure 7.2). At subzero temperatures methane formation, from both CO_2/H_2 and acetate, was observed in native permafrost sediments at temperatures down to −17 °C (20), that is, in the absence of liquid water.

Discussion

The microbial activity detected in cryopegs documents the fact that subzero temperatures themselves do not exclude biochemical reactions. The observed consumption of labeled glucose at these temperatures and salinity provides reason to conclude that in over-cooled water, microbial metabolism occurs, which does not preclude the possibility of microbial reproduction *in situ*. From the astrobiological perspective, mineral-enriched brines provide the only opportunity for free water within the Martian subsurface permafrost (11). Arctic cryopegs, preserved dozens to hundreds of thousands of years and longer, represent the terrestrial model of such exobiological niches with their unique community—a plausible prototype for sub-Martian microbial life.

The results of incorporation of ^{14}C-labeled acetate into lipids can be interpreted as follows. The thawing and mixing of the soil at the beginning of the experiment destroyed the physical structure of the permafrost, including the osmotic gradients in the water films. The rapid freezing of the samples at -20 °C at the beginning of the experiment resulted in the formation of a permanent ice structure within a relatively short time, probably minutes (5), so the channels to sources of nutrients were fully open. Metabolic activity (nutrient uptake) in the log phase and calculated doubling times reflect the physiological growth potential under optimal (e.g., laboratory) conditions, whereas under natural stable permafrost conditions the bacterial population is in a stationary phase. Microbial life in nature is rarely in the log phase of growth (17). Thus, we concluded that measurable metabolic activity of permafrost bacteria is possible at temperatures down to at least -20 °C (Figure 7.3), but in the stationary phase, which is reached less than one year after freezing, the level of activity, if any, is not measurable with present methods.

We suggest that methane formation below and above the freezing point is carried out by different methanogenic communities—psychrophilic or psychrotolerant, respectively. Although currently unexplainable, the most important result of our experiment, nevertheless, is that even under these conditions there was methane generation. It means that the extra low amount of bound water is quite enough to carry out a complete set of such biochemical processes. It is very likely that this strategy at subzero temperature is carried out by specific groups of psychrophilic methanogens that cannot be isolated by routine microbiological techniques. The data show that in a natural dry, deep frozen environment the redox reactions utilized by ancient microorgan-

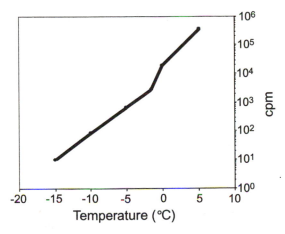

FIGURE 7.3. Incorporation of ^{14}C-labeled acetate by the native microbial population in Siberian permafrost.

isms represent an unknown universal strategy of life maintenance over the eons. For this reason, we contend that this strategy might operate with no time limits. These results were obtained in conditions close to Arctic permafrost with temperatures higher than −15 °C. For cells within Antarctic permafrost with temperatures below −20 °C, additional experiments are needed.

Conclusions

We conclude that viable microorganisms have survived at subzero temperatures for hundreds of thousands to millions of years in permafrost (10), and ice (1), and in these niches retain viability in the absence of free water. Under such conditions, within the native permafrost samples using ^{14}C-labeled substrates both constructive and energetic metabolic activity of ancient microorganisms was detected. From our point of view, viable cells are present in an overcooled, but not a frozen state, within Arctic permafrost at permanent temperatures higher than −15 °C. These cells are surrounded by an unfrozen water film (Figure 7.4 A). In Antarctic permafrost at temperatures below −20 °C, probably, the cells and surrounding water phase are in the frozen state (Figure 7.4 B).

The studied nonsaline loamy soil contains 5 to 6% of unfrozen water at −1.5 °C, and 1 to 2% unfrozen water at temperatures between −15 and −20 °C. In permafrost, thin films of this water cover both soil particles and bacterial cells. Nutrients reach the cells and waste products are eliminated by diffusion through the narrow channels of unfrozen water. In permafrost, therefore, access to nutrients and the ability to eliminate waste materials are limited by the thickness of the unfrozen films of water, which in turn depends on temperature. The thickness of the films decreases from > 8 nm at −1.5 °C to < 0.5 nm at −10 °C (Figure 7.4B, 3). It is significant that the shapes of the curves for the levels of stationary phases, the amounts of unfrozen water in permafrost, and the thickness of its layers were remarkably similar. We suggest that the temperature dependence of the levels of stationary phases is due to the decrease in the thickness of the liquid water layer in permafrost. Thus, it is not the absolute exhaustion of nutrients but the inaccessibility of nutrients due to a diffusion barrier formed at different levels depending on the temperature that limits metabolic activity and results in the asymptotic behavior of the radioactive counts (flattening of the curves) with time.

Our experiments only begin to reveal the capabilities of microorganisms to survive and even to thrive at temperatures below the freezing point of water. We suggest the need for development of more sensitive equipment and methods to continue and indeed expand these kinds of studies of the cryobiosphere. The mechanisms that permit life to exist at subzero temperatures on Earth probably also extend to environments beyond the planet Earth. The results obtained in permafrost sediments give insights for the newly emerging field

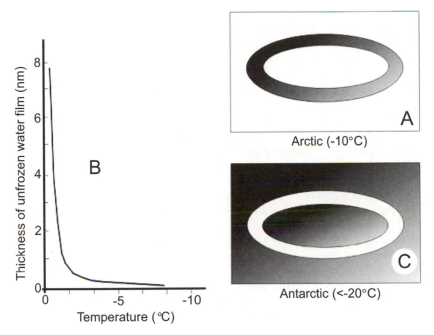

FIGURE 7.4. Unfrozen water and microbial cells in the permafrost environment. (A) Arctic permafrost (t = −10 °C). (B) Unfrozen water thickness in Arctic permafrost. (C) Antarctic permafrost (t = < −20 °C). The white ellipse in (A) represents a super-cooled but unfrozen cell, and the gray area represents an unfrozen water film. The gray ellipse in (C) represents a frozen cell, and the surrounding white shell represents a frozen water film.

of exobiology. Viable chemolithotrophic, psychrotolerant anaerobic microbial communities within permafrost represent an extraterrestrial model for potential life forms with their unique mechanisms to assimilate carbon dioxide and other compounds that might be found in frozen subsurface environments on cryogenic planets without free oxygen, inaccessible organic matter, and a water phase near zero.

Acknowledgment

This work was supported by the Russian Foundation for Basic Research (grants 01–05–65043 and 01–04–49084), the Russian Federal Scientific Program (contract - 43.016.11.1625), and the National Science Foundation (grant LExEn–9978271).

Literature Cited

1. Abyzov, S.S. 1993. Microorganisms in Antarctic ice. pp 265–295. In E.I. Friedmann (ed.), Antarctic Microbiology. New York. Wiley.
2. Alperin, M.J., and W.S. Reeburgh. 1985. Inhibition experiments on anaerobic methane oxidation. Appl. Environ. Microbiol. 50: 940–945.
3. Anderson, D.A. 1967. Ice nucleation and the substrate-ice interface. Nature 216: 563–566.
4. Carpenter, E.J., S. Lin, and D.G. Capone. 2000. Bacterial activity in South Pole snow. Appl. Environ. Microbiol. 66: 4514–4517.
5. Ershov, E.D. 1998. General Geocryology. Cambridge University Press, Cambridge.
6. Finegold, L. 1996. Molecular and biophysical aspects of adaptation of life to temperatures below the freezing point. Adv. Space Res. 18(12): 87–95.
7. Friedmann, E.I. 1994. Permafrost as microbial habitat. pp. 21–26. In D. Gilichinsky (ed.), Viable Microorganisms in Permafrost. Russian Academy of Sciences, Pushchino.
8. Friedmann, E.I., L. Kappen, M.A. Meyer, and J.A. Nienow. 1993. Longterm productivity in the cryptoendolithic microbial community of the Ross Desert, Antarctica. Microbial Ecol. 25: 51–69.
9. Geiges, O. 1996. Microbial processes in frozen food. Adv. Space Res. 18(12): 109–118.
10. Gilichinsky, D. 2002. Permafrost as a microbial habitat. pp. 932–956. In G. Bitton (ed.), Encyclopedia of Environmental Microbiology. Wiley, New York.
11. Gilichinsky, D. 2002. Permafrost model of extraterrestrial habitat. In G. Horneck and C. Baumstark-Khan (eds.), Astrobiology. Springer-Verlag, New York.
12. Gilichinsky, D., E. Vorobyova, L. Erokhina, and D. Fyodorov-Davydov. 1992. Long-term preservation of microbial ecosystems in permafrost. Adv. Space Res. 12: 225–263.
13. Gilichinsky, D., S. Wagener, and T. Vishnivetskaya. 1995. Permafrost microbiology. Permafrost and Periglacial Processes 6: 281–291.
14. Jansen, H., and E. Bock. 1994. Profiles of ammonium, nitrite and nitrate in permafrost soils. pp. 27–36 In D. Gilichinsky (ed.), Viable Microorganisms in Permafrost. Russian Academy of Sciences, Pushchino.
15. Kappen, L.B., B. Schroeter, C. Scheidegger, M. Sommerkorn, and G. Hestmark. 1996. Cold resistance and metabolic activity of lichens below 0 °C. Adv. Space Res. 18(12): 119–128.
16. Laurinavichius, K.S., and S.S. Beljaev. 1978. The rate of microbial methane production determined by the radioisotope technique. Microbiology 47: 1115–1117 (in Russian).
17. Morita, R.Y. 1997. Bacteria in Oligotrophic Environments. Chapman & Hall Microbiology Series, New York.
18. Rivkina, E., E.I. Friedmann, C. McKay, and D. Gilichinsky. 2000. Microbial activity of permafrost bacteria down to −20 °C. Appl. Environ. Microbiol. 66: 3230–3233.

19. Rivkina, E., D. Gilichinsky, S. Wagener, J. Tiedje, and J. McGrath. 1998. Biogeochemical activity of anaerobic microorganisms from buried permafrost sediments. Geomicrobiology 15: 187–193.

20. Rivkina, E., K. Laurinavichius, D. Gilichinsky, and V. Shcherbakova. 2002. Methane generation in permafrost. Doklady Biological Sciences. 383: 179–181.

21. Rivkina, E., D. Gilichinsky, C. McKay, and S. Dallimore. 2001. Methane distribution in permafrost: Evidence for an interpore pressure methane hydrate. pp. 487–
497. In R. Paepe and V. Melnikov (eds.), Permafrost Response on Economic
Development, Environmental Security and Natural Resources. NATO science
series. Kluwer Academic Publisher, Dordrecht/Boston/London.

22. Russell, N.J. 1990. Cold adaptation of microorganisms. Phil. Trans. R. Soc. London
B Biol. Sci. 326: 595–611.

23. Russell, N., and T. Hamamoto. 1998. Psychrophiles. pp. 25–46. In K. Horikoshi
and W. Grant (eds.), Extremophiles: Microbial Life in Extreme Environments.
Wiley, New York.

24. Schroeter, B., T.G.A. Green, L. Kappen, and R.A. Seppelt. 1994. Carbon dioxide
exchange at subzero temperatures: Field measurements on *Umbilicaria aprina*
in Antarctica. Crypt. Bot. 4: 233–241.

25. Soina, V.S., E.V. Lebedeva, O.V. Golyshina, D.G. Fedoriv-Davydov, and D.A. Gilichinsky. 1991. Nitrifying bacteria from permafrost deposits of the Kolyma lowland. Microbiologiya 60: 187–190 (in Russian).

26. Vestal, J.R. 1988. Carbon metabolism of the cryptoendolithic microbiota from the
Antarctic desert. Appl. Environ. Microbiol. 54: 960–965.

27. Wilson, G.S., P. Braddock, S.L. Forman, E.I. Friedmann, E.M. Rivkina, J.P. Chanton, D.A. Gilichinsky, D.G. Fyodorov-Davidov, V.E. Ostroumov, V. Sorokovikov, and M.C. Wizevich. 1998. Coring for microbial records of Antarctic climate. Ant. J.U.S. 1996 Review 31(2): 83–86.

28. Zigert, C. 1987. Greigite and mackinawite in quaternary deposits of central Yakutia. Mineralogical J. 9: 75–80 (in Russian).

8

Yeasts Isolated from Ancient Permafrost

Rushaniya N. Faizutdinova, Nataliya E. Suzina, Vitalyi I. Duda,

Lada E. Petrovskaya, and David A. Gilichinsky

PERMAFROST usually remains frozen, even during interglacial periods. Therefore, permafrost remains unchanged for very long geological time periods.
There is incontrovertible proof that permafrost has never thawed (10). Permanent subzero temperatures, low oxygen concentration, absence of free water,
and high salt concentrations in thin water films that surround organic and mineral particles are features of this habitat.

At the same time, permanently frozen permafrost is a very stable natural
habitat that provides advantages to its inhabitants. Daily, day-to-day, and seasonal fluctuations of many environmental factors to which surface biota must
tolerate or adapt are absent. Therefore, it seems likely that microorganisms
should be able to survive in a stable, although harsh environment like permafrost. Indeed, prokaryotes have been detected in permafrost using microbiological and molecular biological techniques (8, 9, 14, 17, 29–31). But how long can
eukaryotic microorganisms, yeasts, for example, survive in this environment?

Yeasts are common inhabitants of cold regions. Viable yeasts have been
isolated from extremely cold or highly saline habitats, such as Antarctica (2,
3, 4, 6, 12, 19–24) and seawater (15, 28). Eukaryotes have been isolated from
permafrost less than 10,000 to about 40,000 years old (i.e., the Holocene period) (31), but not in older permafrost. Viable yeasts then were detected in
permafrost from 10,000 up to 3 million years old (5, 25).

In this study, we report the isolation of yeasts from permafrost, determination of the number of viable yeast cells, the results of a comparison of various
incubation conditions and medium supplements on the ability to isolate yeasts,
and the results of a preliminary study of their ultrastructure.

Isolation Protocol

We have investigated seven different permafrost samples from the Kolyma
lowland (northeast Siberia, Russia, 68°–72°N; 152°–162°E) (Table 8.1) for the
presence of viable yeasts. The permafrost cores were extracted from different

TABLE 8.1

Characteristics of permafrost samples collected in the Kolyma lowland (northeast Siberia, Russia, 68°–72°N; 152°–162°E), and assayed for the presence of viable yeasts

Sample Number	Borehole/Year/ Depth (m)	Location	Sample Characteristics	Age in Years (×1000)	Viable Yeasts
1	11/89/32.2	Khomus-Yuryakh River, right bank	Fine-grained sand	ca. 3000	+
2	6/90/20.7	Bol'shaya Chukochiya River, right bank	Sandy loam	600–1800	—
3	6/91/4.8	Chukochiya River mouth	Sandy loam with sulphides	5–10	—
4	1/93/3.7	Konjkovaya River floodplain	Silted-up sandy loam	15–30	+
5	2/94/11.0	Chukochiya River, midstream	Silted-up sandy loam	1800–3000	—
6	4/2000/7.0	Chukochy Cape	Loam	40–50	—
7	4/2000/9.1			40–50	—

depths (i.e., ages; see ref. 26 for a detailed discussion concerning the determination of permafrost age). Cores were extracted by dry drilling without fluid, which would contaminate the samples. The corer cuts 20- to 30-cm-long cores. The core surfaces were cleaned by shaving with an ethanol-sterilized knife after removal from the corer. The cores then were split into ca. 5-cm-long segments, placed in sterile metal boxes, and stored in the field in a cave dug in the permafrost at approximately −10 °C, and transported frozen to the laboratory. Detailed protocols for sample selection and sterility control were published earlier (17). Briefly, to monitor the possibility of contamination during the drilling process, we carried out several tests. The drilling barrel was seeded with a pure culture of *Serratia marcescens* for 2 hours before drilling. In a separate test, drilled frozen core segments were seeded with a pure culture of *S.marcescens* for several hours to several months at −10 °C. In tests using the isolation techniques described below, *S. marcescens* was found only on the surface of the frozen sample, never inside the frozen cores (17).

The samples were maintained in a deep-freezer in the laboratory. Material for analysis was aseptically removed from the inner part of a frozen core with a scalpel. A 1 g sample (wet weight) was placed in a preweighed sterile flask, and sterile water added to give a 1:10 dilution. The suspensions then were shaken at room temperature for 10 to 15 minutes. The suspensions were diluted 1:10 once again to give final dilutions of 1:10 and 1:100.

The diluted suspensions from the seven permafrost samples then were plated on malt agar (MA) (6° Balling, 2% [w/v] agar) plates, which were incubated at 4 or 20 °C (five replicates per sample). MA is widely used for the isolation of yeasts (27).

Modified Isolation Conditions for the Quantification of Isolated Yeasts

Cells that remain viable for long periods under extreme (though stable) conditions may be viable but nonculturable, or particularly sensitive to nutrient levels in the cultivation media. Thus, we modified MA with various supplements to quantify viable yeasts. We supplemented MA with sodium pyruvate at 0.1% (MA-SP) because SP is an activator of metabolic processes. MA also was supplemented with lactic acid at 0.4% (MA-LA) to inhibit bacterial growth. The plates were incubated at 20 and 4 °C, as described above.

The number of colony-forming units (CFU) that developed were counted. Light microscopy was used to differentiate the yeast from the bacterial colonies that developed. Cell size and morphology, pigmentation, consistency, budding type, and so on were recorded. Samples of each morphological type were subcultured on fresh plates to obtain monocolonial isolates. The number of distinct yeast isolates (DYI) based on their colony and cell morphology was determined (Table 8.2).

Results

We did not isolate viable yeasts from five of the seven permafrost samples assayed on plain malt agar. The isolation conditions that we used perhaps were not suitable for isolation of yeasts from them. At the same time, the failure to isolate microorganisms from these permafrost samples provides evidence that the sterility precautions we used to handle the permafrost samples were adequate.

We did isolate viable yeasts from two of the permafrost samples. One of these samples is 15,000 to 30,000 years old (Late Pleistocene period) and the other is approximately 3 million years old (Late Pliocene period, Table 8.1). The viable yeasts isolated from these cores are the only known living eukaryotic microorganisms (along with filamentous fungi and green algae, chapters 9 and 10) known to be preserved for this length of time.

Yeasts Isolated from Permafrost

MA without supplements showed the best ratio between CFU count and the number of DYI both at 20 and 4 °C. The addition of SP slightly increased the

TABLE 8.2
The number of colony-forming units (CFU) and distinct isolates of yeasts (DYI) isolated from one g (dry weight) of permafrost sample no. 1 (borehole 11/89, Kolyma lowland, Khomus-Yuryakh River basin, northeast Siberia, Russia, depth 32.2 m, age up to 3 million years, iciness w(i) = 21.5%, total organic carbon 0.29%) diluted 1:10 or 1:100, plated on malt agar alone or with supplements, and incubated at 4 or 20 °C.

		Incubation Temperature				
		+20°C			+4°C	
Media	MA	MA + Sodium Pyruvate	MA + Lactic Acid	MA	MA + Sodium Pyruvate	MA + Lactic Acid
CFU, 1:10	0.3×10^4	0.34×10^4 $0.16x\ 10^{4}*$	$0,2 \times 10^3$	0.42×10^4	0.66×10^4 $0.14x\ 10^{4}*$	0.2×10^3
DYI, 1:10	6	5	3	7	5	5
CFU, 1:100	0.4×10^3	0.31×10^3 $0.19 \times 10^{3}*$	0	0.3×10^3	0.54×10^3 $0.16 \times 10^{3}*$	0.2×10^2
DYI, 1:100	3	3	0	4	3	1

MA = Malt agar.
* Bacteria.

total CFU count because bacterial colonies were isolated, as well as yeasts, at both temperatures. Thus, the addition of SP did not increase the number of DYI isolated or yeast CFU (table 8.2). Bacterial colonies grew concurrently with yeasts on MA-SP. MA-LA, however, reduced both the CFU count and the diversity of yeasts (table 8.2). Incubating plates at 4 °C increased both the CFU count and the DYI compared with incubation at 20 °C (Table 8.2). This effect was especially evident when samples were diluted 1:100, and cultivated on MA-LA (table 8.2). LA apparently inhibited the growth of both bacteria and yeasts. Yeasts were not isolated from MA-LA when samples were diluted 1:100 and incubated at 20 °C. However, a few yeasts were isolated when such plates were incubated at 4 °C. Incubation at 4 °C gave a more complete picture of microbial biodiversity and the number of viable yeasts in ancient permafrost samples than incubation at 20 °C. Metabolic rates are decreased at lower temperatures, which allows the slowly growing yeasts to develop. Incubation at room temperature still may be used to isolate the rapidly growing yeasts.

The relative number of different yeast isolates recovered from the 3 million-year-old permafrost sample when incubated on various media at 20 and 4 °C is shown in Figures 8.1 and 8.2, respectively. The best medium for the isolation of the largest number of yeasts from this sample was MA, on which six different yeast isolates were isolated at 20 °C and seven at 4 °C (Figures 8.1 and 8.2). Some isolates predominated, whereas others although not numerous were consistently present (Figures 8.1 and 8.2). The growth of bacteria was completely inhibited on MA or MA supplemented with lactic acid. However, sodium pyruvate apparently activated bacterial growth on MA (see bar 11 on Figure 8.1, and bar 13 on Figure 8.2).

Almost all of the yeast isolates are basidiomycetous, belonging to the genus *Cryptococcus* or *Rhodotorula* based in part on some physiological and biochemical characteristics: glucose fermentation; formation of starchlike compounds; *myo*-inositol, D-glucuronate, and nitrate assimilation, and urease reaction (Figures 8.1 and 8.2). Nonfermentative basidiomycetous yeasts often are detected in tundra soils (1) and in Antarctica (2, 19–24), where nutrient supplies are very meager. Some isolates were not identified. However, identification of some of the isolated yeasts was confirmed and species assigned by analysis of the large-subunit rDNA D2 domain sequence as *Cryptococcus saitoi, C. albidosimilis, c. victoriae, C. antarcticus*, and *Rhodotorula creatinivora* (data not shown).

Ultrastructure of Ancient Yeasts

For electron microscopy, yeast cells were fixed with 2% glutaraldehyde in 0.05 M cacodylate buffer (pH 7.2) and then in 2% OsO_4 in the same buffer, dehydrated in ethanol, and embedded in EPON 812 epoxy resin. Ultrathin sections were obtained with an Ultracut E (Reichert, Austria) ultramicrotome, contrasted with lead citrate, and examined under a JEM-100B transmission electron microscope.

The ultrastructure of five yeasts freshly isolated from permafrost was investigated. All of the "ancient" yeast isolates examined have a basidiomycetous

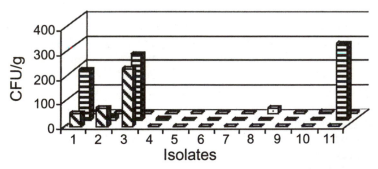

FIGURE 8.1. Relative amounts (CFU/g) of each of ten different yeasts. Isolates 1, 3, and 6 are species of *Rhodotorula*; 2, 7, 8, and 10 are species of *Cryptococcus*; isolates 4, 5, and 9 are unidentified; and isolate 11 is bacteria. See text for details. Yeasts were isolated from 3 million-year-old permafrost samples diluted 1:10, inoculated onto three different culture media [\\= malt agar (MA), = =MA +sodium pyruvate, and . . . = MA + lactic acid], and incubated at +20 °C.

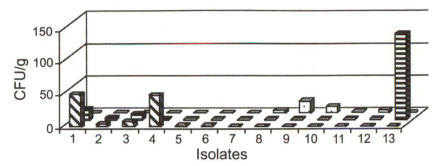

FIGURE 8.2. Relative amounts (CFU/g) of each of twelve different yeasts. Isolates 1, 2, 3, 5, 8, 10, and 11 are species of *Cryptococcus*. The remaining isolates are species of *Rhodotorula*) and bacteria (isolate 13). See text for details. Yeasts were isolated from 3 million-year-old permafrost samples diluted 1:10, inoculated onto three different culture media, [\\ = malt agar (MA), = =MA +sodium pyruvate, and . . . = MA + lactic acid], and incubated at +4 °C.

cell wall. A peculiar ultrastructural feature is the large amount of cellular lipids in both freshly isolated permafrost yeasts and isolates passaged several times on nutrient medium. The important role of endogenous lipids in metabolic regulation of hibernation and estivation is recognized (18). Well-developed capsules also were observed in permafrost isolates (Figure 8.3). The significance of yeast capsules for survival in extreme conditions is well known (11). Freshly isolated yeast strains also exhibit polymorphism (Figure 8.3), which may reflect intensive adaptation processes including possible rearrangements of general metabolism. The polymorphism of newly isolated cells is more common than that observed in passaged cells. Cellular polymorphism also has been described for cancer cells, which are not differentiated (16), for microbial cells without cell walls (i.e., *Mycoplasma*), and for cells that have been carried over from deficient onto enriched media, which is analogous to the present situation.

Conclusions

Several phylogenetic trees for yeasts based on nucleotide sequences have been created (6, 13). Indeed, the complete genomes of some yeasts, for example, *Saccharomyces cerevisiae* and *Schizosaccharomyces pombe*, have been sequenced. It has been estimated, however, that less than 10% of the yeasts that exist in nature have been described. A more complete picture of microbial diversity requires a better understanding of yeast diversity. However, all of the yeast species collected and used for genetic and ecological investigations are

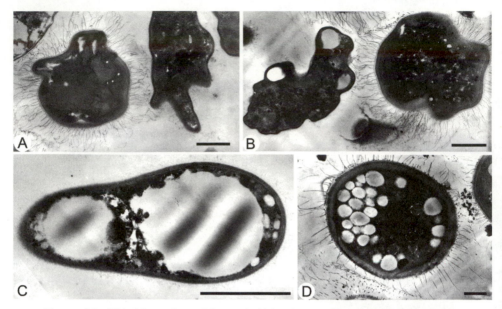

FIGURE 8.3. Ultrathin sections of the newly isolated yeasts from permafrost borehole 11/89 (depth 62.2 m, age ca. 3 million years, iciness, i = 21,5%, total organic carbon 0.29%) showing polymorphism. (A and B) Yeast isolate 11/00 *Cryptococcus antarcticus* var. *circumpolaris*. (C) Yeast isolate 14/00 *Cryptococcus saitoi*. (D) Yeast isolate 15/00 *Rhodotorula creatinovora*. Bars = 1 μm.

"contemporary." The possibility of using "ancient" yeasts, and microorganisms in general, for phylogenetic and other investigations may be very fruitful. The study of yeasts isolated from permafrost may play an important role in understanding eukaryote viability and genome conservation under conditions of prolonged cryopreservation.

Acknowledgments

We thank Prof. W.I. Golubev for his sincere interest and useful recommendations during experimental investigations.

Literature Cited

1. Bab'eva, I.P., and I. Yu. Chernov. 1992. Biology of Yeasts, Moscow University Press, Moscow (in Russian).
2. Bab'eva, I.P., and V.I. Golubev. 1969. Psychrophylic yeasts in the Antarctic oases. Mikrobiologiya 38: 518–524 (in Russian).

3. Baker, J.H. 1970. Yeast, molds, and bacteria from an acid peat on Signy Island. pp. 717–722. In M.W. Holdgate (ed.), Antarctic Ecology, vol. 2. Academic Press, London/New York.

4. Di Menna, M.E. 1966. Yeasts in Antarctic soils. Antonie van Leeuwenhoek 32: 29–38.

5. Dmitriev, V.V., D.A. Gilichinsky, R.N. Faizutdinova, I.N. Shershunov, V.I. Golubev, and V.I. Duda. 1997. Detection of viable yeast in 3-million year-old permafrost soils of Siberia. Mikrobiologiya 66: 655–660 (in Russian).

6. Fell, J.W., T. Boekhout, A. Fonseca, G. Scorzetti, and A. Statzell-Tallman. 2000. Biodiversity and systematics of basidiomycetous yeasts as determined by large subunit rDNA D1/D2 domain sequence analysis. Int. J. System. and Evol. Microbiol. 50: 1351–1371.

7. Fell, J.W., and I.L. Hunter. 1974. *Torulopsis austromarina* sp. nov. A yeast isolated from the Antarctic Ocean. Antonie van Leeuwenhoek 40: 307–310.

8. Gilichinsky, D.A., G.M. Khlebnikova, D.G. Zvyagintsev, D.G. Fedorov-Davydov, and N.N. Kudryavtseva. 1989. Microbial characteristics under cryolythozone sedimentary rocks investigations. Izv. An SSSR, Ser. Geol. 6: 103–115 (in Russian).

9. Gilichinsky, D.A., E.A. Vorob'ova, L.G. Erokhina, D.G. Fedorov-Davydov, and N.R. Chaikovskaya. 1992. Long-term preservation of microbial ecosystems in permafrost. Adv. Space Res. 12: 255–263.

10. Gilichinsky, D.A., S. Wagener, and T.A. Vishnevetskaya 1995. Permafrost microbiology, Permafrost and Periglacial Processes 6: 281–291.

11. Golubev, V.I. 1990 Capsules. pp. 175–198. In A.H. and J.S. Harrison (eds.), The Yeasts Rose, vol. 4. Academic Press, London.

12. Goto, S., J. Sugiyama, and H. Iizuka. 1969. A taxonomic study of Antarctic yeasts. Mycologia 61: 748–774.

13. Hamamoto, M., and T. Nakase. 2000. Phylogenetic relationships among fungi inferred from small subunit ribosomal RNA gene sequences. pp. 57–71. In F.G. Priest and M. Goodfellow (eds.), Applied Microbial Systematics. Kluwer Academic, Dordrecht/Boston/London.

14. Karasev, S.G., L.V. Gurina, E. Yu. Gavrish, V.M. Adanin, D.A. Gilichinsky, and L.I. Evtushenko. 1998. Viable actinobacteria from ancient Siberian permafrost. Kriosfera Zemli II: 69–75 (in Russian).

15. Nakagama, T., M. Hamamoto, T. Nakase, and K. Horikoshi. 1999. *Kluveromyces nonfermentans* sp. nov., a new yeast species isolated from the deep sea. Int. J. System. Bacteriol. 49: 1899–1905.

16. Oberling, Ch., and W. Bernhard. 1961. The morphology of the cancer cells. pp. 405–496. In J. Brachet and A.E. Mirsky, (eds.), The Cell: Biochemistry, Physiology, Morphology, vol. 5, Specialized Cells. Part 2. Academic Press, New York/London.

17. Shi, T., R.H. Reeves, D.A. Gilichinsky, and E.I. Friedman. 1997. Characterization of viable bacteria from Siberian permafrost by 16s rDNA sequencing. Microbial Ecol. 33: 169–179.

18. Storey, K.B. 2001. Turning down the fires of life: Metabolic regulation of hibernation and estivation. pp. 1–21. In K.B. Storey (ed.), Molecular Mechanisms of Metabolic Arrest. BIOS Scientific Publishers, Oxford.

19. Vishniac, H.S. 1985. *Cryptococcus friedmannii*, a new species of yeast from the Antarctic. Mycologia 77: 149–153.
20. Vishniac, H.S. 1985. *Cryptococcus socialis* sp. nov. and *Cryptococcus consortionis* sp. nov., Antarctic basidioblastomycetes. Int. J. Syst. Bacteriol. 35: 119–122.
21. Vishniac, H.S. 1996. Biodiversity of yeasts and filamentous microfungi in terrestrial Antarctic ecosystems. Biodiversity and Conserv. 5: 1365–1378.
22. Vishniac, H.S., and S. Baharaeen. 1982. Five new basidioblastomycetous yeast species segregated from *Cryptococcus vishniacii* emend. auct., an Antarctic yeast species comprising four new varieties. Int. J. Syst. Bacteriol. 32: 437–445.
23. Vishniac, H.S., and W.P. Hempfling. 1979. *Cryptococcus vishniacii* sp.nov., an Antarctic yeast. Int. J. Syst. Bacteriol. 29: 153–158.
24. Vishniac, H.S., and S. Onofri. 2003. *Cryptococcus antarcticus* var. *circumpolaris* var. nov., a basidiomycetous yeast from Antarctica. Antonie van Leewenhoek 83: 231–233.
25. Vorob'ova, E.A., V.S. Soina, M.V. Gorlenko, N.A. Minkovskaya, A.I. Zalinova, A.N. Mamukelashvili, D.A. Gilichinsky, E.M. Rivkina, and T.A. Vishnivetskaya. 1997. The deep cold biosphere: Facts and hypothesis. FEMS Microbiol. Rev. 20: 277–290.
26. Willerslev, E., A.J. Hansen, J. Binladen, T.B. Brand, M.T.P. Gilbert, B. Shapiro, M. Bunce, C. Wiuf, D.A. Gilichinsky, and A. Cooper. 2003. Diverse plant and animal genetic records from Holocene and Pleistocene sediments. Science 300: 791–795. In Supporting Online Material www.sciencemag.org/cgi/content/full/1084114/DC1.
27. Yarrow, D. 1998. Methods for the isolation, maintenance and identification of yeasts. pp. 77–100. In C.P. Kurtzman and J.W. Fell (eds.), The Yeasts. A Taxonomic Study. Elsevier, Amsterdam/Lausanne/New York/Oxford/Shannon/Singapore/Tokyo.
28. Zalar, P., G.S. de Hoog , and N. Gunde-Cimerman. 1999. *Trimmatostroma salinum*, a new species from hypersaline water. Studies in Mycol. 43: 38–49.
29. Zhou, J., M.E. Davey, J.B. Figueras, E.M. Rivkina, D.A. Gilichinsky, and J.M. Tiedje. 1997. Phylogenetic diversity of a bacterial community determined from Siberian tundra soil DNA. Microbiology 143: 3913–3919.
30. Zvyagintsev, D.G., D.A. Gilichinsky, S.A. Blagodatskii, E.A. Vorob'ova, G.M. Khlebnikova, A.A. Arkhangelov, and N.N. Kudryavtseva. 1985. Survival time of microorganisms in permanently frozen sedimentary rocks and buried soils. Mikrobiologiya 54: 155–161 (in Russian).
31. Zvyagintsev, D.G., D.A. Gilichinsky, G.M. Khlebnikova, D.G. Fedorov-Davydov, and N.N. Kudryavtseva. 1990. Comparative characteristics of microbial cenoses isolated from permafrost rocks of different age and genesis. Mikrobiologiya 59: 491–498 (in Russian).

9

Fungi in Ancient Permafrost Sediments
of the Arctic and Antarctic Regions

Nataliya E. Ivanushkina, Galina A. Kochkina,
and Svetlana M. Ozerskaya

IN A CONTEMPORARY MICROBIAL COMMUNITY, microscopic fungi occupy a prominent position by virtue of the large number of species and their role in biogeocenoses. Paleontological studies show that fungi had a prominent role in biocenoses of the past (18). Previous reports of micromycetes isolated from the ancient permafrost of the Arctic and Antarctic regions indicated that viable micromycetes were isolated from permafrost sediments less than 40,000 years old (1, 23). At the same time, viable prokaryotes including actinobacteria, endospore-forming bacteria, proteobacteria, and so on (6, 19, 23) as well as eukaryotic microorganisms (e.g., algae and yeasts) were isolated from permafrost sediments from the entire Pleistocene period (4, 21).

The long-term viability of different groups of microorganisms in nature is determined by a number of factors, including the natural breakdown of biomolecules, inability of cells to reproduce, the impact of radiation, unfavorable temperature, and other physico-chemical conditions (7). Temperature decrease retards cellular metabolic processes, which significantly prolongs microbial viability. Lethal damage that can result from cell supercooling (and freezing) is related to the destruction of cellular structures and changes in biomolecules. Such effects may be caused by the formation of extra- and intracellular ice crystals, cell dehydration, and/or deviations in the pH and the concentration of electrolytes (13, 15). In nature, these processes are greatly weakened by the presence of natural cryoprotectors: sugars, proteins, and similar compounds (8, 13, 15), which are synthesized by the cells themselves or by other organisms that comprise microbial consortia. Films of unfrozen water that surround the organo-mineral particles on which microorganisms are adsorbed can play a role in preserving microbial viability in permafrost (5).

We summarize new data concerning the length of time that microfungi remain in a viable state in ancient permafrost (9, 16). As the cryoconservation of microorganisms under laboratory conditions shows, some of the cells may perish upon thawing of the permafrost. Cell damage can be caused by recrystallization of intracellular ice (at low rates of warming), oxidative and osmotic stresses, as well as the processes connected with phase transitions accompanied by disturbances of cellular structures, particularly the membranes (13, 20). Neutralization of such effects and promoting the repair of cell damage can lead to subsequent cultivation of microorganisms to reveal the total diversity of viable microorganisms present in natural cryoecosystems.

Materials and Methods

Samples of permafrost sediments were collected from the Arctic regions of Russia and Canada, where the average temperature of sandy loam sediments in permafrost is (−7 °C) to (−12 °C); and from the Antarctic regions, where the permafrost temperature drops below −20 °C. At such low temperatures, the sediment pores are filled with ice, and the absence of liquid water and water-bearing horizons precludes downward infiltration and penetration by foreign microorganisms into the frozen strata. The permafrost sediments sampled were from 5000 to 3 million years old (Table 9.1).

Permafrost cores were obtained with a portable gasoline-powered drill that operates without drilling fluid, which would contaminate biological samples. The engine turns and hammers on the drill rods, which are connected to a coring device. The corer cuts 20- to 30-cm-long cores. After removal from the corer, the surfaces of the cores were cleaned by shaving with an alcohol-steri-lized knife. The cores were split into ca. 5-cm-long segments, which were either used immediately for microbiological studies or placed in metal boxes, stored in the field in a cave dug in the permafrost, and transported frozen to the laboratory (while maintaining them below freezing) and were placed in permanent frozen storage at −18 °C. To monitor the possibility of contamination during the drilling process, we carried out several tests. The drilling barrel was seeded with a pure culture of *Serratia marcescens* for 2 hours before drilling. In a separate test, drilled frozen core segments were seeded with a pure culture of *S. marcescens* for several hours to several months at −10 °C. In tests using the isolation techniques described below, *S. marcescens* was found only on the surface of the frozen sample, never inside the frozen cores (19). The samples were stored at −15 to −18 °C.

The material for analysis was removed from the central part of a frozen core using a sterile scalpel under sterile conditions in a laminar flow hood. To isolate fungi, 0.5 g samples were placed in sterile tubes containing 5 ml of sterile water at room temperature (20 °C). Tubes then were heated to 35 °C and 52 °C for one minute. The suspensions then were stirred for 10 minutes at room temperature. Inoculations were made from a 1:10 dilution in triplicate on Czapek's medium and malt extract supplemented with lactic acid (4 ml/l of medium) to inhibit bacterial growth. The isolated fungi were incubated at 4 and 25 °C, and their

TABLE 9.1.

The mean numbers (cells/g air-dried permafrost sample) of micromycetes isolated from permafrost, along with the locations, age, and characteristics of the sediments from which they were isolated.

Number of Samples	Hole/Year	Site Locations (latitude/longitude)	Description of Samples	Depth below Surface (m)	Age (years) (×1000)	Fungal Cells/g Air-dried Permafrost
Kolyma Lowland						
26	6/91	(70°05′N, 159°55′E)	Loamy sand	4.8	5–10	$(0.1–0.7)\,10^2$
27	6/91			12.3	5–10	$(0.1–4.7)\,10^2$
9	4/91		Clay loam,	14.7	100	$(0.2–3.1)\,10^2$
7	4/91		Sea deposit	16.8	100	$(1.3–5.6)\,10^2$
11	5/94	(69°29′N, 156°59′E)	Silt loam	12.6–12.8	200–600	$(0.5–2.4)\,10^2$
35	2/94		Silt loamy sand	11.0–11.5	1800–3000	$(0.1–1.1)\,10^2$
37	2/94		Peat sand	27.5–27.6	1800–3000	$(0.2–1.8)\,10^2$
38	2/94		Peat loamy sand	45.3–45.8	1800–3000	$(1.2–2.9)\,10^4$
12	17/91	(68°57′N, 160°02′E)	Medium sand	3.5	20–30	$(0.7–2.3)\,10^2$
15	17/91			10.2	20–30	$(0.1–1.4)\,10^2$
13	17/91			17.5	20–30	$(0.1–1.5)\,10^2$
21	1/93	(69°23′N, 158°28′E)	Silt loamy sand	4.0	15–30	$(0.1–2.7)\,10^2$
30	1/97		Silt loam	13.0	40	$(0.1–0.3)\,10^2$
24	1/92		Fine sand	48.8	3000	$(0.1–0.4)\,10^2$
Canadian Arctic						
1	Taglu	Mackenzie Delta Region	Silt loamy sand	17.9–18.1	10–2500	$(0.1–0.7)\,10^2$
2	Taglu		Coarse sand	107.2–107.2	10–2500	$(0.1–3.3)\,10^2$
3	Taglu			206.9–206.9	10–2500	$(0.1–2.3)\,10^3$
4	Taglu			303.9–304.0	10–2500	$(2.0–8.9)\,10^2$
5	Unipcat		Silt loamy sand	20.5–20.6	170	$(0.3–1.7)\,10^2$
Antarctic Dry Valleys						
16	4/95	Meyer Valley	Coarse quartz-feldspar sand with rubble and chad	1.4–1.4	30	$(0.1–2.5)\,10^2$
18	4/95			2.0–2.0	30	$(1.8–4.3)\,10^2$
22	3/95	Taylor Valley	with rubble and chad	3.0–3.1	170	$(0.1–0.4)\,10^2$
23	3/95			11.9–12.0	170	$(0.1–1.5)\,10^2$

identity and numbers were determined 30 days postinoculation (dpi) (at 4 °C) and 21 dpi (at 25 °C). The isolates obtained were transferred to malt extract medium and stored at 4 °C. The microfungi were identified on the basis of cultural and morphological features given in the respective manuals (3, 22).

In all experiments, a control inoculation consisting of sterile tap water that was used to prepare the sample suspensions was performed. In addition, open Petri plates containing growth medium for sample inoculation were exposed to the air in the isolation box for 10 minutes, and then incubated at 4 and 25 °C as described above. Fungi were not observed on the control plates.

Quantity of Microfungi Isolated

Viable filamentous fungi were isolated from 21 of 23 samples that ranged in age from 10,000 to 3 million years old (the probability of isolation = 0.95). The quantity varied from 0.1×10^2 to 2.9×10^4 cells per gram of air-dried sample (Table 9.1). The largest quantities usually were detected when samples were cultured on malt extract medium incubated at 4 °C (9). There was not a sharp decrease in the number of fungi isolated from the deeper sediments. On the contrary, in some cases their quantity increased in deeper sediment layers (Table 9.1). Such variations may be completely unrelated to the age of the sediments, but due to the peculiarities of specific soil layers.

Taxonomic Diversity of Microfungi

More than 200 microfungal isolates were recovered and identified, of which 192 were maintained in a collection. These isolates were assigned to 49 species representing 28 genera of the basidiomycota, ascomycota, zygomycota, and anamorphic fungi (Table 9.2). The microfungi most frequently isolated are *Penicillium* (*P. chrysogenum, P. verrucosum, P. rugulosum*), *Cladosporium* (*C. cladosporioides, C. herbarum*), *Aspergillus* (*A. sydowii, A. versicolor*), and unidentified white and dark-colored sterile mycelia. The species mentioned above were isolated from almost 30% of the samples, including the most ancient ones (9). *Geomyces pannorum* var. *pannorum, Geotrichum candidum, Paecilomyces variotii, Botrytis cinerea*, and *Ulocladium botrytis* also were isolated, in addition to those above, but from the most ancient sediment layers (1.8–3 million years old).

An analysis of taxonomic diversity indicates a focal distribution of microfungi in the sediments. The best example is provided by two permafrost samples from different depths of the middle flow of the Chukoch'ya River (Figure 9.1). One of the samples (Figure 9.1 A) showed a sharp increase (ca. 30,000 cells) in the number of fungi, mostly due to the presence of two species, *Penicillium variabile* (in abundance—90.8%) and *Verticillium* sp. (8.8%). Yet, in a sample removed from the same sediment layer but at a more shallow depth, we observed a two-fold increase in species diversity and a more uniform quantitative distribution of fungi. However, the total number was smaller than in the sample discussed above (approximately 200 cells, Figure 9.1 B). The abundance of *Penicillium variabile* decreased to 20%, and new species appeared.

TABLE 9.2.

Species of filamentous microfungi isolated from permafrost

Genus		Species			
Name	Number of Samples*	Name	Number of Isolates	Number of Samples**	Frequency(%)***
Penicillium	17	P. aurantiogriseum Dierckx 1901	3	2	8.7
		P. brevicompactum Dierckx 1901	1	1	4.3
		P. chrysogenum Thom 1910	10	7	30.4
		P. citrinum Thom 1910	1	1	4.3
		P. crustosum Thom 1930	1	1	4.3
		P. decumbens Thom 1910	4	1	4.3
		P. glabrum (Wehmer 1893) Westling 1911	2	2	8.7
		P. minioluteum Dierckx 1911	3	3	13.0
		P. puberulum Bainier 1907	1	1	4.3
		P. purpurogenum Stoll 1904	1	1	4.3
		P. restrictum Gilman et Abbott 1927	2	2	8.7
		P. rugulosum Thom 1910	4	4	17.4
		P. variabile Sopp 1912	5	3	13.0
		P. verrucosum Dierckx 1901	6	5	21.7
Mycelia Sterilia (white)	17		44	17	73.9
Cladosporium	11	C. cladosporioides (Fresenius 1850) de Vries 1952	12	7	30.4
		C. herbarum (Persoon 1794: Fries 1829) Link 1815	14	7	30.4
		C. macrocarpum Preuss 1848	1	1	4.3
		C. sphaerospermum Penzig 1882	1	1	4.3
Aspergillus	10	A. fumigatus Fresenius 1863	2	2	8.7
		A. niger van Tieghem 1867	1	1	4.3
		A. oryzae (Ahlburg 1878) Cohn 1883 (1884)	1	1	4.3
		A. sydowii (Bainier et Sattory 1913) Thom et Church 1926	6	4	17.4
		A. versicolor (Vuillemin 1903) Tiraboschi 1926	5	3	13.0
Mycelia Sterilia (dark)	10		11	11	43.5
Geotrichum	5	G. candidum Link 1809: Fries 1832	5	5	21.7
Genus sp. (Basidiomycetes)	4		5	4	17.4
Alternaria	4	A. alternata (Fries 1832) Keissler 1912	5	3	13.0
Botrytis	3	B. cinerea Persoon 1801: Fries 1832	3	3	13.0
Geomyces	2	G. pannorum var. pannorum (Link 1824: Fries 1832) Sigler et Carmichael 1976	2	2	8.7
Papulaspora	2	Papulaspora sp.	2	2	8.7
Chaetomium	2	C. globosum Kunze 1817: Fries 1832	1	1	4.3
		C. indicum Corda 1840	1	1	4.3

TABLE 9.2. (*cont'd*)
Species of filamentous microfungi isolated from permafrost

Genus		Species			
Name	Number of Samples*	Name	Number of Isolates	Number of Samples**	Frequency(%)***
Arthrinium	2	A. arundinis (Corda 1838) Dyko et Sutton 1979	1	1	4.3
		A. sphaerospermum Fuckel 1873	1	1	4.3
Aureobasidium	1	A. pullulans (de Bary 1866) Arnaud 1910	1	1	4.3
Bispora	1	B. antennata (Persoon 1801: Fries 1832) Mason 1953	2	1	4.3
Chaetophoma	1	Chaetophoma sp.	1	1	4.3
Dipodascus	1	D. aggregatus (Persoon 1801: Fries 1832) Mason 1953	1	1	4.3
Engyodontium	1	E. album (Limber 1940) de Hoog	1	1	4.3
Eurotium	1	E. rubrum Kuoning et al. 1901	1	1	4.3
Fusarium	1	F. oxysporum Schlechtendal 1824 emend. Snyder et Hansen 1940	1	1	4.3
Monodictys	1	M. glauca (Cooke et Harkness) Hughes 1958	1	1	4.3
Mucor	1	M. plumbeus Bonorden 1864	1	1	4.3
Oidiodendron	1	Oidiodendron sp.	1		4.3
Paecilomyces	1	P. variotii Bainier 1907	1	1	4.3
Rhinocladiella	1	R. atrovirens Nannfeldt 1934	1	1	4.3
Sporotrichum	1	S. pruinosum Gilman et Abbott 1927	1	1	4.3
Stachybotrys	1	S. chartarum (Ehrenberg 1818) Hughes 1958	1	1	4.3
Trichoderma	1	T. longibrachiatum Rifai 1969	2	1	4.3
Ulocladium	1	U. botrytis Preuss 1851	1	1	4.3
Verticillium	1	Verticillium sp.	1	1	4.3

* Number of samples in which the genus was detected.
** Number of samples in which the species was detected.
*** Number of samples in which the species was detected as a percentage of the total number of samples tested.

Organic substances enter soil through fractures and spread by water flow to different parts of the soil profile. This process is accompanied by an increase in the number of microorganisms into the particular region in which the organic substrate has been introduced (such regions may be deeply located in the sediments). The majority of the isolated fungi are those that reproduce asexually by producing small, single spores. Such fungi apparently are the most resistant to long-term cryopreservation. Some of these species (*Bispora antennata*, *Alternaria alternata*, and *Ulocladium botrytis*) are multisporous and have melanin in their cell walls. This is not surprising because melanin is believed to

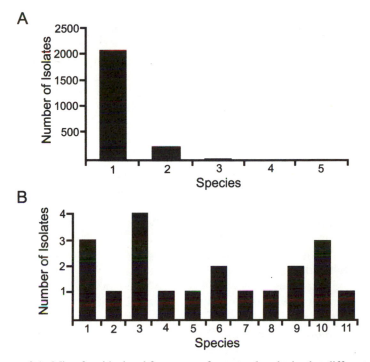

FIGURE 9.1. Microfungi isolated from permafrost samples obtained at different depths from the Chukoch'ya River site in Siberia (A) Sample 38 (hole 2/94, depth 45.3–45.8 m). 1 = *Penicillium variable*, 2 = *Verticillium*, sp., 3 = *Cladosporium cladosporioides*, 4 = *Botrytis cinerea*, and 5 = *Ulocladium botrytis*. (B) Sample 37 (hole 2/94, depth 27.5–27.6 m). 1 = Mycelia sterilia (white), 2 = Mycelia sterilia (dark), 3 = *P. variable*, 4 = *P. chrysogenum*, 5 = *P. verrucosum*, 6 = *Paecilomyces variotii*, 7 = *Geotrichum candidum*, 8 = *Cladosporium herbarum*, 9 = *C. cladosporioides*, 10 = *C. macrocarpum*, 11 = *Aspergillus fumigatus*.

make cells resistant to UV irradiation and extreme temperatures (24). For instance, most fungi found in Antarctica and in the cold Pamirs deserts are melanin pigmented (2).

Some species, for example, e.g., *Arthrinium arundinis, Aureobasidium pullulans, Botrytis cinerea, Dipodascus aggregatus, Fusarium oxysporum, and Geotrichum candidum*, usually are isolated from plant materials and/or are phytopathogens (3, 14, 22). It is possible that vegetative substrates or their derivatives are natural cryoprotectants and provide more favorable conditions for survival of microorganisms when soil horizons become frozen (9, 13). Alternatively, there may be a greater number of microorganisms associated with plants in these permafrost layers because of the presence there of plant materials from which they are derived.

Physiological Peculiarities

The rate of defrosting and warming influences a cell's ability to recover follow-ing deep freezing. In some eukaryotic cells, this ability is enhanced by rapid warming at high temperature (45–52 °C) compared to slow warming at lower temperature (20 °C) (12). Our studies revealed that warming permafrost samples at 35 °C and 52 °C increased the spectrum of the microfungal taxa that was isolated. For example, warming at 35 °C permitted the recovery of *Arthrinium arundinis, Chaetomium indicum,* and *Ulocladium botrytis*, which were otherwise rarely encountered. *Aureobasidium pullulans, Bispora antennata, Oidiodendron* sp., and *Paecilomyces variotii* were isolated from samples thawed at 52 °C.

The incubation temperature also influenced the taxonomic composition and the total number of microfungi isolated. At 4 °C, isolates of *Arthrinium arun-dinis, Aureobasidium pullulans, Bispora antennata, Chaetophoma* sp., *Mono-dictys glauca, Oidiodendron* sp., and *Ulocladium botrytis* were recovered. Mi-crofungi with dark-pigmented mycelia and/or spores made up about 60% of the total number of fungi isolated at 4 °C. The light-colored microfungi were isolated primarily when plates were incubated at 25 °C (9).

Many slow-growing species were isolated at low incubation temperatures: *Aspergillus sydowii, A. versicolor, Eurotium rubrum, Bispora antennata,* and *Stachybotrys chartarum.* Specifically, isolates of *Penicillium* spp. and *Cla-dosporium* spp. were recovered at the highest frequency (Table 9.2).

The growth rates of some of the forty-six isolates of *Penicillium* from per-mafrost sediments did not compare well with those reported by Pitt (17). The ratio of medium cell diameter in several *Penicillium* isolates compared with that reported by Pitt (17) was < one when cultures were incubated at 26 °C and 37 °C, but when incubated at 5 °C the ratio was near 2 (Figure 9.2). The data suggest that permafrost isolates prefer lower temperatures. At 26 °C on MEA and CYA media, isolates of *P. aurantiogriseum* from permafrost grew

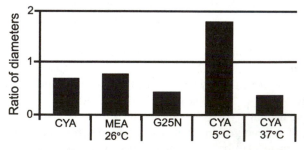

FIGURE 9.2. The average ratio of diameters of *Penicillium* spp. isolated from perma-frost compared with those reported by Pitt (17). CYA = Czapek yeast autolysate agar, MEA = malt extract agar, G 25 N = 25% glycerol nitrate agar.

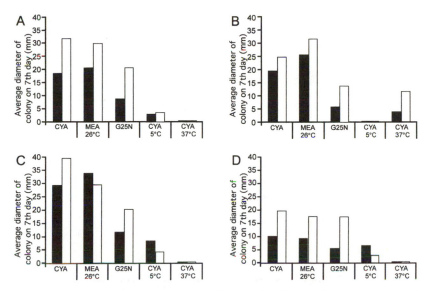

FIGURE 9.3. The growth of various *Penicillium* species isolated from permafrost and cultivated at different temperatures on CYA—(Czapek yeast autolysate agar), MEA (malt extract agar), and G25 N (25% glycerol nitrate agar) compared with their growth under the same conditions as reported by Pitt (17). Dark bars = experimental data, light bars = data from Pitt (17). (A) *P. aurantiogriseum*, (B) *P. decumbens*, (C) *P. chrysogenum*, and (D) *P. verrucosum*.

more slowly, 65% and 94%, respectively, than contemporary isolates on the same media (Figure 9.3 A). Approximately 63% of *P. decumbens* (Figure 9.3 B), *P. restrictum*, *P. rugulosum*, and *P. citrinum* isolates failed to grow at 37 °C, whereas contemporary isolates did grow. At 5 °C, a 1.82-fold increase in growth of isolates of *P. chrysogenum* and *P. verrucosum* was detected compared with literature reports (Figure 9.3 C, D). Six permafrost isolates of *P. crustosum*, *P. minioluteum* and *P. rugulosum* grew at 5 °C, whereas contemporary isolates did not according to the literature. A comparative study of the radial growth rates of *Cladosporium herbarum* and *C. cladosporioides* isolated from permafrost sediments with those isolated from various contemporary habitats showed the distinctions between these populations at different incubation temperatures (Figures 9.4 and 9.5).

Conclusions

Viable microscopic fungi were isolated from permafrost sediments of different origins and ages. In most cases, the quantity of microfungi isolated from more ancient sediments was similar to the quantity isolated from younger sediments. Fungal species with small single-celled spores predominated in per-

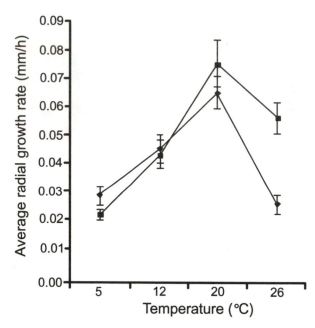

FIGURE 9.4. The average radial growth rate of *Cladosporium herbarum* isolated from permafrost (diamond symbol) and incubated at 5, 12, 20, and 26 °C compared with that of contemporary isolates (square symbol) from our lab.

mafrost sediments, but fungi with large, melanized, multicellular spores also were observed. Many of the isolated microorganisms have adapted to existence at low temperatures. The following observations support this hypothesis:

- a greater number of fungi isolated at 4 °C than at higher temperatures
- the isolation of fungi with large, dark-colored conidia (the genera *Arthrinium, Bispora*, and *Monodictys*) at 4 °C, but not at 25 °C
- the predominance of the genera *Aspergillus* and *Penicillium* among the isolates, as well as many dark-colored hyphomycetes with a slow growth rate, and a temperature range of growth shifted to lower temperatures

These results demonstrate the need to optimize conditions for the resuscitation of frozen ancient microorganisms. We have shown that eukaryotic microorganisms are preserved under conditions of natural cryopreservation for very long periods of time. The preservation of filamentous fungi in permafrost sediments depends on the original composition and density of the ancient microbial consortia, the degree of microbial adaptation to low temperatures, and the protective properties of the environment. All these factors are related to the evolution of biocenoses in ancient times, the geological history of sedi-

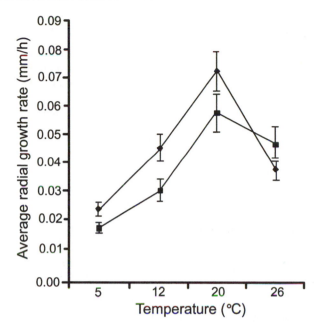

FIGURE 9.5. The average radial growth rate of *Cladosporium cladosporioides* isolated from permafrost (diamond symbol) and incubated at 5, 12, 20, and 26 °C compared with that of contemporary isolates (square symbol) from our lab.

ments, and the conditions of their transition to the permanently frozen state, rather than to their age and, hence, the depth of accommodating sediments.

The isolated fungi are not only of theoretical but also of practical interest because their adaptation to low temperatures might have led to the evolution of enzymes with unique properties. Furthermore, the possibility cannot be excluded that these microorganisms retained the ability to produce some valuable metabolites that are not synthesized by modern microorganisms. We examined the micromycetes isolated from permafrost for new alkaloids. Quinocitrinines A and B were isolated from *Penicillium citrinum* (sample 35, Table 9.1). These compounds are new bioactive representatives of the rare quinoline-type of fungal alkaloids (11). We also reported a new N-acylated derivative of 3,12-dihydroroquefortine from *Penicillium aurantiogriseum* (sample 24, Table 9.1) (10). These two isolates are among the oldest that we have recovered.

Acknowledgment

We thank Dr. D.A. Gilichinskii, Dr. L.I. Evtushenko, and Dr. E.V. Spirina for discussions during the preparation of the manuscript.

Literature Cited

1. Abyzov, S.S., and L.A. Belyakova. 1982. Mycelial fungi from the thickness of the Central Antarctic glacier. Izv. Akad. Nauk SSSR, Ser. Biol. 3: 432–435 (in Russian).
2. Abyzov, S.S., and I.N. Mitskevich. 1993. Microflora of the continental and sea ices of Antarctica. Mikrobiologiya 62(6):582–593 (in English).
3. Carmichael, J.W., W.B. Kendrick, I.L. Conners, and L. Sigler. 1980. Genera of Hyphomycetes. Edmonton, University of Alberta.
4. Dmitriev, V.V, D.A. Gilichinskii, R.N. Faizutdinova, I.N. Shershunov, V.I. Golubev, and V.I. Duda. 1997. Detection of viable yeasts in 3-million-year-old permafrost soils of Siberia. Mikrobiologiya 66:546–550 (in English).
5. Gilichinsky, D., V. Soina, and M. Petrova. 1993. Cryoprotective properties of water in the Earth cryolithosphere and its role in exobiology. Origins of Life and Evolution of the Biosphere 23: 65–75.
6. Karasev, S.G., L.V. Gurina, E. Yu. Gavrish, V.M. Adanin, D.A. Gilichinskii, and L.I. Evtushenko. 1998. Viable actinobacteria from ancient Siberian permafrost. Kriosfera Zemli 2 (2): 69–75 (in Russian).
7. Kennedy, M.J., S.L. Reader and L.M. Swierczynski. 1994. Preservation records of microorganisms: Evidence of the tenacity of life. Microbiology 140: 2513–2529.
8. Kirsop, B.E., and A. Doyle (eds.). 1984. Maintenance of Microorganisms and Cultured Cells: A Manual of Laboratory Methods. Academic Press, London.
9. Kochkina, G.A., N.E. Ivanushkina, S.G. Karasev, E. Yu. Gavrish, L.V. Gurina, L.I. Evtushenko, E.V. Spirina, E.A., Vorob'eva, D.A. Gilichinsky and S.M. Ozerskaya. 2001. Survival of micromycetes and actinobacteria under conditions of long-term natural cryopreservation. Mikrobiologiya 70: 412–420 (in English).
10. Kozlovsky, A.G., V.P. Zhelifonova, T.V. Antipova, V.M. Adanin, S.M. Ozerskaya, N.E. Ivanushkina, F.A. Gollmick, and U. Gräfe. 2003. A new N-carboxylester derivative of 3,12-dihydroroquefortine. Heterocycles 60: 1639–1644 (N7).
11. Kozlovsky, A.G., V.P. Zhelifonova, T.V. Antipova, V.M. Adanin, S.M. Ozerskaya, G.A. Kocshkina, B. Schlegel, H.M. Dahse, F.A. Gollmick, and U. Gräfe. 2003. Quinocitrinines A and B, new quinoline alkaloids from *Penicillium citrinum* Thom 1910 VKM FW-800, a permafrost fungus. J. Antibiotics 56: 188–191 (N5).
12. Lozina-Lozinsky, L. K. 1966. On the capacity of some living systems to endure intracellular crystallization. pp. 33–50. In L.K. Lozina-Lozinsky (ed.), Reactions of Cells and Their Protein Components on the Action of Extreme Factors. Nauka, Moscow. (in Russian).
13. Mazur, P. 1984. Freezing of living cells: Mechanisms and implications. Am. J. Physiol. 247: 125–142.
14. Mel'nik, V.A. 2000. Opredelitel' gribov Rossii: *Hyphomycetes, Dematiaceae*. (The guide of fungi from Russia: Hyphomycetes, Dematiaceae). Nauka, St. Petersburg. 371p (in Russian).
15. Ozerskaya, S.M., N.E. Ivanushkina, S.S. Eremina, K.M. Zaprometova, and G.A. Kochkina. 1996. Cryobank of mycelial fungi in the all-Russian collection of microorganisms (VKM). pp. 145–147. In Materialy soveshchaniya "Kriokonser-

vatsiya geneticheskikh resursov" (Proceedings of the Workshop "Cryopreservation of Genetic Resources"). Pushchino (in Russian).

16. Ozerskaya, S., G. Kochkina, and N. Ivanushkina. 1999. Taxonomic diversity of micromycetes in ancient biocenoses under natural cryoconservation. p. 239. In Proceedings of the IX International Congress of Mycology, Sydney, Australia.

17. Pitt, J.I. 1979. The genus *Penicillium* and its teleomorphic states *Eupenicillium* and *Talaromyces*. Academic Press, London.

18. Popov, P.A. 1967. The micromycetes as object of paleontological investigations. Mikologia i Phitopatologia 1: 158–163 (in Russian).

19. Shi, T., R.H. Reeves, D.A. Gilichinsky, and E.I. Friedmann. 1997. Characterization of viable bacteria from Siberian permafrost by 16S rDNA sequencing. Microbial Ecol. 33: 169–179.

20. Sidyakina, T.M. 1985. Konservatsiya mikroorganizmov (Cryoconservation of Microorganisms). Nauchn. Tsentr Biol. Issled. Akad. Nauk SSSR, Pushchino (in Russian).

21. Vishnivetskaya, T.A., L.G. Erokhina, D.A. Gilichinskii, and E.A. Vorob'eva. 1997. Blue-green and green algac in the Arctic permafrost sediments. Kriosfera Zemli 1: 71–76 (in Russian).

22. Von Arx, J.A. 1981. The genera of fungi sporulating in pure culture (3rd ed.), J. Cramer Vaduz.

23. Vorob'eva, E., V. Soina, et al. 1997. The deep cold biosphere: Facts and hypotheses. FEMS Microbiol. 20: 277–290.

24. Zhdanova, N.N., and A.I. Vasilevskaya. 1982. Ekstremal'naya ekologiya gribov v prirode i eksperimente (Extreme Ecology of Fungi in Nature and Laboratory). Nauk. Dumka, Kiev (in Russian).

10

Viable Phototrophs: Cyanobacteria and Green Algae from the Permafrost Darkness

Tatiana A. Vishnivetskaya, Ludmila G. Erokhina, Elena V. Spirina,

Anastasia V. Shatilovich, Elena A. Vorobyova, Alexander I. Tsapin,

and David A. Gilichinsky

THE WORK reported here is the result of long-standing permafrost studies at the laboratory of Cryology of Soils, Institute for Physicochemical and Biological Problems of Soil Science, Russia. During the last fifteen years, viable prokaryotic and eukaryotic microorganisms have been isolated from buried permafrost soils and sediments up to a depth of 300 m (8, 16, 29, 36, 37). Diverse bacterial communities are dominant inside the permafrost. Indigenous permafrost microorganisms are metabolically active at temperatures as low as −20 °C (14, 27). Active subsurface microbial populations must be chemotrophic, not phototrophic. This leads to the question: Can algae survive at negative temperatures for extended periods of time?

The idea of isolating viable algae from deep subsurface sediments and permafrost is not a new one. The quantitative determination of chlorophyll in sedimentary samples up to 11,000 years old was reported in 1955 (34). Steubing (32) then proved that sedimentary chlorophyll originates from soil algae. Chlorophylls of higher plants degrade rapidly once the leaf dies, for example, tree leaves turn brown when senescing, or chlorophylls are destroyed immediately by fermentation (21) and soil acids (32) once the leaf falls to the ground. The long-term preservation of photosynthetic pigments such as chlorophylls *a* and *b*, and pheophytin under permafrost conditions was shown previously by us (15).

Algae are photosynthetic organisms unlikely to thrive in subsurface darkness. However, there are many examples of the isolation of viable algae from deep subsurface samples. Cameron and Morelli (6) observed many intact cells of green unicellular algae (*Calotrix* sp. and *Chlorococcum* sp.) inside Antarctic permafrost samples from depths up to 68 m. Unfortunately, they did not complete their experiments on the isolation of viable algae. Viable cyanobacteria of the genera *Anacystis*, *Coccochloris*, and *Oscillatoria*, as well as *Protococcus grevillei* were isolated from 0.02 to 0.6 m in Antarctica (5). Viable unicellular green algae, diatoms, and filamentous cyanobacteria were isolated from depths up to 213 m in South Carolina (31). Examples of viable algae isolated from ancient glacial ice are nonexistent to date. There are, however, observations of unicellular algae from different depths of the central Antarctic ice sheet (1) and diatoms within 12,000-year-old ice from Sajama (Bolivia) (7). Environmental clones obtained from 2000- and 4000-year-old ice core samples from Greenland were allocated to the green algal genera *Chlamydomonas* and *Chloromonas* (39).

We isolated viable ancient cyanobacteria and green algae from permafrost sediments that were different in genesis, lithology, and age. Ancient algae have preserved their photosynthetic apparatus in the deep darkness and are able to restore their photosynthetic activity, growth, and development when water and light are made available. After recovery from permafrost, they maintain their morphological and spectral properties and do not differ from typical contemporary strains.

Site Descriptions

The cold polar regions are quite widespread on Earth. Contrary to polar soils, which are characterized by oscillating water availability, long periods of drought, transient freeze-thaw cycles, and repetitive prolonged periods of low temperatures during winter, permafrost soils and sediments represent a constantly frozen, stable, and balanced natural environment. Permafrost soils share similar characteristics: negative temperatures, frozen substrate, thin films of unfrozen water enveloping organic-mineral particles (the thickness of which depends on temperature), higher salt concentration inside water films, increasing water activity, complete darkness, the absence of photochemical reactions, and a very prolonged frozen state.

The permafrost samples were collected during expeditions to the Arctic tundra (Kolyma lowland in northeast Siberia, 67°–70°N, 152–162 °E) and from the polar desert in the Antarctic Dry Valleys (Victoria Land, 77–78°S, 160°–163°E) from 1989 to 1999. Strict protocols for drilling and subsequent handling of the cores were designed to ensure uncontaminated material. Permafrost cores were obtained with a rotary drill that operates without any solutions or drilling fluids. Internal permafrost cores are not melted or contaminated during the drilling process (13). In the field, the handling of permafrost cores was done in a sterile mobile hood with precautions to ensure aseptic manipulation, transportation, and storage of the samples (13, 30). The melted surface material of the extracted core was trimmed away with a sterile knife and the remaining frozen core was divided into sections and immediately placed in presterilized aluminum tins, sealed, and kept frozen during storage and transport. The temperature of extracted cores was never higher than −7 °C. In the

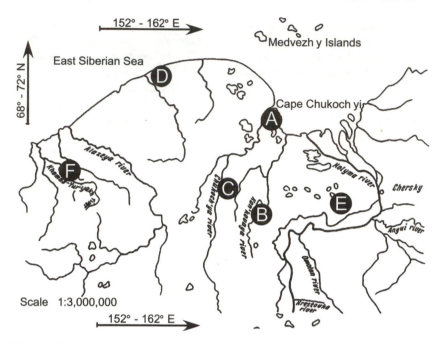

FIGURE 10.1. Location of permafrost sampling sites in the Kolyma lowland of northeast Siberia (marked A–F).

laboratory, the frozen cores were fractured in a class II flow hood with a sterile knife and only internal fragments were taken for microbiological analysis using sterile forceps (29). Altogether, more than 300 permafrost samples were studied. These samples differed in lithology, physico-chemical composition, organic matter content, and age since cryoconservation.

Kolyma Lowland Permafrost Environment

The locations of the Kolyma lowland study sites are shown in Figure 10.1. The permafrost thickness in the tundra ranges from 800 to 1000 m. Annual temperatures at deposition varied from −10 to −12 °C. Late Pleistocene soils with ice wedges up to 30 m thick are the most widespread beneath the sea-sonal thawing upper layers. The thawing starts in June and reaches its maximum in mid-September, so the permafrost table is no lower than 0.3 to 0.5 m in depth. This barrier sharply restricts the influence of external factors and processes. Below this depth, the cross-section is firmly cemented, often to-tally stopped up by ice. In such a closed system, there are no water-bearing horizons or infiltration, and thus the migration downward of matter and moisture is extremely difficult. Hence, under these permafrost conditions, viable microflora isolated from permafrost cannot have originated from the outside but exist *in situ*.

The basic stratigraphic horizons are characterized by mostly hydrocarbonate-calcium freshwater composition, and marine deposits are distinguished by their high sodium chloride salinity. The pH of the medium is nearly neutral and Eh = (+40) − (−250). The organic carbon content varied from 0.5 to 7%, and is independent of bedding depth and age. The tremendous mass of organic carbon not processed by microorganisms is conserved within the frozen sediments. Here, there are 340 to 460 kg of organic carbon per 100 m^3 (18).

Iciness varied from 12 to 20% in sands to the maximum of 40 to 70% when the pores were totally filled with ice, which excludes any migration within the strata. Unfrozen water content is independent of the total iciness of sediments, but is associated with texture: the more dispersed soils have greater unfrozen water content and thicker water films (2). The water films are thinner than the size of the cell, which precludes cell migration along the unfrozen water films. At natural permafrost temperatures, the magnitude of unfrozen water content is 3 to 5% in loams. In the coarse sands it is minimal and tends to be zero. The films of unfrozen water play the major role in the preservation of embedded microorganisms. Unfrozen water films that envelop organo-mineral particles protect viable cells absorbed onto their surface from mechanical destruction by growing ice crystals, and thus serve as the main niches for cells at negative temperatures (14).

The gaseous phase of permafrost pore space is occupied by oxygen, nitrogen, methane, carbon dioxide, and so on. The concentration of oxygen and nitrogen does not differ from their concentration in the air, whereas the concentration of biogenic gases (methane and carbon dioxide) may be appreciably higher than their atmospheric values. These gases, as well as the presence of anaerobic bacteria and the Eh value, suggest anaerobic conditions in permafrost. These gases do not diffuse in the frozen strata (28), and cell migration also is impossible.

Formation of permafrost began at Arctic latitudes in the second half of the Pliocene. Subsequently, prolonged periods of aggradation alternated with degradation. Only in the Kolyma lowland of northeastern Eurasia does the palinospectra solely reflect the tundra and the forest tundra biocenoses, indicating that despite fluctuations, climatic conditions always favored permafrost formation, precluding the degradation of permafrost even during the optima. Syngenetic veins provide the evidence that deposits have never thawed since their initial freezing. The absence of marine sediments indicates that the transgressions did not induce permafrost thawing. The lack of traces of glaciation on the lowland indicates the absence of glaciers under which the permafrost could thaw. Thus, the accumulating and simultaneously freezing deposits since the Pliocene have never thawed under the climatic and geological events in the following epochs, and are preserved until now. These are the most ancient frozen strata in the Arctic, whose age is approximately 3 million years.

Dry Valley Permafrost Environment

Victoria Land occupies almost 2% of the ice-free Antarctic continent (Figure 10.2). This is a polar desert with the most extreme permafrost conditions on Earth. The average annual air temperature is −20 to −35 °C, depending upon elevation and distance from the sea. The annual precipitation in the form of snow is 10 to 45 mm. Even at the negative air temperatures in summer, the soil surface temperature can reach up to 10 °C in an hour (24), although a layer of seasonal thaw does not exist. The upper soil layers are dried and weathered because of snow evaporation and ground ice sublimation. As a result, the soil cover is not formed and the permafrost table coincides with the diurnal surface of large-detrital rocks. Beneath it, in summer a 50 to 70 cm "active layer" is formed, which is not cemented by ice at subzero temperatures. The lining horizons are either the ground ice or the large-grained sorted quartz-feldspar sands that contain an abundance of pebbles and detritus. These are firmly cemented by ice, which has unexpected high contents (25–40% and more) in the subsurface layer, disproving the thesis of the "dry permafrost (< 5% moisture content)" in the Antarctic.

The mean annual temperature of permafrost decreases with air temperature inland and at higher altitudes. At Taylor Valley (50 m above sea level) the sediment temperature is −18.5 °C, at Beacon Valley (1300 m) the temperature drops to −24 °C (in ice and sands), and the value of the down-hole temperature at Mt. Feather (2750 m) is −27 °C. These are the lowest permafrost temperatures on the Earth at which the water-bearing horizons are absent. The permafrost environment of the Dry Valleys differs from that at the Kolyma lowland by (i) being twice as low in sediment temperature, (ii) the absence of surface and ground waters, and (iii) active sublimation that excludes any downward infiltration. For this reason, even more so than in the Arctic, the possibility of penetration of modern microorganisms into the frozen strata is precluded.

The total organic carbon content varies from 0.05 to 0.1% in the aeolian and glacial sands, to 0.11 to 0.25% in the lacustrine sands. The water extract is dominated by sodium ions. The dry residue reflects the freshwater genesis of sediments. The pH of the medium in contrast to Arctic sediments is alkaline. The redox potential (Eh) of the investigated samples varied from +260 to +480, (i.e., these conditions are not as anaerobic as in the Arctic), which is confirmed by the absence of biogenic methane. The unfrozen water values are so small that instruments fail to record them.

The age of the permafrost cores mentioned in Figure 10.2 was estimated as follows. In Miers Valley, the thermoluminescence age of sands from 2.9 m

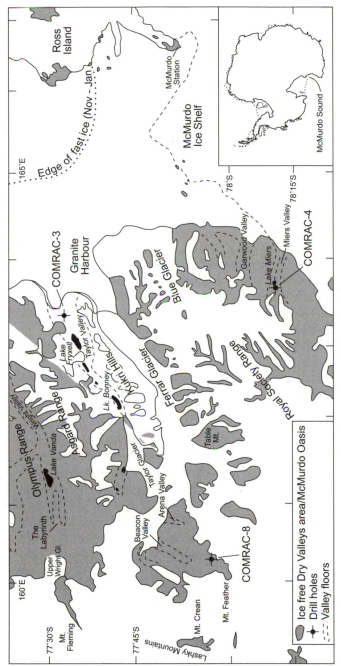

FIGURE 10.2. Location of permafrost sampling sites in the Dry Valleys of Antarctica.

depth in the 956/4 cores is 20,500+/–2,500 and the sands recovered from 4.9 m depth are possibly older than 30,000 years, and were deposited during the last interglacial cycle (40). Chronostratigraphic studies demonstrate that sediments recovered by coring in Taylor Valley (956/3 core) are up to 150,000 years old, by correlation with the magnetostratigraphically dated DVDP-11 core from nearby (10). Coring at Mt. Feather recovered glacigene Sirius Group sediments of at least 2 million years in age (38) and possibly as old as 15 million years (23).

Chlorophyll *a* and Algal Biomass in Permafrost

The *in vivo* content of chlorophyll *a* is a characteristic of algal biomass (9). We had conjectured that permafrost samples, with a high level of total chlorophyll, could contain viable algae. In order to estimate the algal biomass in permafrost and to choose the samples for algae isolation, the total chlorophyll was measured. The permafrost material (one g) was thoroughly shaken with 4 ml of prechilled 80% aqueous acetone in the dark for one hour (32). After filtering the solvent through Whatman No. 50, the absorption and excitation of filtrate was determined with a spectrofluorometer (Hitachi 850, Japan). Concentration of the chlorophyll in the extract as well as the average amount of the total chlorophyll per one g of dry weight sediment was calculated as described (35). The nature of chlorophyll in soil is almost entirely unknown and the chlorophyll degradation products do persist and retain nearly the same absorption maximum as chlorophyll *a* (32). The most likely speculation is that the sedimentary chlorophyll could be derived from higher plants, mosses, different groups of algae, and cyanobacteria. But taking into account that plant chlorophylls are very unstable (32), we assumed that the chlorophyll of the ancient plants was decomposed in permafrost sediments and that measured chlorophyll had its origin in the viable algae and should be related to algal biomass. Despite the difficulties attributed to the fact that the chlorophyll content of plants varies with the type of tissue, the species, and the environment, we applied the method of spectral measuring of chlorophyll to estimate algal biomass in the permafrost samples. The possibility of the preservation of photosynthetic pigments in permafrost, and their indispensability to algae, were discussed in detail previously (15, 35). Many Holocene samples showed high chlorophyll *a* content (Figure 10.3). A sharp decrease of chlorophyll *a* content was observed with increasing permafrost age from Holocene to Pleistocene. The dramatic decrease in chlorophyll *a* content may be related to a reduction in algal biomass possibly caused by cell death or cell damage and the breakdown of chlorophyll *a*. The algae that survive in permafrost are as old as the permafrost itself. Permafrost samples with high fluorescence values were selected for algae isolation because the fluorescence of primary accessory pigments could be a linear function of cell concentration (4).

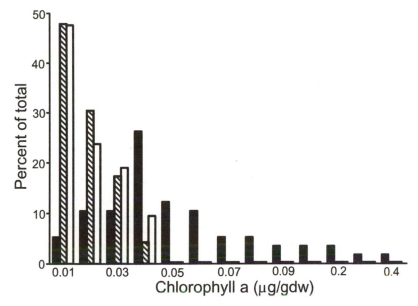

FIGURE 10.3. Algal biomass and age of permafrost sediments. Biomass is expressed as the content of chlorophyll *a* (μg/gram dry weight). The amount of each set of samples is expressed as a percentage of the total. Black bars = Holocene samples, hatched bars = Pleistocene samples, and clear bars = Late Pliocene to Early Pleistocene samples.

Isolation and Distribution of Algae in Permafrost

Permafrost fragments (1–2 g) taken from internal parts of frozen cores were incubated in simple mineral media under continuous illumination from fluorescent white lamps at an intensity of 500 lx for 2 months at both 20 °C and 4 °C to isolate viable green algae and cyanobacteria (35). The precautions and controls taken during the algae isolation are mentioned below. All work was done in a class II positive-flow hood. Tools and glassware before autoclaving were cleaned with 2.5 M HCl or 5% sodium hypochlorite, followed by extensive rinsing in distilled water. All solutions and nutrient media were fresh-made in distilled water and autoclaved. Two types of controls were used to check for possible contamination during sample plating and incubation: (i) sterile control—Petri plates with nutrient media were exposed inside the laminar-flow hood along the work area and no colonies were discovered on them over the 2-month incubation period; (ii) algae *in situ* control—selected permafrost samples were divided, and half of them were autoclaved; later the original and the autoclaved pieces were plated in liquid enrichment medium at the same conditions. No algae were detected in the autoclaved permafrost pieces. To ensure sterile incubation, the Petri plates were covered with parafilm and kept in a hermetically closed sterile glass box. Enrichment cultures were reexamined weekly to assess algal biodiversity. Viable algae growing in enrichment cultures were separated by the classical bacteriological streaking method on Bold's basal medium (BBM) (25) solidified with 1.2% agar.

More than 30% of the Arctic and only 4% of the Antarctic permafrost samples contained culturable green algae and/or cyanobacteria. Green algae and cyanobacteria were isolated frequently from fine lake swamp, lake alluvium, and sandy loam samples. No viable algae were isolated from coarse marine (littoral) sands and channel-fill sands. The occurrence of green algae and cyanobacteria in the Arctic Late Pliocene–Early Pleistocene, as well as in other samples of Antarctic permafrost of different ages, was sporadic. Few Antarctic permafrost samples at a temperature near −23 °C contained viable green algae. Algal survival was best within neutral or slightly alkaline fine alluvium sediments. Soil moisture or ice content did not influence algal survival. Cryoprotective properties of the frozen strata may play a main role in algal survival. Permafrost does contain thin films of unfrozen water (14), the thickness of which depends on temperature and varies from 5 to 75Å, at −10 and −1 °C, respectively (2). The unfrozen water films may act as a cryoprotector, but their cryoprotective properties decrease as their thickness decreases with decreasing temperature. Perhaps Antarctic permafrost possesses less cryoprotective properties than Arctic permafrost on the basis that Antarctic sediments are formed from alkaline coarse sands and at lower temperatures.

Algae-Bacterial Association in the Permafrost

All permafrost samples that contained viable algae also contained viable bacteria (Figure 10.4). Bacteria were widely distributed in all samples with cell amounts that ranged from 10^2 to 9.2×10^5 colony-forming units per gram dry weight (cfu/gdw). All unialgal isolates were nonaxenic. The associated bacteria were nearly always detected by microscopic examination. The bacteria detected belonged to several different types including Gram-positive bacteria (*Microbacterium* sp., *Rhodococcus* sp., *Microbispora* sp., *Paenibacillus* sp.) and Gram-negative bacteria: alpha (*Agrobacterium* sp., *Brevundimonas* sp., *Afipia* sp., *Aminobacter* sp.) and gamma (*Pseudomonas* sp., *Xanthomonas* sp.) subdivisions of the proteobacteria. All attempts to obtain axenic algal cultures were unsuccessful: algal cells did not grow without their associated bacteria. We suggest that microalgal-bacterial permafrost consortial associations are taxonomically complex, metabolically interactive, self-sustaining communities that inhabit extreme subterrestrial environments. These associations may be better able to survive in physically and chemically stressed environments. Consortial members exhibit extensive metabolic diversification, but have remained structurally simple (12).

Algae are not ecologically important while frozen in permafrost, but they probably would play a main role in thawed permafrost ecosystems. After thawing, the insulated permafrost ecosystem can function independently as long as water and light are available. Phototrophic algae and heterotrophs orient along microscale chemical (i.e., oxygen, pH, Eh) gradients to meet and to optimize the biogeochemical processes (carbon, nitrogen, phosphorus cycling) essential for the survival, growth, and maintenance of genetic diversity that are needed to sustain life. New techniques should be applied to the study of permafrost to better understand consortial growth and survival under extreme environmental conditions on Earth. Consortia are ideal model systems to develop a process-

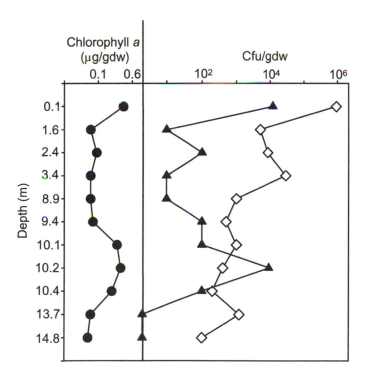

FIGURE 10.4. Correlation between amount of viable algae cells (▲) and algal biomass along the borehole 1/95 was significant with Pearson correlation coefficient (r = 0.85 , P < 0.05, paired t-test). The amount of viable bacteria (◇) was usually higher than the amount of viable algae (▲). The amount of viable cells is expressed as colony forming units per gram dry weight (cfu/gdw), and algal biomass is expressed as the amount of chlorophyll *a* per gram dry weight (Chl *a*, µg/gdw). Permafrost samples were collected from different layers of the same borehole during the 1995 field season. Only the layers in which chlorophyll and algae were observed, are shown (intervals between the layers are not drawn to scale).

based understanding of the structural and functional requirements for life in extreme environments representative of the Earth's earliest biosphere, and possibly other planets.

Algal Diversity in Permafrost

Population size measured as cfu/gdw increases with increasing biomass, measured as chlorophyll *a* content (correlation coefficient r = 0.85, P < 0.05) (Figure 10.4). However, population increase does not necessarily mean an increase in species diversity. Sometimes larger populations were formed by the growth of a single species. Some young Holocene samples contained high amounts (< 10^2-9×10^3 cfu/gdw) of easily recovered green algae, which were isolated by direct plating of soil suspension on agar medium. Many permafrost sediment samples contained low numbers of viable algae that were isolated only by enrichment culture. Permafrost samples contain only limited types of algae; for example, only one kind of algae was isolated from half of all samples. Algal species richness (the number of species in a sample) varied from one to three for every sample tested. Diversity indices for cyanobacteria and green algae in the permafrost sediments were 0.24 and 0.76, respectively. The frequency of viable algae and their biodiversity decreased with increasing permafrost age. The relatively young Holocene permafrost sediments contain a high diversity of viable green algae and cyanobacteria. Algal isolates were identified by morphological criteria (3, 19, 20, 22, 26).

Nonmotile, globular, small cells of *Chlorella* sp. were abundant among the green algae isolates. *Chlorella* sp. was isolated from within permafrost samples of different ages with a diversity index of 0.22. Rod-shaped algae of *Chodatia* sp. and *Stichococcus* sp., as well as unicellular algae of *Chlorococcum* sp., *Pseudococcomyxa* sp., and *Scotiellopsis* sp. were discovered in Late Pleistocene to Holocene lake swamp and alluvium sandy loam samples with a diversity index of 0.04 to 0.06. Unicellular green algae of *Mychonastes* sp. were isolated from both Arctic and Antarctic subsurface samples. Their diversity index in the Arctic permafrost was approximately 0.06. But the unicellular green algae *Mychonastes* sp. and *Pedinomonas* sp. were the only ones isolated from Antarctic permafrost samples.

Filamentous cyanobacteria with narrow, straight, uniseriate trichomes belonging to the order Oscillatoriales were frequently isolated from both young and old Arctic permafrost sediments. Species of the genera of *Oscillatoria* and *Phormidium* were identified and their diversity indices were 0.19 and 0.29, respectively. *Oscillatoria* sp. with wide, straight trichomes was isolated only from Holocene sediments and its diversity index was 0.03. Cyanobacteria of the order Nostocales exhibited diversity indices of 0.14 for *Nostoc* and 0.05 for *Anabaena*. Because of their low potential contamination role and absence

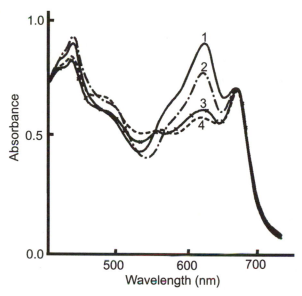

FIGURE 10.5. Absorption spectra of the following cyanobacteria from permafrost: 1) *Oscillatoria*, 2) *Phormidium*, 3) *Nostoc*, and 4) *Anabaena*. The spectra are normalized with respect to the chlorophyll *a* absorption maximum. Absorbance at 440 and 680 nm is determined by the presence of chlorophyll *a*; carotenoids (490–500 nm); C-phycoerythrin (570 nm); and phycocyanin (625 nm).

of any possibility for their migration up and down the profile, the isolation of viable filamentous cyanobacteria confirms that algae exist *in situ*.

Pigment Composition of Ancient Algae

Ancient green algae contained all major groups of photosynthetic pigments—chlorophylls *a* and *b*, and carotenoids (Car)—and their composition did not differ from typical modern strains.

The characteristic and unique property of cyanobacteria is the presence of phycobilin pigments (phycobiliproteins), which form specific supermolecular structures called phycobilisomes. The major function of these pigments is efficient transduction of absorbed light quanta from C-phycoerythrin (PE) through phycocyanin (PC) to allophycocyanin (AP) and allophycocyanin B, and finally to chlorophyll *a* (Chl *a*) (17). The absorption and fluorescence spectra that reflect these molecules were studied for some permafrost cyanobacteria. The absorption spectra of cyanobacterial cells revealed the presence of all known pigments—chlorophyll *a*, carotenoids, and phycobiliproteins (Figure 10.5).

TABLE 10.1.
Photosynthetic pigments in cells of permafrost cyanobacteria

Strain	Car/Chl a^1	PC/Chl a^2	Percentage		Molar Ratio
			PE^3	PC^4	[PE]/[PC]
Oscillatoria sp.	1.26	1.34	0.6	99.4	0.01
Phormidium sp.	1.34	1.29	0	100	0
Nostoc sp.	1.42	0.94	30.2	69.8	0.43
Anabaena sp.	1.37	0.85	23.6	76.4	0.31

[1] Car/Chl a = ratio of carotenoids to chlorophyll a.
[2] PC/Chl a = ratio of phycocyanins to chlorophyll a.
[3] PE = phycoerythrin.
[4] PC = phycocyanin.

Carotenoid content varied slightly among the cyanobacteria studied. PC content varied insignificantly within the same order, but the difference was 30 to 40% between the Oscillatoriales and Nostocales. Significant differences in the content of PE in the cells of cyanobacteria that belonged to different orders also were shown. *Oscillatoria* sp. and *Phormidium* sp. contained only PC, whereas *Nostoc* sp. and *Anabaena* sp. contained both PC and PE, and the PE amount was 30% (*Nostoc* sp.) and 24% (*Anabaena* sp.) of total phycobiliproteins (Table 10.1). PE detected in ancient cyanobacteria suggests their capacity for nitrogen fixation.

Nitrogen Fixation and Chromatic Adaptation in Permafrost Cyanobacteria

The pigment composition of *Nostoc* spp. and *Anabaena* spp., which were expected to be capable of nitrogen fixation, changes under the influence of nitrogen source. Nitrogen fixation by ancient cyanobacteria was discussed in detail earlier (11, 35). Briefly, chlorophyll a and AP content of cells grown in the presence of either nitrate or ammonium did not differ significantly. The source of organic ammonium, however, did lead to slight increases of carotenoids in both genera. In contrast, the PE and PC content of the cells is markedly influenced by the nitrogen source. PE was lower in ammonium- compared to nitrate-grown cells, whereas PC was higher. The molar [PE]/[PC] ratio was higher in nitrate- than in ammonium-grown cells.

An interesting capability of the cyanobacteria is complementary chromatic adaptation. In this process, phycobilisome rods change to achieve better light harvesting that gives a number of advantages to cyanobacteria, allowing them to occupy various environments. Viable permafrost species of *Nostoc* were able to conduct complementary chromatic adaptation, and thus respond to chromatic illumination involving major changes of the specific cellular content of PE and PC (Figure 10.6). The PC content was 17% lower in green, and 35% greater in red than in white light. The specific PE content of the cells was significantly modified after growth in green light (53% more than in white light) and stayed almost the same after growth in red light (Figure 10.6). The [PE (red)/PE (green)] ratio was equal to 0.6 to 0.68, and the [PC (red)/PC (green)] ratio was equal to 1.44 to 1.81. Therefore, ancient species of *Nostoc* show the same response as modern species: green light produces PE, which absorbs green light well, and red light produces PC, which absorbs red light well.

Potassium nitrate did not significantly affect the PE and PC content of the cells incubated under both green and red light, but the organic ammonium source did markedly influence phycobiliprotein content by decreasing PE and increasing PC. This response to nitrogen source during chromatic adaptation was the same as that during growth under white light (Table 10.2). Nitrogen source specifically modulates the composition of the phycobilisome rods by changing the content of the PC or PE within them.

Phylogenetic Analysis of Permafrost Cyanobacteria

Among the nine filamentous cyanobacteria isolated from permafrost sediments, three were heterocystous and six were nonheterocystous forms. All

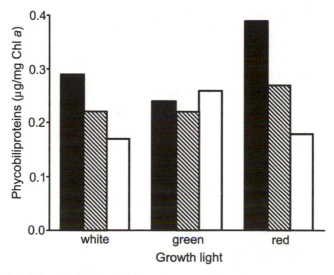

FIGURE 10.6. The cellular phycobiliprotein composition of *Nostoc* cells grown under white, green, and red light. Black bars = phycocyanin, hatched bars = allophycocyanin, and clear bars = phycoerythrin.

TABLE 10.2.
The response of *Nostoc* cells grown under white, green, and red light to a nitrogen source

	White			Green			Red		
Media	*Percentage*		*Molar Ratio*	*Percentage*		*Molar ratio*	*Percentage*		*Molar Ratio*
	PE[1]	*PC[2]*	*[PE]/[PC][3]*	*PE*	*PC*	*[PE]/[PC]*	*PE*	*PC*	*[PE]/[PC]*
Nitrogen-free	46.8	53.2	0.88	52.6	47.4	1.11	31.3	68.7	0.39
KNO₃	45.2	54.8	0.82	51.3	48.7	1.00	27.7	72.3	0.37
Asparagine	39.6	60.4	0.66	46.9	53.1	0.85	20.7	79.3	0.29
Glycine	29.0	71.0	0.41	47.5	52.5	0.56	29.8	70.2	0.16

[1] PE = phycoerythrin content expressed as percentage of total phycobiliprotein (PE × 100/PE + PC).
[2] PC = Phycocyanin content expressed as percentage of total phycobiliprotein (PC × 100/PE + PC).
[3] Concentrations of [PE] and [PC] were determined by using equation (33).

isolates were unicyanobacterial and nonaxenic. A clonal library was created for each isolate. RFLP patterns obtained by amplified rDNA restriction analysis (ARDRA) were compared by Dice coefficient, GelCompar software (version 4.0; Applied Maths, Kortrijk, Belgium). We assumed that clones with 90 to 100% similarity contained a cyanobacterial 16S rDNA gene. Selected clones then were partially or fully sequenced to determine their phylogenetic diversity.

The 16S rDNA sequences of permafrost cyanobacteria were aligned with a selection of cyanobacterial reference sequences. The permafrost cyanobacteria belonged to four separate groups. The most abundant cyanobacterial group belonged to the Lyngbya-Phormidium-Plectonema group. *Oscillatoria* spp. with wide, straight trichomes formed a cluster with *Lyngbya* spp., *Arthrospira* spp., and *Trichodesmium* spp. Heterocystous cyanobacteria formed two different clusters within the genus *Nostoc*.

Conclusions

Although the conditions for photosynthesis do not exist in permafrost, we have shown that permafrost, like the microbial ecosystems of polar soil and other ecosystems, is composed of three parts: producers, reducers, and consumers. The qualitative results obtained from permafrost samples of different ages have shown that subsurface permafrost harbors a viable population of both chemotrophic (different kinds of bacteria and fungi) and phototrophic (green algae and cyanobacteria) microorganisms. In contrast to tundra soil microbial communities, the subsurface permafrost communities are protected from daily and seasonal temperature fluctuations, and do not depend on photosynthesis to provide energy for growth.

Some green algae can be isolated from Holocene samples by direct plating, suggesting that some algae exist in an easily recovered resting state, and that they are active immediately after thawing. Long-time exposure of the resting form of green algae and cyanobacteria to negative temperatures apparently did not influence their capacity to reproduce and grow. However, prolonged cryoconservation within permafrost can lead to (i) decreased cell size, cell shrinkage, and reduction of cell volume; (ii) lowered rates of growth and reproduction; (iii) reduction of cell viability and viable algal populations; iv) reduction of algae diversity and population structure; and (v) reduction of the chlorophyll content of the algae population.

Cyanobacteria and eukaryotic algae are the main primary producers that possess an oxyphototrophic type of photosynthesis. They belong to the most important components of any polar or nonpolar soil microecosystem, because they influence mineral nutrient cycling and energy flow. Understanding the survival of terrestrial microalgae and other microorganisms in the permafrost provides an experimentally tractable analog that can be used to evaluate microbial survival in frozen extraterrestrial environments. Characterization of the biodiversity, structure, and morphology of species that are most frequently isolated from permafrost sediments may well provide important clues to lifestyles that might be encountered on other planets of our solar system.

Acknowledgments

This research was supported by the Russian Foundation for Basic Research, project 98–04–48357. We thank Bonnie Angster for help with English. We are grateful to John Castello for constructive editing and helpful comments.

Literature Cited

1. Abyzov, S.S., I.N. Mitskevich, T.Y. Zhukova, V.A. Kuzhinovskii, and M.N. Poglazova. 1993. Microorganism numbers in deep layers of the central Antarctic ice-sheet. Microbiology 62: 130–135.
2. Anderson, D.M. 1967. Ice nucleation and the substrate-ice interface. Nature 216: 563–566.
3. Andreyeva, V.M. 1998. Terrestrial and aerophilic green algae (Chlorophyta: Tetrasporales, Chlorococcales, Chlorosarcinales). In V.M. Andreyeva (ed.), Soil and Aerophilic Green Algae (Chlorophyta: Tetrasporales.) Nauka, St. Petersburg.
4. Caldwell, D.E. 1977. Accessory pigment fluorescence for quantitation of photosynthetic microbial populations. Can. J. Microbiol. 23: 1594–1597.

5. Cameron, R.E., J. King, and C.N. David. 1970. Microbiology, ecology and micro-climatology of soil sites in Dry Valleys of Southern Victoria Land, Antarctica. Ant. Ecol. 2: 702–716.

6. Cameron, R.E., and F.A. Morelli. 1974. Viable microorganisms from ancient Ross Island and Taylor Valley drill core. Ant. J. U.S. 9: 113–116.

7. Christner, B.C., E. Mosley-Thompson, L.G. Thompson, V. Zagorodnov, K. Sandman, and J.N. Reeve. 2000. Recovery and identification of viable bacteria immured in glacial ice. Icarus 144: 479–485.

8. Dmitriev, V.V., D.A. Gilichinskii, R.N. Faizutdinova, I.N. Shershunov, V.I. Golubev, and V.I. Duda. 1997. Detection of viable yeast in 3-million-year-old permafrost soils of Siberia. Microbiology 66: 546–550.

9. El-Sayed, S. and G. Fryxell. 1993. Phytoplankton. pp. 65–122. In E.I. Friedmann (ed.), Antarctic Microbiology. Wiley, New York.

10. Elston, D.P., and S.L. Bressler. 1981. Magneto-stratigraphic studies in neogene deposits of Taylor Valley and McMurdo Sound, Antarctica. J. Royal Soc. New Zealand 11: 481–486.

11. Erokhina, L.G., E.V. Spirina, and D.A. Gilichinskii. 1999. Accumulation of phycobiliproteins in cells of ancient cyanobacteria from Arctic permafrost as dependent on the nitrogen source for growth. Microbiology 68: 628–631.

12. Friedmann, E.I. 1994. Permafrost as microbial habitat. pp. 21–26. In D. Gilichinsky (ed.), Viable Microorganisms in Permafrost. Russian Academy of Sciences, Pushchino Research Centre, Pushchino.

13. Gilichinsky, D.A., G.M. Khlebnikova, D.G. Zvyagintsev, D.G. Fedorov-Davydov, and N.N. Kudryavtseva. 1989. Microbiological characteristics at the study of sedimentary-rock of cryolite zone. Izvestiya Akademii Nauk SSSR, Seriya Geologicheskaya 6: 103–115.

14. Gilichinsky, D.A., V.S. Soina, and M.A. Petrova. 1993. Cryoprotective properties of water in the Earth cryolithosphere and its role in exobiology. Origins of Life and Evolution of the Biosphere 23: 65–75.

15. Gilichinsky, D.A., E.A. Vorobyova, L.G. Erokhina, D.G. Fyordorov-Dayvdov, and N. Chaikovskaya. 1992. Long-term preservation of microbial ecosystems in permafrost. Adv. Space Res. 12: 255–263.

16. Gilichinsky, D.A., S. Wagener, and T.A. Vishnivetskaya. 1995. Permafrost microbiology. Permafrost and Periglacial Processes 6: 281–291.

17. Glazer, A.N. 1984. Phycobilisome a macromolecular complex optimized for light energy transfer. Biochim. Biophys. Acta 768: 29–51.

18. Glazovskaya, M.A. 1997. Fossilizing functions of the pedosphere in continental cycles of organic carbon. Eurasian Soil Sci. 30: 240–248.

19. Gollerbakh, M.M., E.K. Kosinskaya, and V.I. Polyanskiy. 1953. Blue-green Algae: Classificator of the Fresh-water Algae of the USSR. Ed. M.M. Gollerbakh and V.P. Savich. Sovetskaya Nauka, Moscow.

20. Gollerbakh, M.M., A.M. Matveyenko, I.I. Nikolaev. 1977. The Plant Life: Algae and Lichens. Ed A.A. Fedorov. Prosveshenie, Moscow.

21. Hoyt, P. 1966. Chlorophyll-type compounds in soil: 2. Their decomposition. Plant and Soil 25: 313–328.

22. Komarenko, L.Y., and I.I. Vasil'eva. 1978. The Fresh-Water Green Algae from Reservoirs of Yakutia. L.Y. Komarenko (ed.). Nauka, Moscow.

23. Marchant, D.R., G.H. Denton, C.C. Swisher, and N. Potter. 1996. Late Cenozoic Antarctic paleoclimate reconstructed from volcanic ashes in the Dry Valleys region of southern Victoria Land. Geol. Soc. Am. Bull. 108: 181–194.

24. McKay, C.P., and E.I. Friedmann. 1985. The cryptoendolithic microbial environment in the Antarctic cold desert—temperature-variations in nature. Polar Biol. 4: 19–25.

25. Rippka, R. 1988. Isolation and purification of cyanobacteria. Methods in Enzymol. 167: 3–27.

26. Rippka, R., J. Deruelles, J.B. Waterbury, M. Herdman, and R.Y. Stanier. 1979. Generic assignments, strain histories and properties of pure cultures of cyanobacteria. J. Gen. Microbiol. 111: 1–61.

27. Rivkina, E.M., E.I. Friedmann, C.P. McKay, and D.A. Gilichinsky. 2000. Metabolic activity of permafrost bacteria below the freezing point. Appl. Environ. Microbiol. 66: 3230–3233.

28. Rivkina, E.M., and D.A. Gilichinsky. 1996. Methane as a paleomarker of the genesis and dynamic of the frozen sediments. Litol. and Mineralol. 4: 183–187.

29. Rivkina, E., D. Gilichinsky, S. Wagener, J. Tiedje, and J. McGrath. 1998. Biogeochemical activity of anaerobic microorganisms from buried permafrost sediments. Geomicrobiol. J. 15: 187–193.

30. Shi, T., R.H. Reeves, D.A. Gilichinsky, and E.I. Friedmann. 1997. Characterization of viable bacteria from Siberian permafrost by 16S rDNA sequencing. Microbial Ecol. 33: 169–179.

31. Sinclair, J.L., and W.C. Ghiorse. 1989. Distribution of aerobic-bacteria, protozoa, algae, and fungi in deep subsurface sediments. Geomicrobiol. J. 7: 15–31.

32. Steubing, L. 1973. Soil flora: Studies of the number and activity of microorganisms in woodland soils. Analysis of temperate forest ecosystems. Ecological Studies 42: 131–137.

33. Tandeau de Marsac, N., and J. Houmard. 1988. Complementary chromatic adaptation—physiological conditions and action spectra. Methods in Enzymol. 167: 318–328.

34. Vallentyne, J. 1955. Sedimentary chlorophyll determination as a paleobotanical method. Can. J. Bot. 33: 304–313.

35. Vishnivetskaya, T.A., L.G. Erokhina, E.V. Spirina, A.V. Shatilovich, E.A. Vorobyova, and D.A. Gilichinsky. 2001. Ancient viable green algae and cyanobacteria from permafrost. Nova Hedwigia, Beiheft 123: 427–441.

36. Vishnivetskaya, T., S. Kathariou, J. McGrath, D. Gilichinsky, and J.M. Tiedje. 2000. Low-temperature recovery strategies for the isolation of bacteria from ancient permafrost sediments. Extremophiles 4: 165–173.

37. Vorobyova, E., V. Soina, M. Gorlenko, N. Minkovskaya, N. Zalinova, A. Mamukelashvili, D. Gilichinsky, E. Rivkina, and T. Vishnivetskaya. 1997. The deep cold biosphere: Facts and hypothesis. FEMS Microbiol. Rev. 20: 277–290.

38. Webb, P.N., and D.M. Harwood. 1991. Late Cenozoic glacial history of the Ross Embayment, Antarctica. Quatern. Sci. Rev. 10: 215–223.

39. Willerslev, E., A.J. Hansen, B. Christensen, J.P. Steffensen, and P. Arctander. 1999. Diversity of Holocene life forms in fossil glacier ice. Proc. Nat. Acad. Sci. 96: 8017–8021.

40. Wilson G.S., P. Braddock, S.L. Foreman, E.I. Friedmann, E.M. Rivkina, J.P. Chanton, D.A. Gilichinsky, D.G. Fedorov-Davydov, V.E. Ostroumov, V.A. Sorokovikov, and M.C. Wizevich. 1997. Coring for microbial records of Antarctic climate. Ant. J. U.S. 31: 83–86.

11

The Significance and Implications of the Discovery of Filamentous Fungi in Glacial Ice

Li-Jun Ma, Catherine M. Catranis, William T. Starmer,
and Scott O. Rogers

THE OBJECTIVE of our research was to study fungi entrapped in glaciers. We hypothesized that eukaryotic microorganisms remain viable in glacial ice for very long periods of time. This hypothesis is based on the assumption that airborne microorganisms are deposited in snow on the glacier surface and become immured in the glacier through the continuous thinning and recrystallization of snow. The initial proportion of liquid water in snow is only 10%; the remaining 90% is air and microparticulates, such as soil particles and plant fragments with which the microorganisms are associated. Over time, the weight of the overlying snow-pack compresses the snow, squeezing some of the air to the surface and recrystallizing the snow into granules to form firn. The proportion of liquid water at this stage is 50%. The firn is further compressed to form glacial ice, in which the water proportion is about 90%. This process requires approximately 100 years in the polar regions (58), although this varies with snowfall amounts.

The long chronosequence is one of the most distinctive and important characteristics of glacial cores, and is what makes them so valuable for the study of ancient history. Glacial ice dating systems are well established by using a variety of different techniques, including multiparameter annual layer counting (6, 60) and measurement of oxygen isotope ratios ($\delta^{18}O$ profile) (26, 34, 42). The oldest ice from Antarctica is more than 420,000 years old, which covers four glacial/interglacial cycles (66). The oldest glacial ice in Greenland (GISP 2) is over 100,000 years, which covers the last two glaciations and the last interstadial stage (6, 12, 34, 60, 79). The extensive chronological sequence and availability of glacial ice in both polar regions provide precisely dated ancient materials with detailed geochemical and geophysical information.

Long-term preservation of ancient microorganisms requires a process that retards the breakdown of macromolecules, such as proteins and nucleic acids. Breakdown of DNA is caused mainly by hydrolysis and oxidization when it is hydrated (54). Rapid tissue dehydration and cold temperatures are necessary for the prolonged protection of DNA, and for maintaining the viability of microorganisms (64, 72). The extreme temperatures at the permanently manned South Pole station are −14 to −81 °C, and at Vostok station are −21 to −89 °C (45). Temperatures range from −20 to −32 °C for Greenland (34, 49, 80). Although the temperature increases with depth in the ice sheet due to geothermal heating, the temperature of the ice remains below freezing over thousands of years. Freezing conditions, which limit the amount of free water, combined with sublimation (drying while frozen), which greatly reduces the metabolic rate and arrests normal degradative processes (10), provide a unique protection mechanism for microorganisms and their DNA. Studies of animal corpses entombed in glaciers for thousands of years provide strong support for the argument that macromolecules, including DNA and protein, can be well preserved in glaciers (38, 67, 81).

Microorganisms Immured in Glaciers

The existence of biotic material in glaciers has been known for a long time. Pollen from many different plant species has been recovered from glaciers in Greenland (17, 18), Devon Island (50, 52–53), and Ellesmere Island (16). Some of these pollens have been transported from as far as the Sahara desert (15) and East Asia (13). Wood fragments (e.g., tracheids, vessels, fibers, and parenchyma cells), as well as fungal spores, diatoms, and other microorganisms also have been observed in glacial meltwaters using filtration, electron microscopy, and fluorescent stain methods (1, 5, 33, 52, 53). Plant viral sequences and bacterial viruses were detected in meltwater from glacial ice cores from Greenland (22, 23). Viable bacteria, fungi, and yeasts were recovered from Vostok ice cores up to 400,000 years old (1–5, 55) and from Greenland ice cores over 100,000 years old (24, 55–57).

Although there is not much doubt about the presence of microorganisms and other biotic material in glaciers, that microorganisms or their DNA could survive in a viable or amplifiable form in ice for centuries or millennia is not universally accepted (85). Apparently solid ice at temperatures well below 0 °C could contain an interconnected network of very acidic, liquid water veins (69). These veins contain carbon and other nutrients essential for life. Nutrients in such veins are sufficient to maintain a significant number of metabolically active but nonreproducing cells for more than 10,000 years (69). Viable microorganisms have been detected in the accretion ice above the surface of subglacial Lake Vostok (see ref. 43 and chapters 6, 15, and 16, this volume). Therefore, our report has been independently corroborated by other studies.

We concentrated our efforts on fungi, because, as with other types of microorganisms, fungi respond to stressful environmental conditions by producing several different kinds of resistant structures, for example, dormant spores, sclerotia, chlamydospores, and conidia. Most of these structures have extremely thick cell walls, some with more than four wall layers that comprise more than half of the spore volume (36). The darkly pigmented cell walls of many spores render them less vulnerable to radiation damage (77). For example, spores of *Ulocladium* spp. remain viable after exposure to 210,000 J/m^2 UV irradiation (55). In addition, as relatively simple eukaryotic organisms, fungi follow the same organization and principles as other much more complex organisms, such as plants and animals. They have been widely used as model systems to understand eukaryotic development, cell cycle regulation, genetics, and genomics, and are playing very important roles in understanding evolutionary processes (9, 20, 65, 76).

Microbiologists initially were fascinated by the variety of exquisite and peculiar fungi (both macrofungi and microfungi) that were isolated from the deep-freezing environment of the Antarctic and Arctic (46–48, 50). These fungi live and reproduce in recently deglaciated sites and in snowbeds under poor nutrient conditions (35). The spores of such regional fungi most likely are carried in snow and deposited onto the ice surface. Fungal spores also can be carried by strong winds for hundreds or thousands of miles (39), and deposited onto the glaciers as well.

Detection of Fungi

The greatest challenge for all studies of ancient microorganisms is to avoid contamination with contemporary microorganisms during the isolation and detection processes. Several decontamination procedures have been tested and two—the UV-mechanical (24) and sodium hypochlorite protocols (55, 73)—were used for this study.

Eight ice subcores from the Dye 3 drill site in Greenland and one from the Vostok 5G core in Antarctica were subjected to decontamination using Clorox (5.25% sodium hypochlorite) and assayed for fungi (Table 11.1). Fifteen ice subcores from the GISP 2 and Dye 3 sites in Greenland were subjected to the UV-mechanical decontamination protocol and assayed for the same purpose (Table 11.2). Each subcore was carefully examined on a light table to eliminate those with fractures or fissures. The selected subcores then were transported frozen to a −20 °C room in our laboratory. The depths of each subcore, as measured from the top of the subcore to the glacial surface, ranged from 62 to 3584 m, which corresponded to ages of 300 to 400,000 years before present, respectively (6, 26, 28, 34, 42, 60, 80).

To detect viable fungi, meltwater from each melt shell following Clorox decontamination (see refs. 55, 73 and chapter 2, this volume), or each inner meltwater aliquot following UV-mechanical decontamination (24) was used to inoculate eight different solid culture media (57). Plates were incubated under

TABLE 11.1.
Ice subsections, detection of viable isolates, and fungal sequences after Clorox treatment

Core ID[a]	Depth[b] (m)	Age[c] (ybp)	Number of isolates	PCR[d] Amplification
D3/79–80 (1)	160.65	1 000	5	0
D3/79–80 (2)	160.73	1 000	3	IM 80–2
D3/79–124 (4)	369.23	2 000	1	0
D3/79–522 (4)	598.57	3 500	3	IM 522–4
D3/79–724 (4)	798.86	4 500	1	0
D3/79–939 (4)	1007.46	6 000	1	IM 939–4
D3/79–1150 (4)	1207.62	7 000	2	0
D3/79–1571 (4)	1621.65	11 000	6	0
D3/79–1945 (4)	1994.56	140 000	2	0
Vostok 5G (1)	3584.53	400 000	3	nt

[a] The ice subcores collected from Greenland Dye 3 were designated as D3; those from Antarctic Vostok station were designated as Vostok. The number inside the parentheses was the subsection number of that subcore.

[b] Depth is given from the top of ice core section to the upper glacial surface.

[c] The approximate age of ice was calculated according to references 6, 28, 59, and 80.

[d] The sequences were amplified directly from ice meltwater. nt = not tested; 0 = no amplification.

several different conditions. Controls consisted of four culture plates, one placed in each corner of the sterile hood, collected after all the work had been completed, and incubated alongside the inoculated plates. After more than six months' observation, nothing was found in any of the controls.

Twenty-seven viable fungal isolates were isolated from nine Clorox-treated ice subcores (Table 11.1). In total, 190 fungal isolates were recovered from forty-five inner subsection meltwaters representing fifteen ice subcores following UV-mechanical decontamination (Table 11.2). Viable fungi have been recovered from all ice subcores examined to date, including ice from both Antarctica and Greenland. Three fungal isolates were recovered from the Vostok 5G deep core, which has been dated over 400,000 ybp. These results prove the existence of viable fungi in glaciers, and demonstrate the utility of glaciers as a preservation matrix for ancient microorganisms.

Only some of the fungi entrapped in glaciers may be viable. Even among these, many may not be culturable (8). Nevertheless, the nucleic acids and/ or proteins of such fungi may be detectable utilizing molecular techniques. Meltwaters obtained from treated ice samples were assayed by polymerase

TABLE 11.2.
Ice subsections decontaminated by UV-mechanical decontamination protocol, and
numbers of isolates and DNA sequences amplified from them

Ice core [a]	Depth (m) [b]	Age (ybp) [c]	Number of Isolates	Sequences from Ice[d]
GB63 (1–4)	62.000	300	34	nt
GD131 (1–3)	130.000	500	31	1
GD135 (1–3)	134.000	500	23	1
D3/79–60 (1–3)	140.740	600	8	2
D3/71–49 (1–3)	158.569	700	19	1
D3/71–65 (1–3)	200.770	1000	9	1
GD226 (1–3)	225.000	1250	14	0
D3/71–124 (1–3)	369.113	2000	1	0
D3/71–522 (1–3)	698.504	3500	8	nt
D3/79–724 (1–3)	798.800	4500	8	1
D3/79–939 (1–3)	1007.310	6000	7	0
D3/79–1150 (1–3)	1207.425	7000	12	nt
D3/79–1354 (1–3)	1405.590	8500	6	0
D3/79–1571 (1–3)	1621.499	11 000	5	1
D3/79–1945 (1–3)	1994.480	140 000	5	2

[a] Greenland ice subcores collected from GISP 2 were designated as G and those collected from Dye 3 were designated as D3. The numbers within parentheses are the ice subsections.

[b] Depth is given from the top of ice core section to the upper glacial surface.

[c] The approximate age of ice was calculated according to references 6, 28, 59, and 80.

[d] The sequences were amplified directly from ice meltwater. nt = not tested; 0 = no amplification.

chain reaction (PCR) amplification to detect fungal DNA. To avoid potential contamination, all the equipment and reagents were used exclusively for this project. The amplification reactions were set up in a sterile hood. Four water controls were set up at each corner of the hood and tested by PCR as negative controls. When nested PCR was performed, the primary PCR negative controls were carried through the nested PCR as secondary controls. Through all the tests, none of the negative controls gave PCR product, but thirteen fungal sequences were amplified from meltwaters (Table 11.3) using nested PCR (57).

DNA was amplified from cultured isolates as well as directly from meltwater, and each was sequenced using the dideoxynucleotide chain termination

TABLE 11.3.
Sequences amplified directly from ice meltwater (IM) and results of BLAST search of GenBank

Sequence[a]	GenBank	Closest Taxon [b] (accession number)	Similarity (%)
IM 49–3(D3/71–49)	AF177749	*Tricholoma robustum* (AF062634)	99
IM 49–2 (D3/71–49)	AF177743	Ascomycete	—[c]
IM 60–1 (D3/79–60)	AF177744	Fungi	—[c]
IM 65–3 (D3/71–65)	AF177746	*Torulaspora delbrueckii* (D89599.1)	97
IM 80–2 (D3/79–80)	AF276436	*Torulaspora delbrueckii* (D89599.1)	96
IM 131–2 (GD131)	AF177747	*Phaeosphaeria nodorum* (U77362)	90
IM 135–1 (GD135)	AF177740	Ascomycete	—[c]
IM522–4 (D3/79–522)	AF276437	Basidiomycete	—[c]
IM 724–1 (D3/79–724)	AF177745	*Acremonium alternatum* (AAU57674)	96
IM 939–4 (D3/79–939)	AF276438	Fungi	—[c]
IM 1571–1 (D3/79–1571)	AF177741	*Acremonium alternatum* (AAU57674)	86
IM 1945–1(2) (D3/79–1945)	AF177742	*Acremonium alternatum* (AAU57674)	89
IM 1945–1(1) (D3/79–1945)	AF177748	*Cladosporium herbarum* (L25431)	97

[a] Sequence ID and ice subcores from which the sequences obtained (indicated in parentheses).
[b] Species exhibiting greater than 50% similarity with the sequence amplified from ice are presented.
[c] Sequences that exhibited much less than 50% similarity with contemporary sequences are classified only to division or fungal kingdom.

method (75) in a LONG READIR 4200 automated DNA sequencer (Li-Cor Biotechnology Division, Lincoln, Nebraska). A gapped BLAST search (7) of GenBank (www.ncbi.nlm.nih.gov/cgi-bin/BLAST/) was performed for each sequence obtained to determine the closest available sequence. Sequences were aligned using ClustalW 1.8 (7, 82) (http://searchlauncher.bcm.tmc.edu/

multi-align/multi-align.html) and manually adjusted. The phylogram (Figure 11.1) was conducted using PAUP (78).

Characterization and Phylogenetic Analyses

Among the total 217 viable isolates, sixteen were characterized, and their ribosomal RNA ITS regions were amplified and sequenced (Table 11.4). Eleven were morphologically identified and supported by the molecular data. They include three isolates of *Penicillium* spp., one *Aspergillus* sp. (plectomycetes, Figure 11.2), four *Cladosporium* spp. (three of *C. cladosporioides* and one of *C. herbarum*, Figure 11.3), and one species each of *Fusarium, Ulocladium, and Alternaria* (Fig. 11.4). The remaining five isolates failed to sporulate and could not be identified morphologically. However, based on the comparison of the ITS region to those in GenBank (Table 11.4), two isolates were identified as ascomycetes (*Dactylella lobata*) and three as basidiomycetes (*Multiclavula corynoides, Tricholoma robustum*, and *Pleurotus pulmonarius*). Among these five isolates, GI254 has the lowest similarity (61% in the entire ITS region). The closest contemporary species is *Multiclavula corynoides*.

Nine of the 13 sequences that amplified directly from meltwater were identified as ascomycetes, and 2 were identified as basidiomycetes based on BLAST searches of GenBank (Table 11.3). The remaining 2 amplicons (IM939–4 and IM60–1) had very low similarity to any sequence in GenBank, and thus were excluded from the phylogenetic analyses. Two of the 9 ascomycete sequences (IM49–2 and IM135–1) were distant (20–35% similarity) from any contemporary sequences, and thus were classified only at the divisional level (Table 11.3).

The DNA samples amplified through PCR assays are from either viable or dead fungal materials. Our results demonstrated that it is possible to amplify DNA directly from viable fungal material; for example, the amplicon (IM1945–1–1), amplified directly from meltwater of subcore D3/79–1945, is a perfect match to one sequence amplified from culture GI687, which was isolated from the same subcore. Because the amount of ice meltwater used for culture assays (11 ml per sample) is far larger than the amount used for PCR assays (72.5 μl per sample), we did not see many cases like this. By increasing sample size for PCR assay, we will increase the possibility of detecting the same ancient samples using both direct PCR and conventional culture methods.

In general, sequences amplified directly from meltwaters were more different from contemporary sequences than those amplified from cultures (Tables 11.3 and 11.4), because the majority of the direct amplicons have come from fungal fragments or free DNA fragments embedded in the glaciers. Such sequences may represent fungi that are inherently difficult to culture, and thus

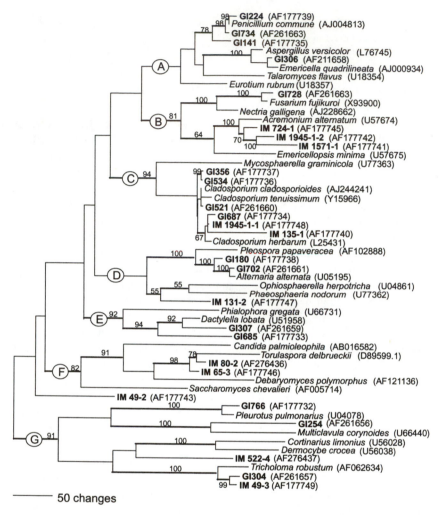

——— 50 changes

FIGURE 11.1. The most parsimonious tree based on sequences of fungal ITS regions
from ice samples and contemporary taxa. GenBank accession numbers are indicated in parentheses. The tree length is 3005. Fungal sequences obtained from ice
specimens are designated as IM (sequenced directly from ice meltwater) and GI
(sequenced from cultured isolates). The tree was generated using heuristic searches
of equally weighted maximum parsimony and midpoint rooting option on PAUP
(78). Bootstrap values (> 50%) are indicated on bold lined branches (500 replications). Gaps were introduced into the sequences and treated as a fifth character in
the parsimony analysis. Consistency index (CI) = 0.4775; retention index = 0.6778.

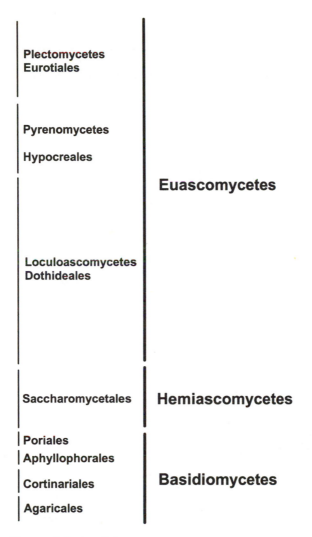

FIGURE 11.1. (*cont'd*)

are under-represented in GenBank and other databases. Alternatively, the organism may be extinct in the contemporary active biosphere. A third possibility is that preservation in ice may not be perfect, especially when viability is lost and the released DNA is damaged, resulting in erroneous PCR amplicons and DNA sequences.

The most parsimonious phylogram includes the 16 sequences from the cultured isolates, the sequences of 11 amplicons amplified directly from meltwater, and 26 contemporary sequences obtained from GenBank. Seven subclades

TABLE 11.4.

Morphological identification and the comparison of ITS sequences (BLAST search of GenBank)

Isolate (ice core)	GenBank Accession Number	Morphological Identification	Closest Taxon (accession number)	BLAST Similarity (%)
GI141 (GD131)	AF177735	*Penicillium* sp.1	*P. kojienum* (AF033489)	94
GI224 (GD131)	AF177739	*Penicillium* sp. 2	*P. commune* (AJ004813)	98
GI734 (D3/79–1354)	AF261663	*Penicillium* sp.	*Penicillium* commune (AJ004813)	98
GI306 (D3/71–124)	AF261658	*Aspergillus* sp.	*Emericella quadrilineata* (AJ000934)	94
GI180 (GD135)	AF177738	*Ulocladium* sp.	*Alternaria raphani* (U05200)	96
GI702 (D3/79–1150)	AF261661	*Alternaria alternata*	*Alternaria alternata* (U05195)	98
GI728 (D3/79–1150)	AF261662	*Fusarium* sp.	*Fusarium fujikuroi* (X93900)	98
GI356 (D3/71–49)	AF177737	*Cladosporium cladosporioides*	*C. cladosporioides* (L25429)	99
GI521 (D3/71–522)	AF261660	*C. cladosporioides*	*C. cladosporioides* (L25429)	97
GI534 (D3/79–724)	AF177736	*C. cladosporioides*	*C. cladosporioides* (L25429)	99
GI687 (D3/79–1945)	AF177734	*C. herbarum*	*C. herbarum* (L25431)	97
GI307 (D3/71–124)	AF261659	None	*Dactylella lobata* (U51958)	94
GI685 (D3/79–1945)	AF177733	None	*D. lobata* (U51958)	95
GI254 (GB63)	AF261656	None (clamp connection present)	*Multiclavula corynoides* (U66440)	61
GI304 (D3/71–124)	AF261657	None	*Tricholoma robustum* (AJ062634)	98
GI766 (D3/79–1945)	AF177732	None	*Pleurotus pulmonarius* (U04078)	97

(A to G) are formed among euascomycetes, hemiascomycetes, and basidiomycetes with 19, 2 and 5 ancient fungal sequences in each, respectively (Figure 11.1). Most sequences identified in this study were those of ascomycetes. Most sequences were resolved at least to the family level, but the sequence of IM 49–2 was distinct, being extremely different from any sequence currently existing on GenBank or in our sequence database. Therefore, its precise position currently is ambiguous.

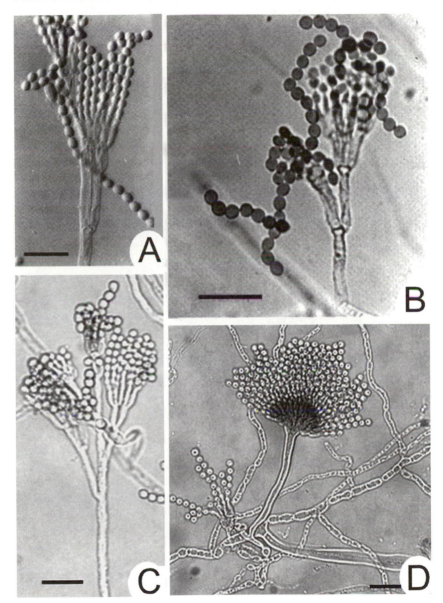

FIGURE 11.2. Isolates GI141, GI224, GI306, and GI734 from glacial ice. (A) GI141: *Penicillium* sp.1, formed brownish-yellow colonies, penicilli biverticillate and symmetric, conidia smooth and in chains. (B) GI224: *Penicillium* sp.2, formed grayish-green colonies, biverticillate and symmetric penicilli with smooth conidia. (C) GI734: *Penicillium* sp.3. (D) GI306: *Aspergillus* sp.1, conidiophores sinuate, smooth walled, 50–70 μm long and about 4 μm wide, bearing vesicle head at the end. Conidia in chain, spherical shaped, and 3–4 μm in diameter. The scale bar is 15 μm.

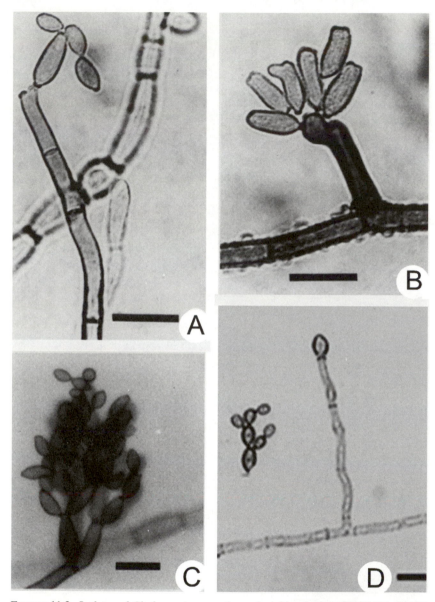

FIGURE 11.3. Isolates of *Cladosporium* spp. from glacial ice. Isolate GI356 (A), GI687 (B), GI534 (C), and GI521 (D). GI356, GI521, and GI534: identified as *C. cladospori-oides* with smooth brown conidiophores around 100 to 200 μm in length, conidiogenous cells terminal on side branches, conidia were in acropetal, branched chains with size ranging from 3–10 μm × 2–5 μm. GI687: identified as *C. herbarum* with dark brown, rough hyphae, terminal and intercalary conidiogenous cells, and bearing acropetal conidia. Conidia were brown and verrucose. The scale bar is 10 μm.

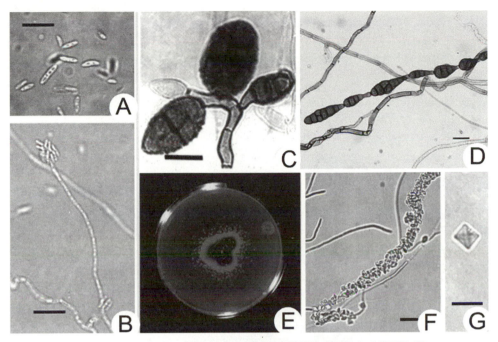

FIGURE 11.4. Morphologies of ice isolates GI728, GI180, GI702, and GI307. Macro-conidia (A) with three to four septa and up to 4 × 12 μm in size and microconidia (B) 1 × 4 μm in size of isolate GI728 identified as *Fusarium* sp. (C) Isolate GI180: *Ulocladium* sp. with smooth golden brown conidiophores. The conidia were ellip-soidal, golden brown, and rough, with two to three transverse and several longitudi-nal septa. (D) Isolate GI702: *Alternaria alternata* with conidiophores unbranched, straight, and up to 60 μm in length. Conidia are arranged in long chains often with short conidial beaks. The mean size was 13 μm × 35 μm, with up to eight trans-verse and several longitudinal septa. Isolate GI307 formed a golden-brown pigment (E) and exuded abundant diamond-shaped crystals (8 × 8 μm, (G) in the culture media that adhered to the hyphae (F). The scale bar is 15 μm.

Significance and Implications

Fungi are the largest, most diverse, and most resilient group of eukaryotic microorganisms. They have the capacity to survive extreme environmental conditions including desiccation and low temperature that makes them the perfect targets to search for in glaciers. Viable fungi were isolated from every ice core we examined (Tables 11.1 and 11.2), irrespective of which decontami-nation protocol was used. The discovery of viable fungi and fungal nucleic acids from both Antarctic and Arctic glaciers provides the direct evidence for the long-term preservation of eukaryotic microorganisms and their nucleic acids in the cold and dry conditions of glacial ice. These results confirm our hypothesis that glacial ice is a very good preservation matrix for the protection of eukaryotic microorganisms and their nucleic acids from hydrolysis, oxidization, UV irradiation, and other destructive agents.

Although obtaining ancient biological materials from glacial ice is still at its early stage, the significance of the research is beyond speculation. We can claim with confidence that the ancient biological materials recovered from glacial ice will have great impacts on biological research and geophysical and geochemical studies, as well as economic impacts.

Impacts on the Study of Evolutionary Biology

Scientific communities now are interested in fungal genomics in part because of work on the genome projects recently published (40, 84). Fungi are comparatively simple eukaryotes. Nonetheless, their cellular and genome structure and organization are similar to those of more complex eukaryotes, including humans, plants, and animals. In general, they have smaller genomes with a higher gene density and relatively low number of repetitive sequences. Therefore, they are unique model organisms for research to understand all aspects of genetics and genomics, including genome organization (9, 20, 68, 71), transcription and translation regulations (61, 83), horizontal gene transfer (74), aging processes (58), and development (14, 44).

In addition, fungal genomics is being used by mycologists to resolve longstanding questions in fungal biology. Most fungal species have both sexual and asexual reproductive structures. The polymorphic and pleomorphic fungi can have several exchangeable phenotypes. Many "orphan genes" (genes without identifiable homologs) have been reported from fungal genomes (20, 62). The broad diversity, great phenotypic complexity, and proportion of "orphan genes" raise interesting questions regarding the evolution of the fungal genome (9, 68, 71). Comparison of the budding yeast *Saccharomyces cereviseae* and the fission yeast *Schizosaccharomyces pombe* reveal that they are as different from each other as either is from animals. Based on this evidence, it is speculated that fungi have evolved faster than either plants or animals (31, 76). What is the evolutionary rate of fungi? What is the guiding principle in this kingdom? How does this compare with the evolutionary rates for plants and animals? Ancient fungi from glacial ice are living fossils that can help to provide us with answers to these kinds of questions through the use of genomic techniques.

Among the fungal isolates we recovered from glaciers, some of them, such as *Penicillium* spp., *Cladosporium* spp., and *Aspergillus* spp., are closely related to contemporary taxa and have been recovered from glaciers at different time points. For example, *Cladosporium cladosporioides* was isolated from three ice cores dated at 700, 3500 and 4500 ybp (Table 11.4). It is feasible to sequence these isolates for genomic comparisons to reveal the differences between contemporary isolates and their ancient siblings. Is the evolutionary rate linear at the genome level? Are there different mutation rates between coding and noncoding regions? What is the principle for the accumulation of mutations among different functional genes? What is the rate of evolution (mutation?) among conserved noncoding regions?

Alternatively, some isolates are distantly related to contemporary species. The closest contemporary taxon to isolate GI254 is *Multiclavula corynoides*, which has only 60% similarity in the ITS region. One possible explanation is that related contemporary species have not yet been studied in detail. It is estimated that only about 3 to 7% of all fungal species have been described. Thus, our understanding of the fungal kingdom is still extremely limited. The information we may obtain from these ancient isolates will expand our knowledge of fungal diversity. Ancient fungi may represent species thought to be extinct. By studying the physiology, genetics, and genomics of these "lost species," we may be able to understand the cause of their extinction.

Impacts on Geophysical and Geochemical Studies

The polar regions offer an immense field laboratory for fundamental research with global implications because of their extremes of climate, habitat and biogeography. Therefore, a survey of fungal populations correlated with specific geophysical and geochemical data is planned. For example, there are low snow accumulations in the polar regions that correspond to cold periods of Earth history (6). But the snow that accumulates during such times contains a high density of wind-blown continental dust (29). We are interested in the biotic profiles of the wind-blown dust in this snow, and would like to correlate biotic diversity with temperature, snow accumulation, dust density, and so on. The electrical conductivity of glacial ice depends on its acidity. Ice formed during cold, dusty periods has a high concentration of alkali, which significantly reduces the conductivity compared to ice formed during warmer and less dusty periods (79). The highly acidic solution that occurs in the veins produced at the triple boundaries of glacial ice provides enough nutrients and the elements necessary for microbial growth (69). Is there a correlation between microbial concentration and the electrical conductivity of the ice? A reasonable hypothesis might be that microbial populations fluctuate with climate change. Higher populations or increased metabolic activity in glaciers would be expected during warmer rather than colder periods.

Obviously, extensive sampling of ancient microorganisms from glaciers at chronosequence will extend the assessment of biodiversity to geologic timeframes and correlate the change with historic environmental events, at each of the time points. These comparisons might disclose the impacts of global climatic changes, geographic events, and anthropogenic impacts on population fluctuation, species extinction, and speciation. For instance, ice core records have shown that volcanic activities had a major impact on the atmosphere (27, 30, 88). Then what is the impact of volcanism on the biosphere? Will we see increased or decreased biomass? What is the effect on the viability? During glacial periods, atmospheric carbon dioxide and methane (greenhouse gas) levels were 30% and 50% (respectively) lower than interglacial levels (11, 25). What are the impacts of the greenhouse gas changes on the fungal population and species based on the century or millennium scale? We would like to determine the impacts of human activities on the fungal community, such as the development of human civilization, increasing population, explorations to the polar regions, industrialization, and deforestation.

Impacts on Pathogenicity

Fungi are major plant pathogens. The study of ancient fungi immured in glaciers will have great impact on the study of fungal pathogenicity. Some of the most devastating epidemics of plant diseases were caused by the invasion of "new pathogens" in established ecosystems or introduction of crops to new regions with different pathogens. Examples include the late blight disease of potato, which caused the Great Irish Potato Famine of the 1840s (32), and the extirpation of mature American chestnut (*Castanea dentata* (Marsh.) Borkh.) trees in the eastern forests of the United States (19), which was caused by the fungus *Cryphonectria parasitica* ((Murr.) Barr.) introduced from Asia at the turn of the twentieth century (21). Pathogens immured in glaciers may not have been active in the biosphere for thousands of years. The reentry of these relatively new pathogens into a uniformly susceptible host population could be a threat to crops, animals, and even humans. According to recent study, some fungal molecules are critical to the outcome of infections (37, 63, 86). These molecules exhibit major differences between pathogens and saprobes (87), and can be used as indicators of the pathogenicity of the organism. Surveying the glacial fungal population and examining the expression of these pathogenicity-related molecules in advance will help us to identify possible virulent isolates and give us the warning of their reentry. In this way, we may reduce some potentially severe damage.

Economic Impacts

Because of the production of various secondary metabolites, fungi have been widely used in industry to produce enzymes (valued at $1.2 billion per year), chemicals ($32 billion per year), and pharmaceutical products including antibiotics ($23 billion per year). Isolates from glaciers may provide new materials to be screened for compounds of commercial interest, for example, drugs (anti-cancer drugs, antibiotics), cold-resistant compounds (such as antifreeze proteins [41]), and so on. See chapter 9 for some novel compounds isolated from ancient permafrost fungi.

The Search for Extraterrestrial Life

The Antarctic environment is believed to be an analogue for conditions on other ice-covered planets and moons, such as the Moon, Mars, Europa, Callisto, Io, and others. Therefore, the techniques developed and employed in the search for life in glacial ice also can be used to search for life and nucleic acids in ice from extraterrestrial sources.

Acknowledgments

We thank Professor C.J.K. Wang for guidance with morphological characterization of the isolates, and J.D. Castello for reviewing the manuscript. This work was supported by a grant from the LExEN Program of the National Science Foundation (#9808676).

Literature Cited

1. Abyzov, S.S. 1993. Microorganisms in the Antarctic ice. pp. 265–295. In E.I. Friedman (ed.), Antarctic Microbiology, Wiley, New York.
2. Abyzov, S.S., I.P. Babeva, V.I. Biryuzova, N.A. Kostrikina, and E.E. Azieva. 1983. Peculiarities of the ultrastructural organization of yeast cells from the interior of an Antarctic glacier. Biol. Bull. Acad. Sci. USSR (Eng. trans. Akademiia nauk SSSR Investia Seriia biologicheskaia) 10: 539–545.
3. Abyzov, S.S., N.E. Bobin, and B.B. Kudriashov. 1979. Microbiological analysis of glacial series of central Antarctica. Biol. Bull. Acad. Sci. USSR (Engl. trans. Akademiia nauk SSSR Investia Seriia biologicheskaia) 6: 828–836.
4. Abyzov, S.S., N.E. Bobin, and B.B. Kudriashov. 1986. Central Antarctica glacier as an object of investigations of prolonged anabiosis of microorganisms in nature. Antarktika: Doklady Komissii 25: 202–208.
5. Abyzov, S.S., I.N. Mitskevich, and M.N. Poglazova. 1998. Microflora of the deep glacier horizons of central Antarctica. Microbiology 67: 451–458.
6. Alley, R.B., D. Meese, C.A. Shuman, A.J. Gow, K. Taylor, M. Ram, E.D. Waddington, and P.A. Mayewski. 1993. Abrupt increase in Greenland snow accumulation at the end of the Younger Dryas event. Nature 362: 527–529.
7. Altschul, S.F., T.L. Madden, A.A. Schaffer, Z. Zhang, W. Miller, and D.J. Lipman. 1997. Gapped BLAST and PSI-BLAST: A new generation of protein database search programs. Nucleic Acids Res. 25: 3389–3402.

8. Amann, R.I., W. Ludwig, and K. Schleifer. 1995. Phylogenetic identification and *in situ* detection of individual microbial cells without cultivation. Microbiol. Rev. 59: 143–169.

9. Arnold, J. 2001. Foreword to the special section on fungal genomics. Genetics 157: 933.

10. Aufderheide, A. 1981. Soft tissue paleopathology: An emerging subspeciality. Hum. Pathol. 12: 865–867.

11. Barnola, J.M., D. Raynaud, Y.S. Korotkevich, and C. Lorius. 1987. Vostok ice core provides 160,000-year record of atmospheric CO_2. Nature 329: 408–414.

12. Bender, M., T. Sowers, M.-L. Dickson, J. Orchardo, P.M. Grootes, P.A. Mayewski, and D. Messe. 1994. Climate connections between Greenland and Antarctica during the last 100,000 years. Nature 372: 663–666.

13. Biscaye, P.E., F.E. Grousset, S. Revel, S. van der Gaast, G.A. Zielinski, and G. Kukla. 1997. Asian provenance of glacial dust (stage 2) in the Greenland Ice Sheet Project 2 ice core, Summit, Greenland. J. Geophys. Res. 102: 26765–26781.

14. Bockmuhl, D.P., and J.F. Ernst. 2001. A potential phosphorylation site for an A-type kinase in the Efg1 regulator protein contributes to hyphal morphogenesis of *Candida albicans*. Genetics 157: 1523–1530.

15. Bortenschlager, S. 1970. Neue pollenanalytische untersuchungen von gletschereis und gletschernahen Mooren in den Ostalpen. Z. Gletscher Glazialgeol 6: 107–118.

16. Bourgeois, J.C. 1986. A pollen record from the Agassiz ice cap, northern Ellesmere Island, Canada. Boreas 15: 345–354.

17. Bourgeois, J.C. 1990. A modern pollen spectrum from Dye 3 south Greenland ice sheet. J. Glaciol. 36: 340–342.

18. Bourgeois, J.C., R.M. Koerner, and B.T. Alt. 1985. Airborne pollen: a unique air mass tracer, its influx to the Canadian High Arctic. Annals Glaciol. 7: 109–116.

19. Braun, E.L. 1950. Deciduous Forests of Eastern North America. Blakiston, Philadelphia.

20. Braun, E.L., A.L. Halpern, M.A. Nelson, and D.O. Natvig. 2000. Large-scale comparison of fungal sequence information: Mechanisms of innovation in *Neurospora crassa* and gene loss in *Saccharomyces cerevisiae*. Genome Res. 10: 426–430.

21. Brewer, L. 1995. Ecology of survival and recovery from blight in American chestnut trees (*Castanea dentata* (Marsh.) Borkh.) in Michigan. Bull. Torrey Bot. Club 122: 40–57.

22. Castello, J.D., S.O. Rogers, W.T. Starmer, C.M. Catranis, L.J. Ma, G.D. Bachand, Y. Zhao, and J.E. Smith. 1999. Detection of tomato mosaic tobamovirus RNA in ancient glacial ice. Polar Biol. 22: 207–212.

23. Castello, J.D., S.O. Rogers, J.E. Smith, W.T. Starmer, and Y. Zhao. 2005. Plant and bacterial viruses in the Greenland ice sheet, chapter 13, this volume.

24. Catranis, C.M., and W.T. Starmer. 1991. Microorganisms entrapped in glacial ice. Ant. J. U.S. 26: 234–236.

25. Chappellaz, J., J.M. Barnola, D. Raynaud, Y.S. Korotkevich, and C. Lorius. 1990. Atmospheric CH_4 record over the last climatic cycle revealed by the Vostok ice core. Nature 345: 127–131.

26. Cuffey, K.M., G.D. Clow, R.B. Alley, M. Stuiver, E.D. Waddington, and R.W. Saltus. 1995. Large Arctic temperature change at the Wisconsin-Holocene glacial transition. Science 270: 455–458.

27. Dai, J., E. Mosley-Thompson, and L.G. Thompson. 1991. Ice core evidence for explosive tropical volcanic eruption 6 years preceding Tambora. J. Geophys. Res. 17 (D9): 17361–17366.

28. Dansgaard, W., H.B. Clausen, N. Gundestrup, S.J. Johnson, and C. Rygner. 1985. Dating, climatic interpretation of two deep Greenland ice cores. pp. 71–76. In C.C.J. Langway, H. Oeschger, and W. Dansgaard (eds.), Greenland Ice Core: Geophysics, Geochemistry, and the Environment, American Geophysical Union, Washington, D.C.

29. Dansgaard, W., J.W.C. White, and S.J. Johnsen. 1989. The abrupt termination of the Younger Dryas climate event. Nature 339: 532–534.

30. Delmas, R., J.S. Kirchner, J.M. Palais, and J.R. Petit. 1992. 1000 years of explosive volcanism recorded at the South Pole. Tellus Ser. B, 44: 335–350.

31. Doolittle, R.F., D.-F. Feng, S. Tsang, G. Cho, and E. Little. 1996. Determining divergence times of the major kingdoms of living organisms with a protein clock. Science 271: 470–477.

32. Foster, R.F. 1989. Modern Ireland 1600–1972. Penguin Books, London.

33. Fredskild, B., and P. Wagner. 1974. Pollen and fragments of plant tissue in core samples from the Greenland ice cap. Boreas 3: 105–108.

34. Grootes, P.M., M. Stuiver, J.W.C. White, S. Johnsen, and J. Jouzel. 1993. Comparison of oxygen isotope records from GISP 2 and GRIP Greenland ice cores. Nature 336: 552–554.

35. Gulden, G., M.K. Jenssen, and J. Stordal. 1988. Arctic and Alpine Fungi, vol. 2. Soppkonsulenten, Oslo, Norway.

36. Hawker, L.W., and M.F. Madelin. 1976. The dormant spore. pp. 1–72. In D.J. Weber and W.M. Hess (eds), The Fungal Spore. Wiley, New York.

37. Hogan, L.H., B.S. Klein, and S.M. Levits. 1996. Virulence factors of medical important fungi. Clin. Microbiol. Rev. 9: 469–488.

38. Höss, M., and S. Pääbo. 1993. DNA extraction from Pleistocene bones by a silica-based purification method. Nucleic Acids Res. 21: 3913–3914.

39. Ingold, C.T. 1953. Dispersal in Fungi. Oxford, London.

40. International Human Genome Sequencing Consortium. 2001. Initial sequencing and analysis of human genome. Nature 409: 860–921.

41. Jia, Z., and P.L. Davies. 2002. Antifreeze proteins: An unusual receptor-ligand interaction. Trends in Biochemical Sciences 27: 101–106.

42. Johnson, S.J., H.B. Clausen, K. Fuhrer, N. Gundestrup, C.U. Hammer, P. Iversen, J. Jouzel, B. Stauffer, and J.P. Steffensen. 1992. Irregular glacial interstadials recorded in a new Greenland ice core. Nature 359: 311–313.

43. Karl, D.M., D.F. Bird, K. Björkman, T. Houlihan, R. Shackelford, and L. Tupas. 1999. Microorganisms in the accreted ice of Lake Vostok, Antarctica. Science 286: 2144–2147.

44. Khalaf, R.A., and R.S. Zitomer. 2001. The DNA binding protein Rfg1 is a repressor of filamentation in *Candida albicans*. Genetics 157: 1503–1512.

45. King, J.C., and J. Turner. 1997. Antarctic Climatology and Meteorology. Cambridge University Press, Cambridge.

46. Kobayasi, Y., N. Hiratsuka, R.P. Korf, K. Tubaki, K. Aoshima, M. Soneda, and J. Sugiyama. 1967. Mycological studies of the Alaskan Arctic. Ann. Rep. Inst. Fermentation Osaka 3: 1–138.

47. Kobayasi, Y., N. Hiratsuka, Y. Otani, K. Tubaki, S. Udagawa, J. Sugiyama, and K. Konno. 1971. Mycological studies of the Angmagssalik Region of Greenland. Bull. Nat. Sci. Mus. Tokyo 14: 1–96.

48. Lange, M. 1955. Macromycetes Part II. Greenland Agaricales. Meddr Grfnland 147 11: 1–69.

49. Langway, C.C.J., H. Oeschger, and W. Dansgaard. 1985. The Greenland ice sheet program in perspective. pp. 1–8. In C.C.J. Langway, H. Oeschger, and W. Dansgaard (eds.), Greenland Ice Core: Geophysics, Geochemistry, and the Environment. American Geophysical Union, Washington, D.C.

50. Laursen, G.A., and J.F. Ammirati (eds.). 1982. Arctic and Alpine Mycology, vol. 1. University of Washington Press, Seattle/London.

51. Lichti-Federovich, S. 1974. Pollen analysis of ice surface snow from the Devon island ice cap. Geological Survey Canada Paper 74–1A:197–199.

52. Lichti-Federovich, S. 1975. Pollen analysis of ice core samples from the Devon island ice cap. Geological Survey Canada Paper 75–1A:441–444.

53. Lichti-Federovich, S. 1975. Pollen analysis of ice core samples from the Devon island ice cap. Geological Survey Canada Paper 75–1B:135–137.

54. Lindahl, T. 1993. Instability and decay of the primary structure of DNA. Nature 362: 709–715.

55. Ma, L.-J. 2000. Ancient Fungi Entrapped in Glaciers. Ph.D. dissertation. SUNY-College of Environmental Science and Forestry, Syracuse.

56. Ma, L.-J., C.M. Catranis, W.T. Starmer, and S.O. Rogers. 1999. Revival and characterization of fungi from ancient polar ice. Mycologist 13: 70–73.

57. Ma, L.-J., S.O. Rogers, C.M. Catranis, and W.T. Starmer. 2000. Detection and characterization of ancient fungi entrapped in glacial ice. Mycologia 92: 286–295.

58. Matthew, R.B., and F.G. Neil. 1996. Glacial Geology: Ice Sheets and Landforms. Wiley, New York.

59. McVey, M., M. Kaeberlein, H.A. Tissenbaum, and L. Guarente. 2001. The short life span of *Saccharomyces cerevisiae* sg1 and srs2 mutations is a composite of normal aging processes and mitotic arrest due to defective recombination. Genetics 157: 1531–1542.

60. Meese, D.A., A.J. Gow, R.B. Alley, P.M. Grootes, P.A. Mayewski, M. Morrison, M. Ram, K.C. Taylor, Q. Yang, and G.A. Zielinski. 1997. The Greenland Ice Sheet Project 2 depth-age scale: Methods and results. J. Geophys. Res. 102: 26411–26423.

61. Munshi, R., K.A. Kaudl, A. Carr-Schmid, J.L. Whitacre, A.E.M. Adams, and T.G. Kinzy. 2001. Overexpression of translation elongation factor 1A affects the organization and function of the actin cytoskeleton in yeast. Genetics 157: 1425–1436.

62. Nelson, M.A., S. Kang, E.L. Braun, M.E. Crawford, P.L. Dolan, P.M. Leonard, J. Mitchell, A.M. Armijo, L. Bean, and E. Blueyes. 1997. Expressed sequences from conidial, mycelial and sexual stages of *Neurospora crassa*. Fungal Genet. Biol. 21: 348–363.

FILAMENTOUS FUNGI IN GLACIAL ICE   **179**

63. Oliver, R., and A. Osbourn. 1995. Molecular dissection of fungal phytopathogenicity. Microbiology 141: 1–9.

64. Pääbo, S., and A.C. Wilson. 1991. Miocene DNA sequences—a dream come true? Curr. Biol. 1: 45–46.

65. Penalva, M.A. 2001. A fungal perspective on human inborn errors on metabolisms: Alkaptonuria and beyond. Fungal Genet. Biol. 34: 1–10.

66. Petit, J.R., I. Badile, A. Leruyuet, D. Raynaud, C. Lorius, J. Jouzel, M. Stievenard, V.Y. Lipenkov, N.I. Barkov, B.B. Kudryashov, M. Davis, E. Saltzman, and V. Kotlyakov. 1997. Four climate cycles in Vostok ice core. Nature 375: 305–308.

67. Poinar, H.N., and B.A. Stankiewicz. 1999. Protein preservation and DNA retrieval from ancient tissues. Proc. Nat. Acad. Sci. 96: 8426–8431.

68. Prade, R.A., J. Griffith, K. Kochut, J. Arnold, and W.E. Timberland. 1997. In vitro reconstruction of the *Aspergillus nidulans* genome. Proc. Nat. Acad. Sci. 94: 14564–14569.

69. Price, P.B. 2000. A habitat for psychrophiles in deep Antarctic ice. Proc. Nat. Acad. Sci. 97: 1247–1251.

70. Priscu, J.C., E.E. Adams, W.B. Lyons, M.A. Voytek, D.W. Mogk, R.L. Brown, C.P. McKay, C.D. Takacs, K.A. Welch, C.F. Wolf, J.D. Krishtein, and R. Avvci. 1999. Geomicrobiology of subglacial ice above Lake Vostok, Antarctica. Science 286: 2141–2143.

71. Radford, A., and J.H. Parish. 1997. The genome and genes of *Neurospora crassa*. Fungal Genet. Biol. 21: 267–277.

72. Rogers, S.O., K. Langenegger, and O. Holdenrieder. 2000. DNA changes in tissues entrapped in plant resins (the precursors of amber). Naturwissenschaften 87: 70–75.

73. Rogers, S.O., L. Ma, Y. Zhao, C.M. Catranis, W.T. Starmer, and J.D. Castello. 2005. Recommendations for elimination of contaminants and authentication of isolates in ancient ice cores. Chapter 2, this volume.

74. Rosewich, U.L., and H.C. Kistler. 2000. Role of horizontal gene transfer in the evolution of fungi. Ann. Rev. Phytopathol. 38: 325–363.

75. Sanger, F., S. Nicklen, and A.R. Coulson. 1977. DNA sequencing with chain-terminating inhibitors. Proc. Nat. Acad. Sci. 74: 5463–5467.

76. Sipiczki, M. 2000. Where does fission yeast sit on the tree of life? Genome Biology 1(2): 1011.1011–1014.

77. Sussman, A.S., and H.O. Halvorson. 1966. Spores: Their Dormancy and Germination. Harper & Row/New York/London.

78. Swofford, D. 1999. PAUP: Phylogenetic analysis using parsimony, Version 4.0b2a (PPC). Sinaurer and Associates, Sunderland, Massachusetts.

79. Taylor, K.C., C.U. Hammer, R.B. Alley, H.B. Clausen, D. Dahl-Jensen, A.J. Gow, N.S. Gundestrup, J. Kipfstuhl, J.C. Moore, and E.D. Waddington. 1993. Electrical conductivity measurements from the GISP 2 and GRIP Greenland ice cores. Nature 366: 549–552.

80. Taylor, K.C., P.A. Mayewski, R.B. Alley, E.J. Brook, A.J. Gow, P.M. Grootes, D.A. Meese, E.S. Saltzman, J.P. Severinghaus, M.S. Twickler, J.W.C. White, S. Whitlow, and G.A. Zielinski. 1997. The Holocene-Younger Dryas transition recorded at summit, Greenland. Science 278: 825–827.

81. Taylor, P.G. 1996. Reproducibility of ancient DNA sequences from extinct Pleisto-
    cene fauna. Mol. Biol. Evol. 13: 283–285.
82. Thompson, L.G., W.L. Hamilton, and C. Bull. 1975. Climatological implications
    of microparticle concentrations in the ice core from Byrd Station, Western Ant-
    arctica. J. Glaciol. 14: 433–444.
83. van Nues, R.W., and J.D. Beggs. 2001. Functional contacts with a range of splicing
    proteins suggest a central role for Brr2p in the dynamic control of the order of
    events in spliceosomes of *Saccharomyces cerevisiae*. Genetics 157: 1451–1467.
84. Venter, J.C., M.D. Adams, E.W. Mayers, P.W. Li, R.J. Mural, G.G. Sutton, H.O.
    Smith, M. Yandell, C.A. Evans, and R.A. Holt. 2001. The sequence of the human
    genome. Science 291: 1304–1351.
85. Willerslev, E. Pers. comm.
86. Yoder, O.C., and B.G. Turgeon. 1996. Molecular-genetic evaluation of fungal mol-
    ecules for roles in pathogenesis to plants. J. Genet. 75: 425–440.
87. Yoder, O.C., and B.G. Turgeon. 2001. Fungal genomics and pathogenicity. Current
    Opinion of Plant Pathol. 4: 315–321.
88. Zielinski, G.A. 1995. Stratospheric loading and optical depth estimates of explo-
    sive volcanism over the last 2100 years derived from the GISP 2 Greenland ice
    core. J. Geophys. Res. 100 (D10): 20, 937–20, 955.

12

Yeasts in the Genus *Rhodotorula* Recovered from the Greenland Ice Sheet

William T. Starmer, Jack W. Fell, Catherine M. Catranis,

Virginia Aberdeen, Li-Jun Ma, Shuang Zhou, and Scott O. Rogers

Rhodotorula species have been recovered from an unusual variety of habitats and sources (Table 12.1), ranging from deep ocean sediments to the upper atmosphere (i.e., 18,000–30,000 m in the stratosphere). These organisms are ubiquitous on leaf surfaces, they commonly occur in soils; fresh, estuarine, and marine waters; indoor and outdoor air; and clouds and fog. In this chapter, we hypothesize that the worldwide distribution of these yeasts may be influenced by ablation of glacial ice. Glaciers and the polar ice caps are the world's greatest reservoirs of water. Microorganisms entrapped in glaciers during accumulation of snow are eventually released to reenter the global population at unknown rates, but the enormous amount of water returning yearly implies a great potential for ice as a source of yeasts found in the oceans, glacial streams, and lakes. In this context, we present evidence on the recovery of *Rhodotorula* species from various depths of the Greenland ice sheet. Our examination relies mainly on one ice core (Dye 3), drilled in 1971 and 1979 (16, 54). We used two different methods in two separate labs to isolate yeasts from water melted from subsections of subcores ranging in depths (ages in years before present) from 158.6 m (ca. 350 ybp) to 1994.3 m (ca. 140,000 ybp). Strains obtained were investigated for their physiological properties, as well as their nucleotide sequences of the ITS and D1/D2 regions of the ribosomal DNA.

Methods

Previous reports of fungi isolated from glacial ice detailed the sources and a mechanical method to obtain interior subcores. These procedures are summarized as follows: (i) the ends of the subcore were sterilized with UV irradiation, (ii) an interior core was mechanically extracted from the sterile ends under sterile conditions, and both the interior core and exterior sheath were melted. Meltwater from these parts were plated and incubated at progressively

TABLE 12.1
Sources of *Rhodotorula* species

Source	References
Surfaces	
Leaves	34, 75, 80
Bark	12, 83
Arid plants	62
Fruit	15
Halophilic plants	42
Human	44, 64
Water	
General	36
Rivers	43, 70, 81
Lakes	49, 51, 67, 78, 82
Brackish	74
Estuaries	37, 38, 88
Coastal seas and oceans	9, 28, 29, 69, 91, 92
Air	
Indoor	3, 45, 76
Outdoor	7, 22, 32, 75
Upper Atmosphere	8
Fog	31
Soil	
General	22, 23
Amazonian	65
Mounds of *Macrotermes*	4
Ice	
Antarctic	2
Greenland	14, 57, 58
Animals	
Drosophilids	68
Mosquitoes	46
Bees	47, 86
Beetles	94
Mollusks and crabs	19
Fresh sea food	48
Trout and turbot	5
Whales	10
Musk oxen	56
Lamb	21
Red meat	20
Milk	39, 77

higher temperatures (8 to 22 °C). Colonies were isolated and cultured. This procedure, referred to as the mechanical method, was used for ice melted in 1991 (13, 14, 57, 58).

A second method developed subsequent to the mechanical method employed surface sterilization with Clorox (sodium hypochlorite) followed by sequential melting of shells of ice from the exterior to the interior (see chapters 2 and 11). This method is referred to as the chemical method. In both cases, each subcore was subsectioned into several parts (two to four) before treatment and melting. The melting and sampling of meltwater was carried out in a sterilized laminar flow hood with open Petri dishes containing media designed to detect airborne contaminants that might enter during the decontamination, melting, and plating procedures.

Ribosomal DNA sequence analysis for the internal transcribed spacer (ITS) and D1/D2 region of the large subunit followed the methods of Fell et al. (30).

Results

A total of 5.691 l of water was obtained by melting ice from the Dye 3 core of the Greenland ice sheet. Some meltwater (470 ml) was saved for other studies (e.g., viral isolation, ref. 13). Another fraction (375 ml) was lost during decontamination of the exterior by the chemical procedure. Of the remainder, about 80% (3.9 l) was plated for yeast isolation and thirty-two colonies of yeasts grew. The yeasts recovered included ascomycetous and basidiomycetous taxa with the majority (twenty-two) in the genus *Rhodotorula*. Table 12.2 lists the *Rhodotorula* species that were recovered and the depth from the surface of the glacier in which the yeasts were found. In addition to the twenty-two *Rhodotorula* colonies isolated, the Dye 3 core yielded ten colonies representing six species or species groups (four *Cryptococcus albidus*, three *Candida* spp., and one each of *Cryptococcus humicola*, *Sporidiobolus salmonicolor*, and *Aureobasidium pullulans*).

Species identification by physiological and nucleotide sequences of the ITS and D1/D2 regions of the ribosomal DNA showed the *Rhodotorula* isolates belong to the *Erythrobasidium* and *Sporidiolobolus* lineages (30). Figure 12.1 shows the position of the isolates in the D1/D2 trees.

The mechanical isolation method revealed three *Rhodotorula* species in the outer shells of the melting procedure. These isolates could be old or recent contaminants of the ice and thus we cannot be certain of their age. Only *R. mucilaginosa* was recovered from a decontaminated subcore at 598.3 and 798.7 m.

The chemical isolation method only detected *R. laryngis*, but this species was frequently found occurring at four depths (160.7, 1007.2, 1621.4, and

TABLE 12.2.

Record of yeasts (Ids in brackets) from the Dye 3 ice core of Greenland

Core[a]	Depth (m)	ss[b]	sh[c]	Isolates (mechanical)	ss[a]	sh[b]	Isolates (chemical)
6	158.6	4	2	None	—	—	
7	200.7	3	2	None	—	—	
8	369.0	2	2	*R. laryngis* (o) [431] *R. mucilaginosa* (o) [435]	1	5	None
9	140.7	1	2	None	—	—	
10	160.7	—	—		2	5	*R. laryngis*(*) [769, 770, 773, 779, 780]
12	598.3	3	2	*R. mucilaginosa* (o,*) [513°, 524*, 528*, 529*] *R. laryngis* (o) [661]	1	5	None
13	798.7	3	2	*R. mucilaginosa* (*) [544] *R.* sp(o) [545]	1	5	None
14	1007.2	3	2	*R. laryngis* (o) [644]	1	5	R. laryngis(*) [784]
15	1207.4	3	2	None	1	5	None
16	1405.5	3	2	*R.* sp (o) [638]	—	—	
17	1621.4	3	2	None	1	5	*R. laryngis*(*) [776, 777, 778, 788]
18	1994.3	3	2	None	1	5	*R. laryngis*(*) [786]

Note: Subcores 6, 7, and 8 were drilled in 1971, while subcores 9 through 18 were drilled in 1979. Each subcore was cut into a number of subsections (ss) for independent isolation. Two isolation methods were used. The mechanical subcoring used in 1991 produced an outer (o) shell of ice and interior (*) subcore. It is presumed that the outer shell contains old and recent contaminants. The chemical method used in 1999 produced a sequence of shells of melted ice. These shells represent isolates from the interior (*) of the ice because the exterior had been decontaminated with Clorox.

[a] The GISP 2 ice core, also obtained from the Greenland ice sheet, was sampled from a subcore at 62 m. Two isolates of *R. mucilaginosa* [129, 405] were recovered from the interior.

[b] Number of subsections cut from subcore.

[c] Number of shells melted from each subsection.

1994.3 m) of the eight depths studied. Two subcores (10 and 17) yielded multiple colonies (five and four, respectively) of this species. All isolates from the chemical isolation method were presumed to be old organisms deposited when the original snow accumulated on the glacier. No contamination was detected by the procedure used to detect airborne contaminants.

Discussion

Two central questions are of concern.
1. Why, among the yeasts, are *Rhodotorula* species common in glacial ice? The answer to this question requires knowledge of the life history, habitats, and dispersal modes of the species and, undoubtedly, will require careful work and attention to all aspects of their adaptations for survival and reproduction.
2. Despite the precautions, controls, and care taken in the isolation of these organisms, could they still be results of recent contamination? They are contaminants in many environments, including health care facilities (3), hospital workers (44), the food industry (20), and the laboratory. The question of possible recent contamination is difficult but important to resolve for investigations to detect organisms in any environment.

Why Rhodotorula?

Davenport (18) noted that basidiomycetes are generally suited to colder habitats with minimum growth temperatures close to 0 °C. In addition, *Rhodotorula* and *Rhodospiridium* species often have wide temperature tolerances for growth. These traits could be a result of several cold-related adaptations such as production of carotenoid pigments, storage of intracellular lipids, high tolerance to osmotic pressure, production of capsules, and a slow-growing life cycle.

Tolerating cold conditions is one facet of the ability to remain viable after prolonged periods. The ability of *Rhodotorula* species to withstand high pressure (40 MPa) indicates these yeasts can live at abyssal depths (33, 55) and this may contribute to their survival in deep ice. Osmotic adjustments to environmental salinity is also a useful adaptation shared by most marine yeasts (35, 40, 41). Furthermore, the relative longevity of *Rhodotorula* species in a variety of aquatic conditions has been reported. For example, *R. rubra*, kept in sterile water after exposure to space flight (*Apollo* 17), was viable after twenty-seven years (95). Mean survival of *R. mucilaginosa* in river water was up to 606 days (71). The utility of water, especially distilled water, for long-term yeast storage has been documented for many fungi and yeasts (61), and periodic bursts of multiplication have been observed (72).

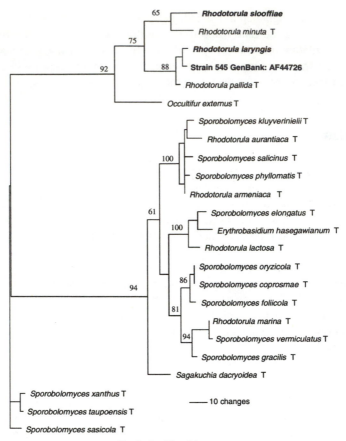

Erythrobasidium Lineage

FIGURE 12.1. *Erythrobasidium* and *Sporidiobolus* lineages of the urediniomy-cetous yeasts. Phylogenetic analysis (PAUP 4.0b10) of the D1/D2 region of the large subunit rDNA to demonstrate the phylogenetic location of the ice core yeasts (in bold). Numbers on branches represent bootstrap percentages (> 50%) from 1000 full heuristic replications. *Erythrobasidium* lineage: the single most parsimonious tree. Number of characters = 634, constant characters = 432, parsi-mony uninformative characters = 26, parsimony informative characters = 176. Tree length 406, consistency index = 0.645, retention index = 0.813. *Sporidiobo-lus* lineage: one of 23 equally parsimonious trees. Number of characters = 629, constant characters = 458, parsimony uninformative characters = 48, parsimony informative characters = 123. Tree length 376, consistency index = 0.566, reten-tion index = 0.794.

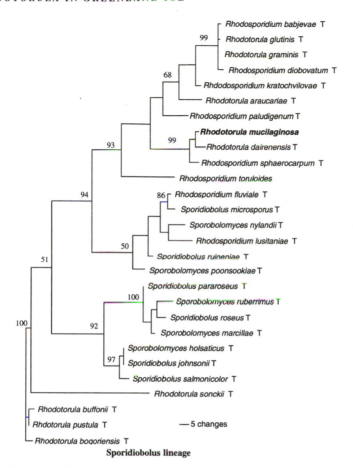

Figure 12.1. (*cont'd*)

Survival and transport in air are critical to the likelihood of organisms becoming entrapped in ice. In this regard, *Rhodotorula* species are frequently found in air samples outdoors (32) and indoors (76), in the upper atmosphere (8) above the clouds in the stratosphere, and deep within permafrost (see chapter 8). Perhaps more relevant is the finding that some yeasts, such as *Rhodotorula*, increase in numbers in fog (31).

Tolerance to cold, survival in water, and distribution in the atmosphere is coupled to growth on surfaces in the terrestrial environment. *Rhodotorula* species are well-known residents of the phyllosphere (17, 24, 25, 53, 59, 73, 79, 96) and surfaces such as tree bark (12, 83) or other plant surfaces (15, 62).

Living on leaf surfaces presents three problems for microorganisms; (i) tolerance to sunlight, air, and exposure to oxidants; (ii) acquisition of resources;

and (iii) adherence. The first problem is partially solved by the light-induced production of carotenoid pigments (87) that may provide protection against damaging UV irradiation as well as serve as antioxidants (50, 66). Resource acquisition is presumed to be facilitated by leaf exudates and the activity of other organisms such as aphids that deposit honey dew (85). *Rhodosporidium toruloides* uses localized mannose residues, possibly in association with glycoproteins, at sites of yeast-bud development to adhere to leaf surfaces (11). This adaptation could explain a necessary component of the *Rhodotorula* phylloplane life history. In addition, *Rhodotorula* species from the phyllosphere produce antibiotics (60) that may provide competitive advantages on surfaces. In a similar manner, *Rhodotorula* species from water or sediment have broad-spectrum killer activity against both ascomycetous and basidiomycetous yeasts (89).

Rhodotorula species are relatively minor components of more conventional ascomycetous yeast habitats (e.g., tree fluxes and insects, 4, 52, 84, 94). The isolation of *Rhodotorula* species from a variety of nonspecific sources (Table 12.1) may be a consequence of their surface dwelling, air- and water-transported lifestyle. These characteristics make them highly likely to be found in glacial ice and to be reintroduced to aquatic environments when glaciers melt (26, 90).

Contamination?

There are several potential sources of recent contamination. Table 12.3 outlines the potential sources of contaminants and the steps used to ensure against them. Concerns regarding contamination should consider the following:

1. Recovery of certain species can occur because those species are adapted to survive long periods of atmospheric transport, and may be contaminants under any condition (ancient or contemporary). This situation makes it difficult to differentiate recent from ancient contamination.

2. Nucleotide sequence homologies between presumed old and contemporary organisms suggest recent contamination. However, those species adapted to recycle through ice (e.g., ice–water–land–atmosphere–ice) might be discriminated against because cycling is likely to mix old and new populations over geological time periods.

3. Repeated recovery of a few forms suggests recent contamination, a conclusion based on the assumption that ice organisms should represent a random sample of airborne forms deposited during snowfall, that is, organisms that occur in the snow simply by chance. However, we also expect that organisms recovered from deep ice were subjected to long-distance transport, cold conditions, and extended time under pressure and therefore might be selectively isolated by these factors. In the present study the isolates of *R. mucilaginosa* and *R. laryngis* were recovered on a regular basis. *R. laryngis* was found using different methods, in different laboratories, and by different investigators. It is thus possible that some species of *Rhodotorula* in the *Erythrobasidium* and

TABLE 12.3.
Potential sources of recent contamination in the isolation of yeasts from ice, and the methods used as precautions against contamination

Potential Source	Control or Method
Drilling	Cores were selected to be free of visible wafering and fractures.
Transport and storage of cores	Exterior surfaces were assumed to be contaminated at this stage.
Exterior surface	Surfaces were decontaminated by two different methods, two different environments by different researchers. Both methods were tested for efficacy by sham ice core experimentation.
Air present during subcore decontamination melting of shells, or plating of melt water	Melting and plating were conducted in sterile laminar flow environments with open control-Petri plates for determination of any potential airborne contaminants.

Sporidiobolus lineages have special adaptive features that account for their repeated recovery. The adaptive features of these groups can and should be explored experimentally. If they do have features that allow them to remain viable for long periods of time in deep ice, then their life history should include this cycling through this medium as an important influence on their evolution and worldwide distribution.

Acknowledgments

This work was supported by grants from the National Science Foundation. We thank André Lachance for constructive comments on an earlier version of the manuscript.

Literature Cited

1. Abranches, J., H.N. Nobrega, P. Valente, L.C. Mendonça-Hagler, and A.N. Hagler. 1998. A preliminary note on yeasts associated with fecal pellets of rodents and marsupials of Atlantic forest fragments in Rio de Janeiro, Brazil. Revista de Microbiologia 29: 170–173.
2. Abyzov, S.S. 1993. Micro-organisms in the Antarctic ice. pp. 265–295. In E.I. Friedmann (ed.), Antarctic Microbiology. Wiley, New York.

3. Aidoo, K.E., A. Anderton, and K.A. Milligan. 1995. A 2-year survey of the airborne mycoflora in a hospital environment. Int. J. Environ. Health Res. 5: 223–228.

4. Amund, O.O., M.N. Onumonu, T.A. Onawale, and S.L.O. Malaka. 1988. A study of microbial composition and lignocellulose degradation in the mound soil of *Macrotermes bellicosus* Smeathman. Microbios Letters 37: 69–74.

5. Andlid, T., R.V. Juarez, and L. Gustafsson. 1995. Yeast colonizing the intestine of rainbow trout (*Salmo gairdneri*) and turbot (*Scophtalmus maximus*). Microbial Ecol. 30: 321–334.

6. Ba, A.S., Sh. A. Phillips Jr., and J.T. Anderson. 2000. Yeasts in mound soil of the red imported fire ant. Mycol. Res. 104: 969–973.

7. Barreto de Oliveira, M.T., R.F. Dos Santos Braz, and M.A. Guerra Ribeiro. 1993. Airborne fungi isolated from Natal, State of Rio Grande do Norte-Brazil. Revista de Microbiologia 24: 198–202.

8. Bruch, C.W. 1967. Microbes in the upper atmosphere and beyond. pp. 345–374. In P.H. Gregory and J.L. Monteith (eds.), Airborne Microbes. Cambridge University Press, London.

9. Bruni, V., R.B. Lo Curto, R. Patane, and D. Russo. 1983. Yeasts in the Strait of Messina. Memorie di Biologia Marina e di Oceanografia 13: 65–78.

10. Buck, J.D. 1984. Microbiological observations on two stranded live whales. J. Wildlife Dis. 20: 148–153.

11. Buck, J.W., M.-A. Lachance, and J.A. Traquair. 1998. Mycoflora of peach bark population dynamics and composition. Can. J. Bot. 76: 345–354.

12. Buck, J.W., and J.H. Andrews. 1999. Attachment of the yeast *Rhodosporidium toruloides* is mediated by adhesives localized at sites of bud cell development. Appl. Environ. Microbiol. 65: 465–471.

13. Castello, J.D., S.O. Rogers, W.T. Starmer, C.M. Catranis, L. Ma, G.D. Bachand, Y. Zhao, and J.E. Smith. 1999. Detection of tomato mosaic tobamovirus RNA in ancient glacial ice. Polar Biol. 22: 207–212.

14. Catranis, C.M., and W.T. Starmer. 1991. Microorganisms entrapped in glacial ice. Ant. J. U.S. 26: 234–236.

15. Chand-Goyal, T., and R.A. Spotts. 1996. Enumeration of bacterial and yeast colonists of apple fruits and identification of epiphytic yeasts on pear fruits in the Pacific Northwest United States. Microbiol. Res. 151: 427–432.

16. Dansgaard W., H.B. Clausen, N. Gunderstrup, S.J. Johnson, and C. Rygner. 1985. Dating and climatic interpretation of two deep Greenland ice cores. pp. 71–76. In C.C.J. Langway, H. Oeschger, and W. Dansgaard (eds.), Greenland Ice Core: Geophysics, Geochemistry, and the Environment. American Geophysical Union, Washington, D.C.

17. Davenport, R.R. 1973. Micro-ecology of yeasts and yeast-like organisms in a vineyard. pp. 163–165. In Report of the Long Ashton Agricultural and Horticultural Research Station.

18. Davenport, R.R. 1980. Cold-tolerant yeasts and yeast-like organisms. pp. 215–230. In F.A. Skinner, S.M. Passmore, and R.R. Davenport (eds.), Biology and Activities of Yeasts. Academic Press, New York.

19. De Araujo, F.V., C.A.G. Soares, A.N. Hagler, and L.C. Mendonça-Hagler. 1995. Ascomycetous yeast communities of marine invertebrates in a Southeast Brazilian mangrove ecosystem. Antonie van Leeuwenhoek 68: 91–99.

20. Dillon, V.M., and R.G. Board. 1991. Yeasts associated with red meats. J. Appl. Bacteriol. 71: 93–108.

21. Dillon, V.M., R.R. Davenport, and R.G. Board. 1991. Yeasts associated with lamb. Mycol. Res. 95: 57–63.

22. di Menna, M.E. 1955. A quantitative study of air-borne fungus spores in Dunedin, New Zealand. Trans. British Mycol. Soc. 38: 119–129.

23. di Menna, M.E. 1957. The isolation of yeasts from soil. J. Gen. Microbiol., 17: 678–688.

24. di Menna, M.E. 1959. Yeasts from leaves of pasture plants. N. Z. J. Agric. Res. 2: 394–405.

25. di Menna, M.E. 1971. The mycoflora of leaves of pasture plants in *New Zealand*. pp. 159–174. In T.F. Preece and C.H. Dickinson (eds.), Ecology of Leaf Surface Micro-organisms. Academic Press, London.

26. Ellis-Evans, J.C., and D.D. Wynn-Williams. 1985. The interaction of soil and lake microflora at Signy Island. pp. 663–668. In W.R. Siegfried, P.R. Condy, and R.M. Laws (eds.), Antarctic Nutrient Cycles and Food Webs. SCAR Symposium on Antarctica Biology, Wilderness (South Africa), 12–16 September.

27. Estermann, R. 1994. Biological functions of carotenoids. Aquaculture 124: 219.

28. Fell, J.W. 1976. Yeasts in oceanic regions, pp 93–124. In E.B.G. Jones (ed.), Recent Advances in Aquatic Mycology. Elek Science, London.

29. Fell, J.W., D.G. Ahearn, S.P. Meyers, and F.J. Roth Jr. 1960. Isolation of yeasts from Biscayne Bay, Florida and adjacent benthic areas. Limnol. and Oceanogr. 5: 366–371.

30. Fell, J.W., T. Boekhout, A. Fonseca, G. Scorzetti, and A. Statzell-Tallman. 2000. Biodiversity and systematics of basidiomycetous yeasts as determined by large-subunit rDNA D1/D2 domain sequence analysis. Int. J. System. Evol. Microbiol. 50: 1351–1371.

31. Fuzzi, S., P. Mandrioli, and A. Perfetto. 1997. Fog droplets—an atmospheric source of secondary biological aerosol particles. Atmospheric Environment 31: 287–290.

32. Gambale, W., A. Purchio, and C.R. Paula. 1981. Daily periodicity of airborne fungi from Sao Paulo City, Brazil. Revista de Microbiologia 12: 176–181.

33. Gross, M., I.J. Kosmowsky, R. Lorenz, H.P. Molitoris, and R. Jaenicke. 1994. Response of bacteria and fungi to high-pressure stress as investigated by two-dimensional polyacrylamide gel electrophoresis. Electrophoresis 15: 1559–1569.

34. Gunasekera, T.S., N.D. Paul, and P.G. Ayres. 1997. Responses of phylloplane yeasts to UV-B (290–320 nm) radiation. Interspecific differences in sensitivity. Mycol. Res. 101: 779–785.

35. Gupta, R. 1996. Growth of marine yeast on different strength of stress solutes. Proc. second workshop on scientific results of FORV Sagar Sampa., Department of Ocean Development, New Delhi, pp. 91–95.

36. Hagler, A.N., R.B. De Oliveira, and L.C.M. Hagler. 1982. Yeasts in the intertidal sediments of a polluted estuary in Rio de Janeiro, Brazil. Antonie van Leeuwenhoek 48: 53–56.

37. Hagler, A.N., and L.C.M. Mendonça-Hagler. 1981. Yeasts from marine and estua-
    rine waters with different levels of pollution in the State of Rio De Janeiro,
    Brazil. Appl. Environ. Microbiol. 41: 173–178.
38. Hagler, A.N., and D.G. Ahearn. 1987. Ecology of aquatic yeasts. pp. 181–205. In
    A.H. Rose and J.S. Harrison (eds.), The Yeasts, vol. 1, 2nd ed. Academic Press,
    London.
39. Haridy, M.S.A. 1992. Yeast flora of raw milk in El-Minia City, Egypt. Cryptogami.
    Mycol. 13: 321–326.
40. Hernandez-Saavedra, N.Y., J.L. Ochoa, and R. Vazquez-Duhalt. 1994. Effect of
    salinity on the growth of the marine yeast Rhodotorula rubra. Microbios 80:
    99–106.
41. Hernandez-Saavedra, N.Y., J.L. Ochoa, and R. Vazquez-Dulhalt. 1995. Osmotic
    adjustment in marine yeast. J. Plankton Res. 17: 59–69.
42. Hinzelin, F., P. Lectard, and J.M. Pelt. 1979. Yeast ecology in both fluvial and
    saline continental ecosystems. Rev. Mycol. 43: 149–168.
43. Hinzelin, F., and P. Lectard. 1983. Microflora of the phyllosphere of some halo-
    philic plants. Cryptogam. Mycol. 4: 333–345.
44. Huang, Y.C., T.Y. Lin, H.S. Leu, J.L. Wu, and J.H. Wu. 1998. Yeast carriage on
    hands of hospital personnel working in intensive care units. J. Hospital Infection
    39: 47–51.
45. Hung, L-L., C.S. Yang, F.A. Lewis, and F.A. Zampiello. 1994. Quarterly monitor-
    ing and seasonal variation of microorganisms in air handling units—a two-year
    study. pp. 481–484. In L. Morawska, N.D. Bofinger, and M. Maroni (eds.), In-
    door Air: An Integrated Approach. Elsevier Science, New York.
46. Ignatova, E.A., S.S. Nagornaya, T.N. Povazhnaya, and G.S. Yanishevskaya. 1996.
    Yeast flora of blood-sucking mosquitoes. Mikrobiologichnyi Zhurnal 58: 12–15.
47. Inglis, G.D., L. Sigler, and M.S. Goettel. 1993. Aerobic microorganisms associated
    with alfalfa leafcutter bees (Megachile rotundata). Microbial Ecol. 26: 125–143.
48. Kobatake, M., N.J.W. Kreger-van Rij, M.T.L.A. Placido, and N. van Uden. 1992.
    Isolation of proteolytic psychrotrophic yeasts from fresh raw seafoods. Letters
    in Appl. Microbiol. 14: 37–42.
49. Kolesnitskaya, G.N., and E.A. Maximova. 1982. The composition of yeast species
    in the water of the southern Baikal Lake. Mikrobiologiya 51: 501–505.
50. Krinsky, N.I. 1989. Antioxidant functions of carotenoids. Free Radical Biology
    and Medicine 7: 617–635.
51. Kwasniewska, K. 1988. Horizontal distribution and density of yeasts and filamen-
    tous fungi in Lake St. Clair water. J. Great Lakes Res. 14: 438–443.
52. Lachance, M.A., B.J. Metcalf, and W.T. Starmer. 1982. Yeasts from exudates of
    Quercus, Ulmus, Populus, and Pseudotsuga: New isolations and elucidation of
    some factors affecting ecological specificity. Microbial Ecol. 8: 191–198.
53. Lamb, R.J., and J.F. Brown. 1970. Non-parasitic microflora on leaf surfaces of
    Paspalum dilatatum, Salix babylonica and Eucalyptus stellulata. Trans. British
    Mycol. Soc. 55: 383–390.
54. Langway, C.C.J., H. Oeschger, and W. Dansgaard. 1985. The Greenland ice sheet
    program in perspective. pp. 1–8. In C.C.J. Langway, H. Oeschger, and W. Dans-
    gaard (eds.), Greenland Ice Core: Geophysics, Geochemistry, and the Environ-
    ment. American Geophysical Union, Washington, D.C.

55. Lorenz, R., and H.P. Molitoris. 1997. Cultivation of fungi under simulated deep sea conditions. Mycol. Res. 101: 1355–1365.
56. Lund, A. 1980. Yeasts in the rumen contents of Musk Oxen. J. Gen. Microbiol. 121: 273–276.
57. Ma, L-J., C.M. Catranis, W.T. Starmer, and S.O. Rogers. 2000. Revival and characterization of fungi from ancient polar ice. Mycologist 13: 70–73.
58. Ma, L-J., S.O. Rogers, C.M. Catranis, and W.T. Starmer. 2000. Detection and characterization of ancient fungi entrapped in glacial ice. Mycologia 92: 286–295.
59. McBride, R.P., and A.J. Hayes. 1977. Phylloplane of European larch. Trans. British Mycol. Soc. 69: 39–46.
60. McCormack, P.J., H.G. Wildman, and P. Jeffries. 1994. Production of antibacterial compounds by phylloplane-inhabiting yeasts and yeastlike fungi. Appl. Environ. Microbiol. 60: 927–931.
61. McGinnis, M.R., A.A. Padhye, and L. Ajello. 1974. Storage of stock cultures of filamentous fungi, yeasts, and some aerobic actinomycetes in sterile distilled water. Appl. Microbiol. 28: 218–222.
62. Middelhoven, W.J. 1997. Identity and biodegradative abilities of yeasts isolated from plants growing in an arid climate. Antonie van Leeuwenhoek 72: 81–89.
63. Middelhoven, W.J., and F. Spaaij. 1997. *Rhodotorula cresolica* sp. nov., a cresol-assimilating yeast species isolated from soil. Int. J. System. Bacteriol. 47: 324–327.
64. Mok, W.Y., and M.S. Barreto da Silva. 1984. Mycoflora of the human dermal surfaces. Can. J. Microbiol. 30: 1205–1209.
65. Mok, W.Y., R.C.C. Luizao, M.S. Barreto da Silva, M.F.S. Teixeira, and E.G. Muniz. 1984. Ecology of pathogenic yeasts in Amazonian soil. Appl. Environ. Microbiol. 47: 390–394.
66. Moore, M.M., M.W. Breedveld, and A.P. Autor. 1989. The role of carotenoids in preventing oxidative damage in the pigmented yeast, *Rhodotorula mucilaginosa*. Archiv. Biochem. Biophys. 270: 419–431.
67. Morais, P.B., M.A. Resende, C.A. Rosa, and F.A.R. Barbosa. 1996. Occurrence and diel distribution of yeasts in a paleo-karstic lake of southeastern Brazil. Revista de Microbiologia 27: 182–188.
68. Morais, P.B., C.A. Rosa, J. Abranches, L.C. Mendonça-Hagler, and A.N. Hagler. 1996. Yeasts vectored by *Drosophila quadrum* (Calloptera group) in tropical rain forests. Revista de Microbiologia 27: 87–91.
69. Paula, C.R. de, A. Puricho, and W. Gambale. 1983. Yeasts from beaches in the southern area of Sao Paulo State "Baixada Santista," Brazil. Revista de Microbiologia 14: 136–143.
70. Peçanha, M.P., F.C. Pagnocca, C.A. Rugani, and F.A. Neves. 1996. Yeasts and other parameters of pollution of the Ribeirao Claro stream in Rio Claro, Sao Paulo. Revista de Microbiologia 27: 177–181.
71. Peter, M., and Z. Peter. 1988. Experimentelle Studien bezueglich der Persistenz einiger fakultativ pathogener Pilze in flusswasser. Zentralbl. Mikrobiol. 143: 523–528.
72. Peter, Z., and M. Peter. 1989. Experimental studies on the persistence in distilled water of certain conditional pathogenic fungi. Arch. Roum. Pathol. Exp. Microbiol. 48: 275–282.

73. Pollard, A. 1969. The flora of a vineyard. pp. 120–121. In Report of the Long Ashton Agricultural and Horticultural Research Station.

74. Prabhakaran, N., and P. Sivada. 1995. Hydrocarbon degrading yeasts from Cochin Backwater. J. Marine Biol. Assoc. India 37: 226–230.

75. Robbs, P.G., A.N. Hagler, and L.C. Mendonça-Hagler. 1989. Yeasts associated with a pineapple plantation in Rio de Janeiro, Brazil. Yeast as a Main Protagonist of Biotechnology, S485–S-489, Yeast, 2, 7. International Symposium on Yeasts, Perugia, Italy.

76. Rogers, S.A. 1984. A 13-month work-leisure-sleep environment fungal survey. Annals of Allergy 52: 338–341.

77. Rosa, C.A., F.R. Novak, J.A.G. de Almeida, L.C. Mendonça-Hagler, and A.N. Hagler. 1990. Yeasts from human milk collected in Rio de Janeiro, Brazil. Revista de Microbiologia 21: 361–363.

78. Rosa, C.A., M.A. Resende, F.A.R. Barbosa, P.B. Morais, and S.P. Franzot. 1995. Yeast diversity in a mesotrophic lake on the karstic plateau of Lagoa Santa, MG-Brazil. Hydrobiologia 308: 103–108.

79. Ruinen, J. 1963. The phyllosphere. II. Yeasts from the phyllosphere of tropical foliage. Antonie van Leeuwenhoek 29: 425–438.

80. Shivas, R.G., and J.F. Brown. 1984. Identification and enumeration of yeasts on *Banksia collina* and *Callistemon viminalis* leaves. Trans. British Mycol. Soc. 83: 687–689.

81. Slavikova, E., and R. Vadkertiova. 1997. Seasonal occurrence of yeasts and yeast-like organisms in the River Danube. Antonie van Leeuwenhoek 72: 77–80.

82. Slavikova, E., R. Vadkertiova, and A. Kockova-Kratochvilova. 1992. Yeasts isolated from artificial lake waters. Can. J. Microbiol. 38: 1206–1209.

83. Sorenson, W.G., J. Simpson, and J. Dutkiewicz. 1991. Yeasts and yeast-like fungi in stored timber. Int. Biodeterior. 27: 373–382.

84. Spencer, D.M., J.F.T. Spencer, E. Fengler, and L.I. De Figueroa. 1995. Yeasts associated with algarrobo trees (*Prosopis* spp.) in northwest Argentina: A preliminary report. J. Ind. Microbiol. 14: 472–474.

85. Stadler. B., and T. Müller. 1996. Aphid honeydew and its effect on the phyllosphere microflora of *Picea abies* (L.) Karst. Oecologia 108: 771–776.

86. Suciu, M., G. Boltasu, A. Vladescu, and B. Murg. 1981. Identification of some strains of yeasts in the intestine of the working bee (*Apis melifera* L.) from Romania. Arch. Roum. Pathol. Exp. Microbiol. 40: 163–172.

87. Tada, M., and M. Shiroishi. 1982. Mechanism of photoregulated carotenogenesis in *Rhodotorula minuta* I. Photocontrol of carotenoid production. Plant and Cell Physiol. 23: 541–548.

88. Taysi, I., and N. van Uden. 1964. Occurrence and population densities of yeast species in an estuarine-marine area. Limnol. and Oceanogr. 9: 42–45.

89. Vadkertiova, R., and E. Slavikova. 1995. Killer activity of yeasts isolated from the water environment. Can. J. Microbiol. 41: 759–766.

90. Vaeaetaenen, P. 1982. Effects of freshwater outflows on microbial populations in the Tvaerminne archipelago, southern Finland. Holarctic Ecol. 5: 61–66.

91. van Uden, N., and R.C. Branco. 1963. Distribution and population densities of yeast species in Pacific water, air, animals, and kelp off southern California. Limnol. and Oceanogr. 8: 323–329.

92. Velegraki-Abel, A., U. Marselou-Kinti, and C. Richardson. 1987. Incidence of yeasts in coastal sea water of the Attica Peninsula, Greece. Water Res. 21: 1363–1369.

93. Vishniac, H.S. 1987. Psychrophily and the systematics of yeast-like fungi. Pp. 389–402. In G.S. de Hoog, M.Th. Smith, and A.C.M. Weijman (eds.), The Expanding Realm of Yeast-like Fungi. Elsevier, Amsterdam.

94. Vishniac, H.S., and D.T. Johnson. 1990. Development of a yeast flora in the adult green june beetle (*Cotinis nitida*, Scarabaeidae). Mycologia 2: 471–479.

95. Volz, P.A., and S.L. Parent. 1998. Space flight micro-fungi after 27 years storage in water and in continuous culture. Microbios 96: 111–125.

96. Warren, R.C. 1976. Microbes associated with buds and leaves: Some recent investigations on deciduous trees. pp. 361–374. In C.H. Dickinson and T.F. Preece. (eds.), Microbiology of Aerial Plant Surfaces. Academic Press, London.

13

Plant and Bacterial Viruses in the Greenland Ice Sheet

John D. Castello, Scott O. Rogers, James E. Smith, William T. Starmer, and Yinghao Zhao

MANY MICROORGANISMS, including filamentous fungi, yeasts, algae, cyanobacteria, and bacteria, have been isolated from polar ice and/or permafrost and identified (refs. 1, 14, and see chapters 6, 7, 8, 9, 10, 11, 12, 13, 15, 16, and 19, this volume). Nevertheless, only a few investigators are searching for viruses in ice. Should we even expect to find viruses in ice? If so, which ones and how would they have gotten there? Viruses are obligate parasites that in their simplest form consist of one or more molecules of DNA or RNA surrounded by a protein coat that serves to protect the viral nucleic acid from degradation, among other functions. There is an enormous diversity of them. Every known group of living organisms, including plants, animals, fungi, protists, bacteria, and archaea, are hosts of viruses. Currently there are approximately 3600 described virus species (4), and an estimated 130,000 species that have yet to be described (8). Consequently, we might expect to find some stable viruses in this frozen and dehydrated, yet stable environment. We hypothesize that they are variously waterborne, soilborne, or airborne, with wide host ranges and/or geographic distributions. Ice also might encapsulate varying numbers of temperate viruses of common soil, aquatic, and marine procaryotes. Why search for viruses in polar ice? What are the implications if we find them? This chapter, the following chapter, and the summary chapter (chapter 20) deal with these questions.

We began our research to assess microbial diversity in ancient polar ice in 1998 with a grant from the LExEN (Life in Extreme Environments) Program of the National Science Foundation. Our objective was to identify and characterize the fungi, bacteria, archaea, and viruses entrapped in polar ice. In this chapter, we discuss the bacteria and the bacterial and plant viruses that we have detected in the Greenland ice sheet to date.

Isolation and Identification of Bacteria

In 1991, Catranis and Starmer isolated bacteria, fungi, and algae directly from meltwater obtained from 12 ice subcores from the GISP 2 and Dye 3 drill sites in Greenland (7). The subcores were surface decontaminated using the UV irradiation protocol (chapters 2 and 11), and then aseptically melted as described (7). Ten different media were inoculated with one ml meltwater aliquots, and then incubated at 8 °C for 6 weeks, 15 °C for 2 weeks, and finally 20 °C for an additional week. Approximately 200 bacterial isolates were obtained from these meltwaters (7). The isolates were subcultured and stored at 4 °C. In 1998, we selected 13 of them for identification because they represented a chronosequence of isolates from "young" (500 years before present [ybp]) to "old" (> 100,000 ybp) ice (Table 13.1).

The 13 cultures were identified by molecular analysis. DNA was liberated from the cells by incubating a cell suspension in 1% Triton X-100 at 100 °C for 5 minutes. Then, aliquots of 10 μl each were subjected to PCR amplification (10 μM Tris-[HCl [pH 8.3]; 5 pmol each dATP, dCTP, dGTP, dTTP; 1.5 mM MgCl$_2$; 50 mM KCl; 0.001% gelatin) using 25 pmol of each primer EUB16S1 (CGGTGGCGAAGGCGGCTCTC) and EUB16S3 (CATGGTGT-GACGGGCGGTGTG), with the following temperature regime: 94 °C for 10 minutes, 65 °C for 2 minutes, 72 °C for 1 minute, followed by 30 cycles of 94 °C for 1 min and 72 °C. Subsequently, there was a final elongation step of 72 °C for 10 min. PCR products were subjected to electrophoresis on 1.5% low melting point agarose gels (NuSieve GTG, FMC, Rockland, Maine), eluted from the gels, and rehydrated in water. Approximately 30 ng of this DNA was added to a reaction with the Terminator Ready Reaction Kit (Applied Biosystems, Foster City, California). The cycling program was: 1 minute at 94 °C, then 30 cycles of 94 °C for 10 seconds, 50 °C for 30 seconds, and 60 °C for 4 minutes. Following purification on a Centri-Sep column (Princeton Separations Inc., Adelphia, New Jersey), sequence determination was performed on an Applied Biosystems 373A DNA sequencer (Cornell University, Ithaca, New York).

The sequences were compared to those on GenBank using a BLAST search (www.ncbi.nlm.nih.gov/cgi-bin/BLAST/). The sequences exhibiting the highest similarity were aligned with the glacial sequences using CLUSTALW (http://searchlauncher.bcm.tcm.edu/multi-align/multi-align.html). Phylogenetic trees were determined with PAUP (19), using a heuristic search with 500 bootstrap replications.

A phylogram depicting the systematic relationships of the bacterial isolates from ice with the most closely related contemporary taxa (sequences deposited in GenBank) is depicted in Figure 13.1. Bacterial identifications were corroborated by examination of morphology, Gram stain reaction, endospore stain, and metabolic reactions on selected media (Figure 13.1 and Table 13.1). Eleven of the thirteen isolates were most closely related to *Bacillus subtilis* or related genera and species. Of the remaining two, one was most closely related to *Rhodococcus erythreus*, an actinobacterium, and the other to *Tatumella ptyseos*, a member of the enterobacteriaceae of the gamma proteobacteria (Table 13.1). Gram stain and endospore stain reactions, as well as growth on

TABLE 13.1

Characteristics of the thirteen bacterial isolates from the GISP 2 and Dye 3 drill sites on the Greenland ice sheet

Isolate Number	Drill Sites	Core Number	Depth (m)	Age (ybp)	Isolate ID	Cell Shape	Gram Stain Rxn[1]	Endo-spores[2]	V-P 2 hr[3]	Arabinose[4]	Xylose	Mannitol	Casein[5]	Gelatin	Starch	Killer Activity	PBSY-like Phage
331	GISP2D	131	130.1	<500	B. subtilis	Rods	+	+	+	+	+	−	+	+	+	+	+
332	GISP2D	131	130.1	<500	B. subtilis	Rods	+	+	+	−	+	+	−	−	−	+	+
334	Dye3-71	49	158.6	<500	B. subtilis	Rods	+	+	+	+	+	−	+	+	+	+	+
338	Dye3-71	49	158.6	<500	B. subtilis	Rods	+	+	+	+	+	+	+	+	+	+	+
338B	Dye3-71	49	158.6	<500	B. subtilis	Rods	+	−	+	+	+	−	+	+	+	+	+
451	Dye3-71	65	200.8	600	Rhodococcus erythreus	Rods & cocci	+	−	−	−	−	nt[7]	+	−	+	+	+
523	Dye3-79	522	598.5	1400	Tatumella ptyseos	Rods	−	−	−	−	+	+	+	+	+	+	−
526	Dye3-79	522	598.5	1400	Paenibacillus sp.	Rods	+	+	+	+	−	−	+	−	+	+	+
560	Dye3-79	724	798.8	2200	B. subtilis	Rods	+	+	+	+	+	+	−	+	+	+	+
646	Dye3-79	1945	1994.5	140K	B. subtilis	Rods	+	+	+	−	+	+	−	−	−	+	+
656	Dye3-79	1354	1405.7	5500	B. subtilis	Rods	+	+	+	+	+	+	+	−	−	−	−
733	Dye3-79	1354	1405.7	5500	B. subtilis	Rods	+	+	−	−	+	−	+	+	+	+	+
744	Dye3-79	1571	1621.5	7000	B. subtilis	Rods	+	+	+	−	−	nt	+	−	−	−	−
					B. subtilis SP02	Rods	+	+	+	+	+	−	+	+	+	+	nt
					B. subtilis SB11	Rods	+	+	+	−	−	−	−	−	+	nt	nt
					B. subtilis NR 1297 M+S	Rods	+	+	+	+	+	+	+	−	+	nt	nt
					Negative control[6]		−	−	−	−	−	−	−	−	−	−	−

[1] Gram stain reaction.

[2] Presence or absence of endospores.

[3] Reaction in Voges Proskauer medium after 2 hours' incubation.

[4] Acid from sugars (arabinose, xylose, and mannitol) after 48 hours' incubation at 37 °C.

[5] Hydrolysis of casein, gelatin, and starch after 24 hours' incubation at 37 °C.

[6] Negative controls = blank media.

[7] Not tested.

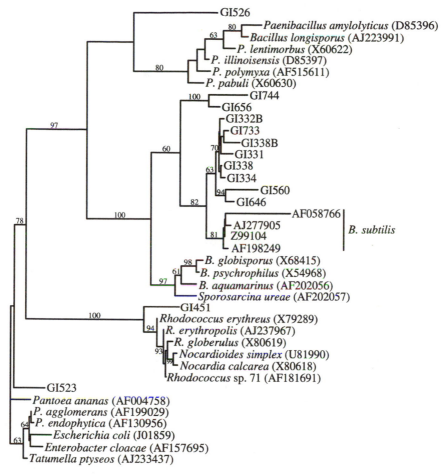

FIGURE 13.1. Results of phylogenetic analysis based on comparison of a 617 bp region within the eubacterial 16S rDNA SSU region (corresponding to 565 nucleotides of *E. coli*) of each of 39 bacterial taxa, including 13 bacteria from ice and 26 contemporary taxa. Gaps were used as a fifth character state. The phylogram was generated using PAUP with the maximum parsimony method and the general heuristic option. Steps = 1015 with a total of 242 parsimony-informative characters. Bootstrap support (shown for 500 replications) for the consensus tree was determined by the fast stepwise addition protocol.

TABLE 13.2.
Properties of the defective phage of *Bacillus subtilis*[a]

Type	Diameter of the Head (Nm)	Tail Length (Nm)	Diameter of the Tail (Nm)	Number of Cross-Striations
PBS X	45	200	20	53–54
PBS Y	45	265	20	69–70
PBS Z	45	255	20	65–66
PBS V	45	185	20	49–50
PBS W	45	225	20	58–59

[a]From reference 17.

selected media, confirm the tentative identities of the isolates as determined by phylogenetic analyses.

Bacterial Viruses in Ice

The thirteen bacterial isolates were tested for latent bacteriophage because, to our knowledge, bacteria isolated from glacial ice have not been examined for viruses. Bacteria can be infected with lytic or temperate viruses (phage). Lytic phage replicate within the bacterial cell and release infectious progeny virions that usually cause lysis and death of the infected cell. Temperate phage, on the other hand, have both latent and lytic phases. During the initial latent phase, the phage DNA is incorporated into the bacterial chromosome, where it remains transcriptionally silent and is passed along to daughter cells as a prophage. The lytic phase can be induced during periods of stress to the host cell. At this time the prophage is activated, replication begins, progeny virions are synthesized, and the cell lyses to release them.

Bacillus subtilis is susceptible to a wide variety of temperate phage, some of which are defective and cannot productively infect certain new host cell strains even though they inject their DNA efficiently. The viral DNA is integrated within the host chromosomal DNA, and is thus passed on to daughter cells. Although unable to replicate, the defective phage may prevent future cell divisions and even lyse the cell; hence the name *killer particles*. There are five groups of temperate, defective phage of *B. subtilis*, which are very common as mixtures in the same cell. They are widespread in this and related species. These phage are differentiated on the basis of morphological features, for example, size of head and tail, and number of cross-striations on the tail (Table 13.2 from ref. 17).

We hypothesized that ancient bacteria in ice would be lysogenized wih temperate phage. Therefore, the bacterial isolates from the Greenland ice sheet were screened for latent phage using stress induction followed by spot test for killing activity. The bacteria were subcultured, irradiated with ultraviolet light, or treated with mitomycin-C to induce temperate phage to become lytic, and subjected to a "spot test" for killing activity (3). Because eleven of the thirteen isolates from ice are closely related to *B. subtilis*, we spot-tested all of the induced cultures onto five different *B. subtilis* clones as well as *B. natto, B. pumilis*, and *B. licheniformis* clones (provided by H.E. Hemphill, Syracuse University, Syracuse, New York). Production of a large, clear zone or plaque (small, clear lesions) on the indicator culture constitutes a positive spot test, and suggests the test culture has killer activity, which can be caused by phage, bacteriocins, antibiotics, or killer particles. Aliquots of induced and noninduced subcultures of all isolates were subjected to low-speed and high-speed centrifugation to pellet any viruses. The pellet suspensions then were examined by TEM for viruses. Subcultures of two isolates (GI 332 and 733) were subsequently concentrated by ultracentrifugation and a ten-fold serial dilution of the suspended high-speed pellets prepared and spot tested against three indicator species: *B. subtilis* (clone SB 11), *B. natto* (R.A. Slepecky clone), and *B. pumilis* (clone NRS 576). The plates were observed for the presence of single, small plaques indicative of infectious phage.

All except the two isolates GI 656 and 744 produced large, clear zones on all eight *Bacillus* indicators, revealing that they possessed killer activity. Phage particles typical of PBSY group defective *B. subtilis* phage were observed in all isolates that showed killer activity except for isolate GI 523 (Table 13.1). Possible icosahedral, tailless phagelike particles were observed in cultures of GI 523, but these have not yet been characterized or identified. Isolates GI 656 and 744 did not show killer activity and phage were not observed in these cultures (Table 13.1). Representative phage particles detected in B. subtilis isolates 332 (Figure 13.2. A, B, C, D, E, and F) and 733 (Figure 13.2 H) are shown. Phage particles with smaller heads also were detected in the noninduced subculture of isolate 332 (Figure 13.2 B, G, and I). The particles depicted in Figure 13.2 A, C, D, E, and F measure 41 nm, $22 \pm 2 \times 252 \pm 28$ nm for head diameter, and tail width and length, respectively. The particles have sixty-nine to seventy cross striations. Based on these measurements, the phage are tentative members of the PBSY group of defective *B. subtilis* phage. The PBSY-like phage were not infectious on the indicator organisms tested, suggesting that they are defective and supporting their placement within the PBSY group. Isolate GI 451, identified as an actinobacterium most closely related to *Rhodococcus erythreus*, also showed killer activity and PBSY-like phage were observed in induced cultures (Table 13.1).

The particles depicted in Figure 13.2. B, G, and I have head and tail dimensions of 30 ± 3 nm, and $14 \times 243 \pm 14$ nm, respectively, sixty-eight cross-striations, and four apparent tail fibers (Figure 13.2 B and I). Thus, these particles have smaller heads and thinner tails than the particles described above.

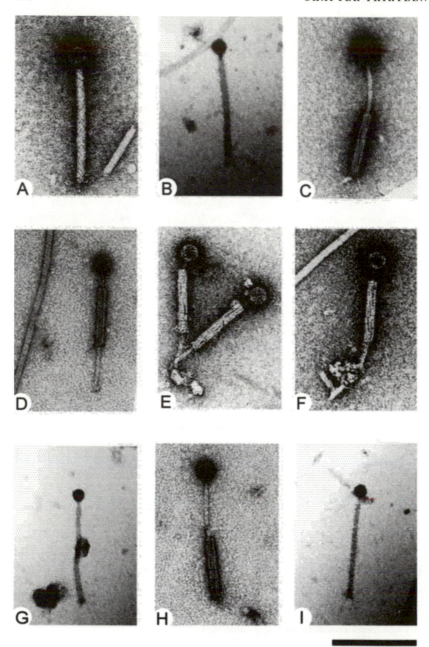

FIGURE 13.2. A, C, D, E, and (F) Phage particles detected in a UV-irradiated culture of *B. subtilis* isolate GI 332 from the Greenland ice sheet. (B, G, and I) Phage particles detected in a non-UV-irradiated culture of isolate GI 332. (H) Phage particle detected in a UV-irradiated culture of *B. subtilis* isolate GI 733. Bar = 200 nm.

To our knowledge, these phage (Figure 13.2 B, G, and I) are not similar to any known group of lytic or temperate phage of *B. subtilis*. These smaller phage have not yet been identified or isolated in pure culture.

Tomato Mosaic Tobamovirus in Ice

Tomato mosaic virus (ToMV) is a member of the tobamovirus group of plant viruses. Tobamoviruses possess all of the characteristics mentioned above to make them likely to be present in ancient ice. They are extraordinarily stable, and indeed, are arguably the most stable nucleoproteins known on Earth. Most of them, including ToMV, have very wide host ranges within many plant families, and many have a worldwide geographic distribution (12). In addition, ToMV is soilborne (11), waterborne (13), and airborne (10), and has been detected in surface water in North America (13), Europe (5), and New Zealand (JDC unpublished data). For these reasons, we hypothesized that this virus would be present in the polar ice sheets.

To date, 30 meltwaters from 11 subcores of the Dye 3 and GISP 2 Greenland cores have been assayed for ToMV using a seminested RT-PCR protocol that amplifies a 347 bp amplicon within the coat protein open reading frame (ORF) of the virus as described by Castello et al. (6). The amplicons were cycle sequenced and a 244 bp DNA sequence of 15 ice isolates was compared with 9 contemporary isolates of ToMV using neighbor-joining analysis (6).

In total, ToMV was detected in 17 of the 30 meltwaters, but not in any water controls (Table 13.3, which is Table 1 from ref. 6). Detection sensitivity of the RT-PCR assay was 10 fg (= 150 virions). The virus was present in Greenland ice < 500 to 140,000 ybp. Based on the neighbor-joining analysis (Figure 13.3, which is Figure 2 from ref. 6), 4 subcores (49, 724, 1354, and 1945) contained > 1 ToMV genotype, and 5 subcores (131, 226, 724, 1150, and 1945) contained sequences nearly identical to those of extant ToMV genotypes.

All meltwaters need to be assayed for other viruses likely to be present in ice. These viruses include the potexviruses, tombusviruses, and tobacco necrosis virus among the plant viruses; the caliciviruses and parvoviruses among the animal viruses; and the enteroviruses and influenza viruses among the human viruses. The following chapter discusses the animal and human viruses that we expect to detect in glacial ice.

Conclusions

In summary, Gram-positive endospore-forming bacteria, actinobacteria, and Gram-negative bacteria are present in the Greenland ice sheet (Table 13.1 and Figure 13.1). *B. subtilis* was most frequently isolated among the thirteen isolates

TABLE 13.3

Location, depth, and approximate age of ice subcore meltwaters tested for tomato mosaic virus (ToMV) by RT-PCR, and the sequence differences (by base pair and % difference) of ice virus isolates (IV) from the type strain of ToMV (TMV-L, GenBank accession # X02144).

Sample Number[1]	Drill Site Core	Depth[2] (m)	Approximate Age (ybp)[3]	RT-PCR Results	Sequence Difference (bp and %)	GenBank Accession Number
IV49-3A1297	Dye-3-71	158	< 500	Positive	7, 3.2%	AF067246
IV49-3B1097	Dye-3-71			Positive	15, 7.3%	AF067252
63-2A1297	GISP2-B	62	< 500	Negative	—	—
63-2B1297	GISP2-B			Negative	—	—
131-1A796	GISP2-D	130	< 500	Negative	—	—
IV131-1B796	GISP2-D			Positive	Not sequenced	—
IV131-2A1297	GISP2-D			Positive	Not sequenced	—
IV131-2B1297	GISP2-D			Positive	4, 1.8%	AF067233
IV131-3A1297	GISP2-D			Positive	4, 1.8%	AF067245
131-2B1097	GISP2-D			Negative	—	—
226-2A1297	GISP2-D	225	1000	Negative	—	—
IV226-2B1097	GISP2-D			Positive	3, 1.4%	AF067237
522-2A1297	Dye 3-79-81	598	1400	Negative	—	—
522-2B1097	Dye 3-79-81			Negative	—	—
IV724-1A796	Dye 3-79-81	798	2200	Positive	3, 1.3%	AF067229
724-1B796	Dye 3-79-81			Negative	—	—
IV724-2A1097	Dye 3-79-81			Positive	3, 1.4%	AF067232
IV724-2B1097	Dye 3-79-81			Positive	3, 1.4%	AF067239
IV724-2B1097RS	Dye 3-79-81			—	11, 5.1%	AF067243
IV939-2A1297	Dye 3-79-81	1007	3000	Positive	6, 2.7%	AF067244
IV939-2B1297	Dye 3-79-81			Positive	Not sequenced	—
1150-1A1097	Dye 3-79-81	1207	4000	Negative	—	—
IV1150-2B1097	Dye 3-79-81			Positive	3, 1.4%	AF067238
IV1354-2A1297	Dye 3-79-81	1405	5500	Positive	6, 2.9%	AF067247
IV1354-2B1297	Dye 3-79-81			Positive	11, 5.3%	AF067250
1571-2A1297	Dye 3-79-81	1621	7000	Negative	—	—
1571-2B1297	Dye 3-79-81			Negative	—	—
1945-1A796	Dye 3-79-81	1994	140,000	Negative	—	—
1945-1B796	Dye 3-79-81			Negative	—	—
IV1945-2A1097	Dye 3-79-81			Positive	Not sequenced	—
IV1945-2B1097	Dye 3-79-81			Positive	3, 1.4%	AF067240
IV1945-2B1097RS	Dye 3-79-81			—	24, 12.5%	AF067251

Table from reference 6.

[1] The ice virus (IV) isolates are identified by subcore (preceding the hyphen), followed by core subsection number, prior to (B) or subsequent to (A) meltwater filtration through a 0.45 µm pore size filter, and the month and year of amplification. RS denotes the sequence of an amplification product subjected to repeat sequencing and analysis.

[2] Depth corresponds to the depth from the surface of the ice core subsections from which the internal subcores were aseptically removed and meltwaters obtained.

[3] GISP 2 subcore ages estimated from Taylor et al. (18). Dye 3 subcore ages estimated from Beer et al. (2), Dansgaard et al. (9), and Reeh et al. (15).

that we identified from both relatively young and old ice. All except two showed killer activity against *Bacillus* tester strains, and all except these two *B. subtilis* isolates showed PBSY-like phage when examined by TEM (Table 13.1).

These PBSY-like phage were detected in *B. subtilis* isolates recovered from ice < 500 up to 140,000 years old, as well as in an isolate of *Rhodococcus* sp., but not in a Gram-negative isolate related to *T. pytseos*. In addition, an unidentified group of *B. subtilis* phage was observed, but has not yet been

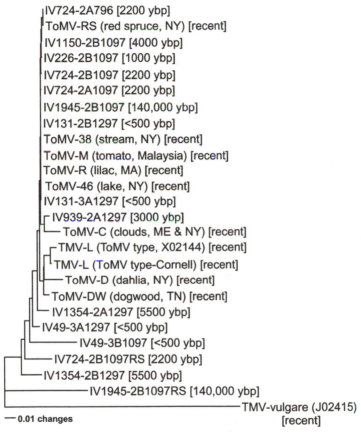

FIGURE 13.3. Neighbor-joining tree of ice (IV) and extant (ToMV) isolates of ToMV based on 244 bp of the coat protein gene (6). The Minimum Evolution score was 0.59. Ice virus and extant sequences originated from the subcores and plant or environmental sources indicated. The IV isolates are identified by subcore (preceding the hyphen), followed by core subsection number, prior to (B) or subsequent to (A) meltwater filtration through a 0.45 μm pore size filter, and the month and year of amplification. RS denotes the sequence of an amplification product subjected to repeat sequencing and analysis. Sequences of TMV-L (ToMV type strain) and tobacco mosaic virus (TMV-Vulgare) from GenBank are included for comparison.

isolated in pure culture or characterized in detail. Phage may serve to regulate bacterial numbers in ice and permafrost, especially in the absence of protozoan herbivory, which has not yet been demonstrated in ice to our knowledge. Similarly, viruses may serve to transfer genes among bacteria in ice via the mechanism of transduction.

Similar bacteria have been isolated from ice by other researchers. Isolates of *Bacillus* and *Paenibacillus* species, as well as various proteobacteria and actinobacteria, were among the most frequently isolated by Christner et al. in both polar and nonpolar ancient ice (see chapter 15). Abyzov also isolated many species of the endospore-forming genus *Bacillus* including *B. subtilis*, actinobacteria of the genera *Nocardia, Nocardiopsis*, and *Streptomyces*, and γ-proteobacteria of the genus *Pseudomonas* from the deep Vostok core in Antarctica (see ref. 1 and chapter 16, this volume). Viable bacteria also have been isolated from 3 million-year-old Siberian permafrost (see ref. 16 and chapters 7, 8, and 10, this volume). Among the species isolated were endospore-forming bacilli including *B. subtilis* and *B. psychrophilus*, β- and γ-proteobacteria including species in the Enterobacteriaceae, and the actinobacterium *Rhodococcus fascians*. Apparently, many different bacteria, most of which are not true psychrophiles, are capable of prolonged survival in the cryosphere.

Isolates of ToMV both similar to and distinct from contemporary ones are present in the ice, and ToMV RNA is present in young (< 500 ybp) and very old (> 100,000 ybp) ice of the Greenland ice sheet. However, we do not yet know if intact, infectious particles are present in the ice. *Bacillus subtilis*, the PBSY-like phage, and ToMV each has been isolated from different layers, hence ages, of the glacial ice sheet. Several ice cores already have been demonstrated to contain genotypes of ToMV identical to extant ones, which lends support to the concept of the recycling of microbial genotypes.

Literature Cited

1. Abyzov, S.S. 1993. Microorganisms in the Antarctic ice. pp. 265–295. In E.I. Friedmann (ed.), Antarctic Mirobiology. Wiley, New York.
2. Beer, J., M. Andree, H. Oeschger, B. Stauffer, R. Balzer, G. Bonani, Ch. Stoller, M. Suter, W. Wolfli, and R.C. Finkel. 1985. [10]Be variations in polar ice cores. In C.C. Langway Jr., H. Oeschger, and W. Dansgaard (eds.), Greenland Ice Core: Geophysics, Geochemistry and the Environment. Geophys. Monogr. 33: 66–70.
3. Bradley, D.E. 1965. The isolation and morphology of some new bacteriophages specific for *Bacillus* and *Acetobacter* species. J. Gen. Microbiol. 41: 233–241.
4. Buchen-Oswald, C., L. Blaine, and M.C. Horzinek. 2000. The Universal Virus Database of ICTV (ICTV dB). pp. 19–24. In M.H.V. van Regenmortel (ed.), Seventh International Committee on the Taxonomy of Viruses Report. Academic Press, New York.

5. Büttner, C., and F. Nienhaus. 1989. Virus contamination of waters in two forest districts of the Rhineland area (FRG). Eur. J. For. Pathol. 19: 206–211.

6. Castello, J.D., S.O. Rogers, W.T. Starmer, C.M. Catranis, L. Ma, G.D. Bachand, Y. Zhao, and J.E. Smith. 1999. Detection of tomato mosaic tobamovirus RNA in ancient glacial ice. Polar Biol. 22: 207–212.

7. Catranis, C.M., and W.T. Starmer. 1991. Microorganisms entrapped in glacial ice. Ant. J. U.S. 26: 234–236.

8. Cowan, D.A. 2000. Microbial genomes-the untapped resource. Tibtech 18: 14–16.

9. Dansgaard, W., H.B. Clausen, N. Gundestrup, S.J. Johnsen, and C. Rygner. 1985. Dating and climatic interpretation of two deep Greenland ice cores. In C.C. Langway Jr., H. Oeschger, and W. Dansgaard (eds.), Greenland Ice Core: Geophysics, Geochemistry, and the Environment. Geophys. Monogr. 33: 71–76.

10. Fillhart, R.C., G.D. Bachand, and J.D. Castello. 1997. Airborne transmission of tomato mosaic tobamovirus and its occurrence in red spruce in the northeastern United States. Can. J. For. Res. 27: 1176–1181.

11. Fillhart, R.F., G.D. Bachand, and J.D. Castello. 1998. Detection of infectious tobamoviruses in forest soils. Appl. Environ. Microbiol. 64: 1430–1435.

12. Gibbs, A.J. 1977. Tobamovirus group. No. 184 in Descriptions of Plant Viruses. Commonwealth Mycological Institute / Association Applied Biologists Wm. Culcross & Son, Ltd. Scotland.

13. Jacobi, V., and J.D. Castello. 1991. Isolation of tomato mosaic virus from waters draining forest stands in New York State. Phytopathology 81: 1112–1117.

14. Ma, L., S.O. Rogers, C.M. Catranis, and W.T. Starmer. 2000. Detection and characterization of ancient fungi entrapped in glacial ice. Mycologia 92: 286–295.

15. Reeh, N., S.J. Johnsen, and D. Dahl-Jensen. 1985. Dating the Dye 3 deep ice core by flow model calculation. In C.C. Langway Jr., H. Oeschger, and W. Dansgaard (eds.), Greenland Ice Core: Geophysics, Geochemistry, and the Environment. Geophys. Monogr. 33: 57–65.

16. Shi, T., R.H. Reeves, D.A. Gilichinsky, and E.I. Friedmann. 1997. Characterization of viable bacteria from Siberian permafrost by 16S rDNA sequencing. Microbial Ecol. 33: 169–179.

17. Steensma, H.Y., L.A. Robertson, and J.D. van Elsas. 1978. The occurrence and taxonomic value of PBSX-like defective phages in the genus *Bacillus*. Antonie van Leeuwenhoek 44: 353–366.

18. Taylor, K.C., P.A. Mayewski, R.B. Alley, E.J. Brook, A.J. Gow, P.M. Grootes, D.A. Meese, E.S. Saltzman, J.P. Severinghaus, M.S. Twickler, J.W.C. White, S. Whitlow, and G.A. Zielinski. 1997. The Holocene-Younger Dryas transition recorded at Summit, Greenland. Science 278: 825–827.

19. Zhao, Y. 2001. Identification of selected bacteria and their phage from Greenland glacial ice. MS thesis, State University of New York, College of Environmental Science and Forestry, Syracuse.

14

Viral Pathogens of Humans Likely to Be Preserved in Natural Ice

Dany Shoham

VIRUSES probably have been in existence for about 3.5 billion years. Human viruses most likely have been in existence since the rise of the first hominid, *Australopithecus afarensis*, nearly 3.6 million years ago. However, human technology permitted the first isolation of a human virus only very recently, a little over 100 years ago. It was in 1901 that the pathogen causing yellow fever was proved to be a filterable virus carried by mosquitoes. There was considerable progress made in our understanding of human viruses during the twentieth century.

The evolution of human viruses probably occurred in the following ways: solely within human hosts, from animals to humans (the zoonotic human viruses), simultaneously in man and animals, and from man to animals (which is less likely). Although more than one evolutionary route could have taken place, in actuality, the origin of human viruses remains unknown. RNA viruses may have followed evolutionary pathways paralleling the evolution of cellular RNA genomes during the so-called RNA world (from 4.5 to 3.5 billion years ago). Therefore, self-replicating nucleic acid molecules may have become the progenitors of RNA viruses. Later, during the emergence of cells with DNA genomes, DNA viruses may have developed in parallel to the development of these cellular DNA genomes. Thus, cells and viruses are likely to have coevolved (7). A wide variety of viruses with the capacity to infect humans has evolved, and many of these likely occurred long before the emergence of ancestral hominids.

Several human viruses still constitute eminent threats of particular concern to mankind, including the viruses that cause AIDS, influenza, hepatitis, hemorrhagic fevers, and, potentially, smallpox, not to mention the slow and oncogenic viruses. These viruses are powerful killers that readily transgress the boundaries of time and space. The devastating smallpox epidemics in the Middle Ages, the "Spanish Flu" pandemic in the first half of the twentieth century and the Ebola and AIDS viruses in the second half of the twentieth century are the most tangible examples of the threat posed by viruses. Hence, the fear of a current or resurrected virus possessing pandemic potency is not an unfounded one, and the prospects for such a reemergence are reasonable. The natural reappearance of a Spanish Flu–like virus, or a smallpox virus, for instance, is dependent on their reemergence from perennial ice that may contain them in a still viable state. Alternatively, if preserved in ice in a nonviable state, these viruses nevertheless would provide a wealth of valuable scientific knowledge that ought to be pursued.

The adaptation of human viruses to the body temperature of man, around 37 °C, does not reduce their survival under cryogenic conditions. Thus, while multiplying optimally at 37 °C, they can be stored at temperatures far below freezing for many years. This attribute might be a key factor in the evolutionary strategy of human viruses (plus mammalian and avian viruses in general) because it permits them to survive in two totally distinct spheres: in the body of their homeothermic host and in the ice formed when virus-containing water freezes. The spectrum of human viruses is remarkably large. Biologically, they include viruses that are found uniquely in human beings, as well as viruses that are parasites of both man and animals, including vertebrates and invertebrates. At the same time, human viruses can endure in various abiotic environments. Excluding ice, human viruses exist in five major natural environments: air (airborne viruses), water (waterborne viruses), arthropods (arboviruses), nonhuman vertebrates, and man.

Annual and Perennial Ice

Two types of ice can be distinguished: annual ice and perennial ice. The geographic separation between them depends on whether summer temperatures in a given year at a specified location rise above freezing or not. The geographic line that separates the areas marked by annual ice and those areas that are totally free of ice depends on whether winter temperatures drop below freezing or not. Thus, three zones are found over the Northern and Southern Hemispheres: one without ice, one that bears annual ice, and one that bears perennial ice.

Almost any water source could be seeded with human viruses prior to freezing. Consequently, any ice may release human viruses upon thawing with a fair prospect to reinfect a susceptible host and to proliferate. But space (boundaries of the geographical zones discussed above) and time (duration of the perennially frozen state) are two important variables in this complex ecological model.

The chances of human viruses immured in ice relate to the presence of nearby past and present human communities. In the lack of close physical proximity to humans, various vectors such as currents of water and air, as well as migrating animals, may serve to bridge the transmission gap. Several milestones in human history have facilitated the spread of human viruses: (i) the tremendous expansion of human communities, accompanied by greatly increased density; (ii) the domestication of various animals, which facilitated the evolution and proliferation of viral pathogens common to man and animals; (iii) the formation of wild synanthropic species (birds and mammals), including migratory ones, that distribute human viruses over all parts of the world; and (iv) The establishment of sewage systems, thereby bringing about the input of human virus-containing-effluents into inland natural water systems as well as the oceans.

Presently, the no ice and the annual ice zones are entirely occupied by man. The influx of human viruses into prefreezing water of these zones is thus assured. Less likely is the deposition of human viruses into prefreezing water of the perennial ice zone. However, it does occur because these zones are occupied by wandering aquatic birds and mammals, which harbor zoonotic human viruses. Yet, while the recirculation of zoonotic human viruses released into natural water systems upon ice thaw in these zones is feasible, it may constitute a dead end in terms of human infections, because these viruses may not reach man. Nevertheless, regardless of their fate after thawing, the very existence of human viruses of any type in perennial ice comprising such a vast age range is a phenomenon of potentially supreme scientific value.

Evolutionary Features and Constraints

The evolutionary survival of human viruses is chiefly a function of the multifactorial interface between them and their human hosts. First and foremost is the herd immunity factor. Most human viruses induce an immune response and cytopathic effect in the body of their host. This capacity is not an optimal way to evolve parasitism. Although some human viruses multiply in arthropods and in certain vertebrate hosts without causing cytopathic effects, very few can do so in man. Also, human viruses must be able to cope with the main human defense mechanism, which is herd immunity. They do so in several ways, but the primary mechanism is antigenic mutagenicity (i.e., constantly mutating their antigenic sites to avoid herd immunity). However, since the twentieth century, the use of antiviral vaccines has made a tremendous contribution to herd immunity. Nevertheless, herd immunity is at best a transient nongenetic factor that disappears within one or two host generations as long as there is no virus contact with its host population during that time. On the other hand, avoiding contact with the host population for many host generations would be unfavorable for the virus because genetically dictated host traits may disappear or corrode during that time, specifically resulting in diminished virus avidity and infectivity. Yet, genetically dictated host traits may change unfavorably in an inverse manner, such that widened viral histotropism and increased virulence may bring about elevated host mortality, beyond the level that enables virus sustainability within its host population. In that sense, the coevolution of pathogenic viruses and human hosts would take place in an optimally gradual and simultaneous way. The capacity to induce an immune response and a cytopathic effect is thus a prominent disadvantage for the parasitic virus; therefore, the optimal way to evolve parasitism is the absence or detainment of that capacity.

Theoretically, then, virus dormancy that lasts for two host generations is an optimal period of perennial human virus endurance. In reality, though, the dormancy period may be much shorter or enormously longer because there are two totally different ways for viral dormancy to take place: biotic dormancy in a host and abiotic dormancy in ice. Biotic dormancy actually is prolonged latency, whereby the virus may persist in an inactive state within host cells that could last until the host dies. During biotic dormancy virus reactivation may occur at any time, but viral avoidance of the human immune system may continue for tens of years. Furthermore, the duration of biotic dormancy may be extended through vertical transmission of the latent infection. Prolonged biotic dormancy of certain human viruses may take place within animal hosts of notable longevity, particularly reptilian species. Vertical transmission of latent human viruses may occur within animal hosts as well.

Nonetheless, the contribution of biotic dormancy to the evolutionary survival of human viruses is uncertain because there are no data concerning the number of successive host generations that may support congenital latent infection, allowing viral dormancy to persist. In all probability it would not be more than two, except for oncogenic viruses that are fully integrated within the human host chromosome, (e.g., leukuviruses), or viruses that give rise to fully immunologically tolerated—yet at times productive—infections (like intrauterine prefetal infections). Such human viruses are indeed rare. One may therefore assume that the duration of perennial biotic dormancy would not exceed several human lifetimes, even if the case involves, at the extreme, a human virus of the zoonotic type, dormant within a reptilian host of remarkable longevity. The range of human viruses known to regularly undergo a biotic latent phase is quite narrow, including mainly *Herpes simplex* viruses, measles virus, certain adenoviruses, and AIDS viruses. Congenital infections by human viruses include cytomegalovirus, rubella virus and certain enteroviruses, but usually take place in the form of active rather than latent infection. All in all, it looks as if only a minority of human viruses is capable of undergoing meaningful biotic dormancy. The principles underlying perennial biotic dormancy of those human viruses are fundamentally valid, and to a certain degree influence the evolutionary survival of some human viruses. Yet, loss of virus viability during biotic dormancy would be an evolutionary dead end, although such viruses would still be of immense scientific value in terms of the preserved viral DNA or RNA. Conversely, biotic dormancy may viably last for numbered host generations, as long as the primary individual host or his congenitally infected descendants survive. Abiotic dormancy, however, would persist for much longer periods of time, if the host dies and his body or parts of it are somehow retained in permafrost, snow, or ice (e.g., the preserved Alps Iceman). Evolutionarily, then, those conditions imply that abiotic dormancy in the frozen state is the prime mechanism of prolonged human viral dormancy.

This leads to a discussion of abiotic viral dormancy mechanisms at large. Abiotic viral dormancy takes place primarily in air and water. The duration of dormancy is variable, from minutes when very susceptible human viruses are exposed to UV in the air, to months within liquid water, and perhaps millennia or longer in perennial ice. Water comprises about 70 percent of the upper layer of Earth. Temperature fluctuations around 0 °C allow dormancy-reemergence cycles of a wide spectrum of life forms, including human viruses. Such dormancy-reemergence cycles are usually of an annual periodicity in the sub-Arctic and of a perennial periodicity in the Arctic. These cycles may underlie a very fundamental natural strategy that represents an important and productive interface between abiotic and biotic ecosystems. In the Arctic this natural strategy may extend the persistence of various microorganisms and viruses throughout complete geological eras, a previously unthinkable timeframe. The domain of natural cryobiology may thus gain a totally unique dimension.

At a global level, then, water appears to play a double role, in that it perpetuates life forms during both their reproductive and dormant phases. The duration of the dormant, frozen phase may immeasurably exceed that of the reproductive one. For viruses, as obligate parasites, water is only a temporary medium in its liquid state, yet the ultimate medium for perennial dormancy in its solid state. The more simple the life form, the greater are its prospects for retaining viability while in the dormant phase. Viruses might thus have the best chances for long-term perennial dormancy. The time course of complete abiotic dormancy may be divided into five stages: the presence of virions in prefrozen water, freezing, the frozen state, thawing, and infection of susceptible hosts. This sequence is identical with respect to both annual and perennial viral dormancy, even if the latter lasts millions of years. Virions may survive all those stages and consequently reemerge, thus, genetically conserved, to face a significantly altered environment.

Stages of Abiotic Dormancy

Presence of Virions in Prefreezing Water

Water is a supportive medium for viruses, particularly compared to air, in that it prevents dehydration and reduces exposure to UV radiation. The presence of human virions in prefreezing water may result from both accidental and regular events. Accidental events include the chance presence in prefreezing water of human virus-infected carcasses of arthropods, birds and mammals, and man himself. The occurrence of human viruses in water due to such accidental events is rare, and is not regarded to be a basis for a subsequent systematic virus reemergence, but the possibility is noted. Of more importance, then, are regular events. Such events generate the current influx of human viruses into Arctic or Antarctic freshwater or seawater reservoirs that later freeze (entirely or partially). The following biotic and abiotic mechanisms are probably most important in that respect: the circulation of human viruses within aquatic vertebrates (e.g., migratory waterfowl and marine mammals) and invertebrates (e.g., oysters); transovarial and trans-stadial nonpathogenic transmission of human viruses within mosquitoes and water mites (human viruses affiliated with the arthropod-borne viruses); virus-containing human discharges of alimentary, urogenital, or respiratory origin (feces, urine, catarrh, and phlegm) that are routinely found in effluent, sewage, and river waters, which constitute a major route of viral contamination of prefreezing lakes and seawater. Virus-containing fecal material is obviously of crucial importance in that concern, as well as the settling of airborne human viruses onto prefreezing water or thickening ice and people who lost their lives due to viral infections, and were buried or engulfed in the permafrost or by snow.

Arthropod-borne human viruses (arboviruses) are a very large, diverse, and cosmopolitan group of viruses. They reach extremely remote sites all over the globe because they infect many migratory bird populations, including Arctic and Antarctic ones. Many arboviruses are transmitted by mosquitoes. Female mosquitoes lay their eggs in water and many species overwinter as diapausing eggs or larvae. Although infected females do not shed virus into water, they do seed water with human arboviruses through their deposited eggs, which often bear apathogenic viral contagion, both internally and externally. Subsequently infected mosquito larvae and pupae do not shed virus into the water (as far as is known), but they harbor the viral contagion and flourish in prefreezing freshwater until freezing occurs.

Waterborne human viruses include a remarkable variety of enteric viruses, as well as other human pathogens like reovirus and adenovirus. All are found in effluents, rivers, lakes, and oceans in the form of free virions, feces-bound virions, or within aquatic animals, like oysters and clams. They reach lakes and oceans through sewage systems connected directly with them, or in contaminated river water. The durability of these viruses is temperature-dependent; the lower the water temperature, the higher their durability. The colder waters of the polar oceans and high-latitude rivers are likely to contain abundant, infectious human waterborne viruses. Thus, the presence of waterborne human viruses in prefreezing water is expected. Yet while the recirculation of zoonotic human viruses released into natural waters upon ice thaw is feasible and probable, nonzoonotic and therefore exclusively human viruses may constitute an evolutionary dead end if they do not infect humans subsequent to their release from ice.

Freezing

The process of freezing is a biophysical one, subject to regular thermodynamic laws. The ability of water to expand while freezing does not harm the viruses contained within it. Although cryogenic temperatures are perfectly tolerated by human viruses, the speed of the freezing process affects virus durability. Slow freezing is best because rapid freezing can disrupt virus particles. A gradual drop in temperature from room temperature to 4 °C, 0 °C, −20 °C, and finally to −80 °C is the best way to preserve viruses in the laboratory. This process most likely simulates that which occurs in nature. Because viruses are routinely stored in the laboratory in this manner, and they retain viability once thawed, it is obviously at least feasible that intact viral particles can be preserved in a viable state in natural ice. The rate of survival can be measured following repeated freeze-thaw cycles.

Frozen State

The frozen state is the essence of perennial viral endurance. During that stage, only two variables are subject to fluctuation: temperature and radiation (mainly solar). Thermal fluctuations within ice, though possibly extreme, are insignificant unless such fluctuations are too rapid. Solar or cosmic radiation of various types, some of which penetrate ice, are potentially damaging to human viruses. However, damaging UV radiation is extremely reduced by ice. No other known factor has any significant influence on virus dormancy once they are frozen, unless time itself has some intrinsic deleterious but unknown effect. Assuming that temperature and radiation do not significantly decrease the viability of human viruses, it is highly likely that their durability in the frozen state is practically unlimited. Experimentally, the persistence of various human viruses in the frozen state has been widely demonstrated, and most human viruses can be preserved frozen for prolonged periods of time.

Thawing

Annual thawing of ice is a common and regular phenomenon. Thawing of perennial ice would occur during those rare summers marked by unusually elevated temperatures that bring about ice melting. The speed of the process may be influential with respect to virus viability. The more gradual the thawing the higher the survivability of the preserved virus. Particle-bound virions, like enteric viruses attached to fecal material, are better protected from deleterious factors than free virions, and that is true also with respect to thawing.

Virus-infected mosquito eggs can reemerge in a viable state upon thawing and proceed with normal development, including trans-stadial transmission of viral contagion. Even if they reemerge from ice in nonviable or infertile condition, they may decompose and release viable virions into the water. The same pertains to infected larvae and pupae. Upon thawing, larvae and pupae may revive or remain lifeless, but the viruses harbored by them would retain viability.

Infection of Susceptible Hosts by Virions Released from Ice

Biologically, all of the above-mentioned stages are meaningless unless infection of susceptible hosts occurs following thawing. In the springtime and summertime during ice thawing there are good prospects for human beings, intermediate animal hosts, and arthropods to be in close proximity to reemerging waterborne viruses. The marginal ice zone (MIZ) by way of its melting ice serves as a major congregational area for various invertebrates and vertebrates, including migratory ones. Man himself frequently sojourns to such sites while seeking food. If renewed infection of susceptible hosts by reemerging viruses does not take place, the viruses may enter another dormancy phase when the water refreezes.

Relevant Viruses

Enteric Viruses: Polio, Coxsackie, Norwalk, Hepatitis A, and Rota-Viruses

Human enteric viruses (RNA viruses) replicate in the intestines and are excreted in feces. They are abundant in sewage system water. Runoff water and runoff cycles still play a major role in carrying enteric viruses, adsorbed to feces, into natural waters. The worldwide construction of central sewage systems put high concentrations of enteric viruses into rivers, lakes, seas, and oceans, despite various effluent decontamination treatments. Various human enteric viruses reach, then, prefreezing water all over the world, and are frozen. Presumably, some of them may endure throughout the entire period of freezing and reemerge upon thawing. Moreover, mollusks, oysters, mussels, and clams often harbor human enteric viruses and are significantly involved in their circulation. Three human enteric viruses affiliated with the enterovirus group, (polio, Coxsackie, and hepatitis A viruses), an enteric virus affiliated with the Caliciviridae family (Norwalk virus), and finally the Reoviridae family (rotavirus), will be discussed. Other human enteric viruses also may be equally relevant. Unless otherwise noted, the following findings pertain to viably recovered viruses.

Six river water samples in Switzerland were assayed for viruses (15). Enteroviruses (rotavirus) were detected in all of them. In Galveston Bay, Texas, a comprehensive study dealing with enteroviruses alone included samples collected sequentially from water, suspended solids, fluffy sediments (uppermost layer of bottom sediments), and compact sediment (33). In that study, a total of 103 samples were examined, of which 27 (26%) were positive. Polio viruses were recovered most often, followed by Coxsackie B viruses and echoviruses (7, 29). Virus was found most often attached to suspended solids: 72% of the samples were positive, whereas only 14% of water samples without solids yielded virus. Fluffy sediments yielded virus in 47% of the samples, whereas only 5% of compact bottom-sediment samples were positive. Polio virus and rotavirus retained infectivity for 9 days when freely suspended in seawater, and up to 19 days when associated with solids.

Virus persistence is significantly increased, however, by aquatic invertebrate hosts. Various human enteric viruses are abundant within shellfish (22). This was demonstrated by applying reverse transcription-PCR and hybridization in shellfish during a three-year study. Infection percentages in oysters were found to be 27% for rotavirus, 23% for Norwalk, 19% for enterovirus, and 17% for astrovirus; and rather higher in mussels: 52% for rotavirus, 50% for astrovirus, 45% for enterovirus, 35% for Norwalk, and 13% for hepatitis A virus.

Polio virus is a typical human virus that has been controlled using vaccination with live, circulating vaccine strains even though an ongoing interplay with wild type strains is still taking place. The wild type virus is one of the most stable viruses. Its durability sharply increases as the temperature drops. It survives for many months in aqueous suspensions of human feces at 4 °C, and can be preserved for many months or years at −20 °C or −70 °C. Both wild type and vaccine strains are prevalent in prefreezing water. Intermediate strains also can be isolated at times. Man is the only host of the virus, and although the disease may be fatal, there are many infected but asymptomatic persons who excrete the virus in feces and sustain its ongoing circulation. Ancient mummies showing clinical symptoms of polio have been found, implying that the virus is an ancient one. Taking into consideration that this virus is an old killer causing an easily recognized disease, it might be possible to recover it from frozen bodies of polio victims, if available, or from ice, particularly near sites of lakes or seas connected to sewage systems.

Coxsackie virus is usually a mild pathogen that causes diarrhea in man. Due to its mildness, unlike poliovirus, there is no involvement of vaccine strains and, as a result, the ecology of this virus is influenced, first and foremost, by the hydrobiological impact of sewage systems. The wild type virus, with about thirty clinical strains, is predominant in sewage water. Further, it has been detected specifically in lake water in North America (11) and in river water in Germany (42). Coxsackie viruses are found in prefreezing water and undergo freezing, retaining their viability for long periods. The virus moves in water currents although some variants also are airborne.

Norwalk virus is an enterovirus with pronounced ecoepidemiological traits. Unlike polio and Coxsackie viruses, it multiplies in and injures the intestines, whereas the former ones propagate benignly in the intestines but injure other organs. Norwalk virus circulation is basically confined to the alimentary tract and water, but shellfish involvement is not rare. It is a major cause of extensive waterborne gastroenteritis outbreaks, as has been observed in the United States (Georgia) (18) and in Europe (Finland) (20). Notably, this hydrophilic virus has been the causal agent of several outbreaks resulting from the serial consumption of commercial ice, made from Norwalk-virus-contaminated drinking water (9,19, 23). It is considered to be a virus typically residing in water, and liable to endure in the frozen state.

Hepatitis A virus is an enterovirus of paramount importance. It has caused infectious hepatitis outbreaks due to the consumption of both treated and untreated water in many communities (25). The persistence of the virus in experimentally seeded water (e.g., distilled, tap, waste, and seawater) was demonstrated, while high recovery of infectious virus was obtained from all water types tested (8). Frozen food contaminated with hepatitis A virus also has been implicated as a source of an infectious hepatitis outbreak (31), indicating the cryostability of the virus. Harbored and excreted exclusively by man, this virus probably reaches prefreezing water in lakes and seas that receive sewage water. We presume that it would be viably preserved in perennial ice.

Rotavirus is an enteric zoonotic reovirus of increasing ecoepidemiological significance. Its durability in raw and treated river water has been experimentally demonstrated over periods of several days at 20 °C and more than 64 days at 4 °C (34). In Spain, marine sediments collected in an area receiving fecal pollution contained rotavirus at frequencies of 57 to 63% (17). Several freeze-thaw cycles did not impair the durability of the virus (12). The presence of rotavirus in aquatic ecosystems is considerably amplified by various farm animals including birds that harbor the virus. Migratory birds may move this virus to remote northerly aquatic habitats. It is therefore reasonable to conclude that this virus exists in prefreezing water and survives in ice.

Influenza Type A Virus

This RNA virus is an ecologically complex virus. It is probably the most likely human virus to be preserved in ice. It is a common respiratory virus of man and a few other mammals. This member of the genus *Orthomyxovirus* is foremost an asymptomatic enteric virus of numerous avian species. Aquatic birds may be the archaic source of all influenza A viruses in other species. The virus is an ancient pathogen of humans and animals. The principal hosts of influenza A viruses in the wild are migratory waterfowl that occupy inland and coastal sub-Arctic and Arctic aquatic biotopes as well as nomadic marine mammals (e.g., seals and whales) that occupy huge oceanic areas over the sub-Arctic and Arctic ecosystems. The strong affinity of influenza A virus to cryoaquatic environments is obvious. These polar or high-latitude biotopes may indeed be extremely remote, thus enhancing the likelihood of continuous viral dormancy in ice. One of the potential marine mammals that may harbor human influenza A viruses is the Greenland ballena whale. This animal is fully and perfectly adapted to the marginal ice zone (MIZ) habitat. It wanders regularly between 55 and 78° N latitude. The recovery rate of influenza A viruses subsequent to prolonged freezing and thawing under experimental conditions also is very high (39).

The presence and persistence of influenza A virus in freshwater has been clearly demonstrated. In particular, it is excreted in abundance in the feces of feral ducks, which are most probably the ancestral host of this virus. These ducks are fully tolerant of infection, and they harbor infection by multiple influenza A strains, including virulent epidemic and pandemic ones. During migration, they and other avian species disseminate the virus all over the world. They also constantly contribute viral genetic material to the gene pool of human influenza viruses through their ongoing aquatic interface with farm animals. This is especially true with respect to Asia, where most epidemic and pandemic influenza A strains originate. However, Antarctic birds are sero-positive for influenza A virus, indicating its presence in the southern polar regions as well (4). Vaccine strains bring about a considerable worldwide amplification of herd immunity in human populations, but are a marginal factor in relation to the enormous overall gene pool and derived wild antigenic variants of this virus.

Current mutational genetic drift, involving both adaptive and neutral mutations, has been quantitatively and meticulously determined for influenza A virus genes, to establish a reliable evolutionary biomolecular clock (1, 36). The rate of this drift is steady and remarkably high. In contrast, there are cases of the reemergence of dormant influenza A virus genes and genomes that represent the opposite phenomenon of multiyear preservation, which takes place through some unknown abiotic mechanism. A salient example of this phenomenon is the unexplained reappearance after twenty-seven years of the A/USSR/1977 pandemic strain (30). This and other examples illustrate this phenomenon, which is usually described as "frozen evolution," without any concrete evidence to support it. Nevertheless, all epidemiological, ecological, and genetic data corroborate the concept that influenza A virus has an inherent capacity for perennial endurance within and subsequent reemergence from ice (37, 38).

The influenza strain of greatest concern is the agent of the 1918 "Spanish Flu" pandemic, which killed approximately 30 million people worldwide. This pathogen has never been isolated alive. There is a fear that it might reemerge, and once again inflict a colossal strike. Attempts were made to isolate this virus from frozen victims of the "Spanish Flu" pandemic, buried in permafrost in North America (Alaska) and Europe (Norway).

Four victims buried in a mass grave in Alaskan permafrost were exhumed and frozen lung tissues biopsied *in situ* from each one. One of those victims tested positive for influenza A RNA. In addition, two of seventy-eight formalin-fixed, paraffin-embedded lung tissue samples from 1918 autopsy cases tested positive for influenza A virus. The complete sequence of the hemagglutinin (HA) gene of the 1918 virus was then determined from these cases. The complete coding sequence of the A/South Carolina/1/18 HA gene was obtained. The HA1 domain sequence was confirmed by using the two additional isolates (A/New York/1/18 and A/Brevig Mission/1/18). The sequences show little variation. Phylogenetic analyses suggest that the 1918 virus HA gene, although more closely related to avian strains than any other mammalian sequence, is nonetheless mammalian, and may have been adapting in humans before 1918 (35).

Based on archival records, seven men were identified who died of influenza in 1918 and were interred in Longyearbyen, Svalbard, Norway, 1300 km from the North Pole. Ground penetrating radar (GPR) was used to locate a large excavation with seven coffins near the existing seven grave markers. The GPR indicated that the ground was disturbed to 2 m depth, and was frozen below one m. The coffins were buried less than one m deep, and the frozen ground was 1.2 m deep where the coffins were located. Virologic investigations are planned (10).

The "Spanish Flu" strain is certainly but one of an immeasurable variety of influenza A strains, any of which may be present in ice. Preliminary results in the laboratory of S.O. Rogers indicate that influenza virus is present within ice cores as well as Arctic lakes (S.O. Rogers, pers. comm.). They have found a low concentration of influenza in glacial ice, and much higher concentrations in Siberian lakes that are visited by migratory birds. The sequences so far obtained are closest to the H1N1, H1N2, and H1N5 influenza A subtypes.

Smallpox Virus

This DNA virus is indeed one of the most dreadful viral pathogens of mankind. It has been eradicated by strict global vaccination programs. There is no known natural reservoir to sustain this virus for long because man is its sole host. The last case of smallpox was in 1979; consequently, vaccination practically terminated all over the world, and the few wild strains still held in laboratories were intended to be killed. Lately, there has been concern that this virus might somehow reemerge naturally from the wild. Indeed, the virus that causes smallpox (*Variola major*) is very stable and survives in exudates from patients for many months. The virus is unlikely to survive in dried crusts for more than a year (3). Yet, it can be preserved in sealed ampules at 4 °C for many years, and indefinitely by freeze drying. This leads to the presumption that the virus still may be preserved alive within bodies of victims buried in permafrost. The prospects of free virions or viral DNA in perennial ice are low, although not zero, because it is less likely that free smallpox virions can travel in air currents and settle on prefreezing water or thickening glaciers.

Ramses V, who lived some 3000 years ago, is the earliest known victim of smallpox, based upon an analysis of his mummy (16). Many virus-like particles were revealed by electron microscopy, and identified serologically as smallpox virus in a 400-year-old mummy from Italy. The antigenic structure of the particles was well preserved (13). The virions in the mummy's skin had lost their viability. Viral antigen was not detected by (enzyme immunoassay) (EIA) or reverse passive haemagglutination (RPHA), and its DNA was not detected by molecular hybridization (26). Attempts to recover the virus from the frozen bodies of persons who died of smallpox continue in Arctic sites in Siberia (32) and in Canada (24).

Arthropod-Borne Viruses: California Encephalitis, Northway, West Nile, and Others

Two groups of aquatic arthropod vectors, mosquitoes and water mites, combined with various migratory holarctic birds and several mammals are responsible for the worldwide circulation of many human arboviruses (RNA viruses). The ecosystems affected include multiple Arctic and sub-Arctic aquatic habitats. Several different human arboviruses are harbored and maintained within these habitats. Subclinically and chronically infected birds that arrive at northern lakes infect local female mosquitoes that feed on bird blood. The viruses readily propagate in the mosquitoes without causing pathogenic effects. The viruses often are transmitted vertically to eggs, larvae, pupae, and maturing progeny, in which they proliferate harmlessly. Premature stages of insect development take place in freshwater throughout the summer. Transmission of virus from transovarially infected males to uninfected females amplifies the overall infection rate of deposited eggs. At the end of the summer, prefreezing water contains virus-bearing eggs and premature stages. They all freeze and are preserved in annual or perennial ice until thawing occurs.

In North America, mosquito-borne viruses are prevalent throughout sub-Arctic and Arctic regions of Canada and Alaska, principally in the boreal forest between 53 and 66° N latitude. They have been identified in tundra regions as far north as 70 ° N (27). The viruses involved are primarily bunyaviruses, specifically California encephalitis (CE) virus (also found in northern Eurasia) and Northway virus. Mosquito vectors include several *Aedes* species and Culiseta inornata, all of which support replication of CE virus at incubation at 13 °C or lower. Where the boreal forest merges into prairie grassland around 53 °N, *Culex tarsalis* mosquitoes become prevalent, and a mosquito-borne alphavirus, western equine encephalitis (WEE) virus, is detected much more frequently than CE and Northway viruses (27). CE virus was isolated from adult *Ae. melanimon* reared from transovarially infected eggs that were stored for up to nineteen months and exposed to repeated freeze-thaw cycles. Neither time since oviposition nor storage conditions affected infection rates in surviving embryos. Survival rates were highest in eggs stored at 4 °C. Transovarial infection with CE virus did not affect survival of embryos, larvae, or adults (41).

Northway virus replication was detected in the salivary glands of wild *C. inornata* and *Aedes communis* mosquitoes from the Canadian Arctic inoculated with the virus by oral ingestion or intrathoracic injection after incubation at 4 °C for nine to eleven months postinoculation (28). WEE virus showed natural vertical transmission in larvae of *Aedes dorsalis* collected from a salt marsh in a coastal area (14). These findings support a perennial ice dormancy mechanism for human arboviruses. In the Eastern hemisphere, CE viruses were isolated from three *Aedes* species in Arctic Norway (40).

West Nile (WN) virus (flavivirus), a mosquito-borne human virus of the Eastern Hemisphere, was recently introduced to the Western Hemisphere, probably by migrating birds, which generated a dreadful disease outbreak in New York City. The virus was isolated from two species of mosquitoes (*Culex pipiens* and *Aedes vexans*) and from brain tissues of twenty-eight American crows (2). Vertical transmission of WN virus was demonstrated in the wild by *Culex* sp. in Kenya (29) and experimentally by *Aedes* sp. and *Culex* sp. (5). Though very thoroughly investigated, the provenance and course of the New York WN virus have not been deciphered, the only positive findings conclusively pointing at involved migratory birds and mosquito-associated mechanisms (43). The virus is native to Africa, Asia, and southeastern Europe; its remarkable spread westward by migrating birds suggests, however, that certain birds may harbor the virus in sub-Arctic and Arctic habitats. North-south seasonal bird migrations are much more likely than east-west migrations. Therefore, virus traveling from the Old to the New World might have been initiated in a remote Arctic biotope that serves as a congregating site for migrant birds from both hemispheres. Thereupon, local mosquitoes may transmit viruses from one bird species to another. This mechanism could explain the presence of CE and WN viruses in both hemispheres, and likewise supports the concept of the preservation of human arboviruses in ice.

Finally, in Siberia, water mites (Hydrachnidae) were found to be associated with the tick-borne encephalitis virus complex (21). These insects, therefore, also likely play a role in the circulation of human arboviruses, in conjunction with water and ice.

Conclusions

The perpetuation of human viruses within ice has two crucial implications, one proximal and one ultimate. In the short run, it rescues human viruses from perishing in liquid water. Viable human virus particles in sewage systems, rivers, lakes, and seawater will either contact a susceptible host, die, or be preserved in ice. Therefore, freezing often is merely a mechanism of prompt viral endurance. Viral endurance in the frozen state may last from several months to many thousands of years. This variability is linear in space, that is, ascending northerly, southerly, and upward, including both inland (sewage systems, rivers, lakes, glaciers) and marine aquatic ecosystems. Dormancy in ice provides a mechanism representing, conceivably, a natural strategy of paramount importance that would in essence immortalize a virus. It is a design that enables the total and complete avoidance of the normally inevitable effects of time. This ice-encapsulated state of arrested life antagonizes the intrinsic course of time and allows for an evolutionary cutoff.

Human viruses are young in terms of evolutionary age, except for zoonotic viruses that are primarily animal-adapted. Apart from affording proximal rescue for human viruses, dormancy in ice may constitute an ultimate mechanism to overcome lasting unfavorable conditions. Abiotic perennial dormancy in ice may thus be a way for viruses to avoid herd immunity as well as to maintain established viral traits compatible with stable host traits (6). From a viral evolutionary perspective, the shorter the periods of time throughout which virus biotic dormancy is feasible, the greater the need for perennial viral dormancy in ice. Stable host properties do change but usually much more slowly than herd immunity. Compatible viral coevolution then is favored. The gradient relating to the duration of viral persistence in ice serves to input into the contemporary biosphere reemerging preserved viral strains that are chronologically and genetically stratified. The ice-related variety of viral strains addresses a need for an ongoing interface between human viruses and their hosts, including men, animals, and arthropods, so that coevolution may readily take place. As long as herd immunity is prevalent (e.g., smallpox, polio, and influenza viruses)—there is a relative advantage for viral dormancy to persist. Otherwise, ice-released viral strains that gave rise to the prevailing herd immunity are a priori doomed under such circumstances. If, however, vastly diversified viral strains are available in the dormant state (i.e., influenza virus) they may reemerge successfully and proliferate. If not, viral reproliferation will commence when herd immunity ceases or appreciably decreases.

The interplay between human viruses and their hosts in the context of viral adaptation toward stable host properties prevails concomitantly with viral adaptation to unstable host properties discussed above, but within an enormously prolonged temporal dimension. It occurs via the spatially alternating but basically random process of ice thawing such that the influx of recurrent viral strains includes ancient, recent, and middle-aged genotypes altogether, in any possible combination. Thus, if the existing parasite-host (virus-human) relationship is evolutionarily an optimal one, it merely has to be maintained (apart from confronting herd immunity). But if the current relationship is evolutionarily suboptimal, virus adaptation to its human host—as well as to contemporary animal hosts—would persist, thanks to ongoing current genotypic modifications that ordinarily take place within active, replicating virions. Conversely, dormancy in ice may confer an evolutionary advantage in that it permits passive viral adaptation toward current genotypic modifications within host populations (in the long run), for example, increased or decreased histotropism toward certain human tissues. Histotropism is a cardinal trait that affects both viral virulence and transmissibility. The suboptimal parasite-host relationship may become closer to optimal through genetic changes in the host itself, while at the same time the parasitic virus is frozen. Notably, such a mechanism certainly may be at times counterproductive, when the virus-host relationship already is optimal, or even suboptimal, bringing about increased incompatibilities between a given dormant virus and its changing host.

An outstanding feature of perennial dormancy in ice, then, is the remarkable built-in ability to move temporally onward and practically backward at the same time, thus permitting the ice-released virions a chance to thrive and to evolve. The longer the entire timeframe of perennial dormancy in ice, the wider is the complete historical diversity and dispersal of the viral gene pool. Hence, the more feasible is perennial preservation and eventual resurrection. This paradigm, although largely theoretical for now, is coherent and is substantiated by an appreciable body of empirical data. Even if complex in theory, it nevertheless relies on sound, plain principles that may reflect some pristine essentials of the globe, and possibly of the universe.

In summary, a dynamic, evolutionarily significant interface between human viruses and ice has existed for a long time. Its current and anticipated ongoing persistence shapes the evolution of viruses. Assuming that water constitutes the very fundamental life-supporting substance wherever life may have existed all over the universe, it is inferred that ice is the most plausible cosmic perpetual preserver of primordial life forms, chiefly viruses. The rise and fall of various alternating ice eras over the Earth, and elsewhere, may signify such an inherent and perpetual natural course. The following terms are corollaries of that interface: ice-borne viruses, perennial ice-sheltered/ice-shielded viral dormancy, and ice-dictated noncontinuous virus evolution.

Literature Cited

1. Air, G.M. 1981. Sequence relationships among the hemagglutinin genes of 12 subtypes of influenza A viruses. Proc. Nat. Acad. Sci. 78: 7639–7643.
2. Anderson, J.F. T.G. Andreadis, C.R. Vossbrinck, S. Tirrell, E.M. Wakem, R.A. French, A.E. Garmendia, and H.J. van Kruiningen. 1999. Isolation of West Nile virus from mosquitoes, crows, and a Cooper's hawk in Connecticut. Science 286(5448): 2331–2333.
3. Arita, I. 1980. Can we stop smallpox vaccination? World Health, May 1980: 27–29.

4. Austin F.J and R.G. Webster. 1993. Evidence of ortho- and paramyxoviruses in fauna from Antarctica. J. Wildlife Dis. October. 29(4): 568–571.

5. Baqar S. 1993. Vertical transmission of West Nile virus by Culex and Aedes species mosquitoes. Am. J. Trop. Med. Hyg. 48: 757–766.

6. Baranowski, E., C.M. Ruiz-Jarabo, and E. Domingo. 2001. Evolution of cell recognition by viruses. Science 292: 1102–1105.

7. Becker, Y. 1996. Molecular evolution of viruses: An interim summary. Virus Genes: 11: 299–302.

8. Biziagos, E., J. Passagot, J. Crance, and R. Deloince. 1989. Hepatitis A virus concentration from experimentally contaminated distilled, tap, waste and seawater. Water Sci. Tech. 21: 255–258.

9. Cannon, R.O, J.R. Poliner, R.B. Hirschhorn, D.C. Rodeheaver, P.R. Silverman, E.A. Brown, G.H. Talbot, S.E. Stine, S.S. Monroe, and D.T. Dennis. 1991. A multistate outbreak of Norwalk virus gastroenteritis associated with consumption of commercial ice. J. Infect. Dis. 164: 860–863.

10. Davis, J.L., J.A. Heginbottom, A.P. Annan, R.S. Daniels, B.P. Berdal, T. Bergan, K.E. Duncan, P.K. Lewin, J.S. Oxford, N. Roberts, J.J. Skehel, and C.R. Smith. 2000. Ground penetrating radar surveys to locate 1918 Spanish flu victims in permafrost. J. Forensic Sci. 45: 68–76.

11. Denis, F.A., E. Blanchouin, A. Lignieres, and P. Flamen. 1974. Letter: Coxsackie A16 infection from lake water. JAMA 288: 1370.

12. Estes, M.K., D.Y. Graham, E.M. Smith, and C.P. Gerba. 1979. Rotavirus stability and activation. J. Gen. Virol. 43: 403–409.

13. Fornaciari, G., and A. Marcetti. 1986. Intact smallpox virus particles in an Italian mummy of 16th century. The Lancet 2 (8507): 625.

14. Fulhorst, C.F., J.L. Hardy, B.F. Eldrige, S.B. Presser, and W.C. Reeves. 1994. Natural vertical transmission of western equine encephalitis virus in mosquitoes. Science 263: 276–278.

15. Gilgen, M., D. Germann, J. Luthy, and P. Hubner. 1997. Three-step isolation method for sensitive detection of enterovirus, rotavirus, hepatitis A virus, and small round structured viruses in water samples. Int. J. Food Microbiol. 37: 189–199.

16. Hopkins, D.R. 1980. Ramses V. Earliest known victim? World Health 5: 22.

17. Jofre J., A. Blasi, and F. Lucena. 1989. Occurrence of bacteriophages and other viruses in polluted marine sediments. Water Sci. Technol. 21: 15–19.

18. Kaplan, J.E, R.A. Goodman, L.B. Schonberger, E.C. Lippy, and G.W. Gary. 1982. Gastroenteritis due to Norwalk virus: An outbreak associated with a municipal water system. J. Infect. Dis. 146: 190–197.

19. Khan, A.S., C.L. Moe, R.I. Glass, S.S. Monroe, M.K. Estes, L.E. Chapman, X. Jiang, C. Humphrey, E. Pon, and J.K. Iskander. 1994. Norwalk virus-associated gastroenteritis traced to ice consumption aboard a cruise ship in Hawaii: Comparison and application of molecular method-based assays. J. Clin. Microbiol. 32: 318–22.

20. Kukkula, M., P. Arstila, M.L. Klossner, L. Maunula, C.H. Bonsdorff, and P. Jaatinen. 1997. Waterborne outbreak of viral gastroenteritis. Scand. J. Infect. Dis. 29: 415–418.

21. Lakimenko, V.V., P.V. Tuzovskii, O.B. Kalmin, I.I. Bogdanov, D.A. Drokin, and A.A. Tagil'tsev. 1997. The Hydrachnidae water mites of southwestern Siberia in relation to their participation in arbovirus circulation. Parazitologiia 31: 414–426.

22. Le Guyader, F., L. Haugarreau, L. Miossec, E. Dubois, and M. Pommepuy. 2000. Three-year study to assess human enteric viruses in shellfish. Appl. Environ. Microbiol. 66: 3241–3248.

23. Levine, W.C., W.T. Stephenson, and G.F. Craun. 1990. Waterborne disease outbreaks, 1986–1988. Mor. Mortal. Wkly. Rep. CDC Surveill. Summ. 39: 1–13.

24. Lewin, P.K. 1985. Mummy's riddles unravelled. Microsc. Soc. Can. Bull. 12: 4–8.

25. Lippy, E.C. and S.C .Waltrip. 1984. Waterborne disease outbreaks—1946–1980: A thirty-five-year perspective. J. Am. Wat. Wks. Assoc. 76: 791–793.

26. Marennikova, S.S., E.M. Shelukhina, O.A. Zhukova, N.N. Yanova, and V.N. Loparev. 1990. Smallpox diagnosed 400 years later: Results of skin lesions examination of 16th century Italian mummy. J. Hyg. Epidemiol. Microbiol. Immunol. 34: 227–231.

27. Mclean, D.M. 1975. Mosquito-borne arboviruses in arctic America. Med. Biol. 53: 264–270.

28. McLean, D.M., P.N. Grass, B.D. Judd, and K.J. Stolz. 1979. Bunyavirus development in Arctic mosquitoes as revealed by glucose oxidase staining and immunofluorescence. Arch. Virol. 62: 313–322.

29. Miller, B.R., R.S. Nasci, M.S. Godsey, H.M. Savage, J.J. Lutwama, R.S. Lanciotti, and C.J. Peters. 2000. First field evidence for natural vertical transmission of West Nile virus in *Culex univittatus* complex mosquitoes from Rift Valley province, Kenya. Am. J. Trop. Med. Hyg. 62: 240–246.

30. Nakajima, K., U. Desselberger, and P. Palese. 1978. Recent human influenza A (H1N1) viruses are closely related genetically to strains isolated in 1950. Nature 274: 334–339.

31. Niu, M.T., L.B. Polish, B.H. Robertson, B.K. Khanna, B.A. Woodruff, C.N. Shapiro, M.A. Miller, J.D. Smith, J.K. Gedrose, and M.J. Alter. 1992. Multistate outbreak of hepatitis A associated with frozen strawberries. J. Infect. Dis.166: 518–524.

32. Orent, W. 1998. Escape from Moscow. The Sciences 38: 26–31.

33. Rao, V.C., K.M. Seidel, S.M. Goyal, T.G. Metcalf, and J.L. Melnick. 1984. Isolation of enteroviruses from water, suspended solids, and sediments from Galveston Bay: Survival of poliovirus and rotavirus adsorbed to sediments. Appl. Environ. Microbiol. 48: 404–409.

34. Raphael, R.A., S.A. Sattar, and V.S. Springthorpe. 1985. Long-term survival of human rotavirus in raw and treated river water. Can. J. Microbiol. 31: 124–128.

35. Reid, A.H., T.G. Fanning, J.V. Hultin, and J.K. Taubenberger. 1999. Origin and evolution of the 1918 "Spanish" influenza virus hemagglutinin gene. Proc. Nat. Acad. Sci. 96: 1651–1656.

36. Saitou, N., and M. Nei. 1986. Polymorphism and evolution of influenza A virus genes. Mol. Biol. Evol. 3: 57–74.

37. Shoham, D. 1985. Studies on the Ecology of Influenza Type A Virus in Israel. Ph.D. thesis, Tel Aviv University.

38. Shoham, D. 1993. Biotic-abiotic mechanisms for long term preservation and reemergence of influenza type A virus genes. Prog. Med. Virol. 40: 178–192.

39. Smith, T.F., and L. Reichrath. 1974. Comparative recovery of 1972–1973 influenza virus isolates in embryonated eggs and primary rhesus monkey cell cultures after one freeze-thaw cycle. Am. J. Clin. Pathol. 61: 579–584.

40. Traavik T., R. Mehl, and R. Wiger. 1978. California encephalitis group viruses isolated from mosquitoes collected in Southern and Arctic Norway, Acta Pathol. Microbiol. Scand. [B], 86B: 335–341.

41. Turell, M.J., W.C. Reeves, and J.L. Hardy. 1982. Transovarial and trans-stadial transmission of California encephalitis virus in *Aedes dorsalis* and *Aedes melanimon*. Am. J. Trop. Med. Hyg., September 31(5): 1021–1029.

42. Walter R., H. Kaupa, M. Johl, J. Durkop, U. Kramer, and W. Macht. 1989. Viruses in river water and health risk assessment. Water Sci. Technol. 21: 21–26.

43. White, D.J., and Morse, D.L. (eds.). 2001. West Nile virus—detection, surveillance and control. Annals of the New York Academy of Sciences, 951.

15

Classification of Bacteria from Polar and Nonpolar Glacial Ice

Brent C. Christner, Ellen Mosley-Thompson,

Lonnie G. Thompson, and John N. Reeve

SNOWFALL ACCUMULATES as glacial ice at both poles and globally at high altitudes in nonpolar regions. Archived chronologically within these glaciers are samples of the atmospheric constituents at the time of snow deposition, including particulates of inorganic and biological origin deposited originally on the surface of the snow, often by attachment to snowflakes. Studies of ice cores have established past climate changes and geological events, both globally and regionally, but rarely have these results been correlated with the insects, plant fragments, seeds, pollen grains, fungal spores, and bacteria, that also are present, and very few attempts have been made to determine the diversity and longevity of viable species entombed in such glacial ice. Fungi, algae, protists, bacteria, and viruses have been detected and recovered from polar ice cores (1, 2, 3, 10, 15, 41), but there are very few similar reports describing the microorganisms preserved in nonpolar glacial ice of different age and from different locations. Fortunately, for such studies, we have access to ice cores archived at the Byrd Polar Research Center (BPRC) at the Ohio State University. These ice cores have been collected over many years from globally distributed sites, and many have already been subjected to extensive physical and chemical analyses. These, therefore, provide the opportunity to isolate and to characterize microorganisms from glacial ice formed at defined dates, under known climate conditions, at geographically very different locations (Figure 15.1). To avoid problems of surface contamination, we constructed an ice core sampling system that melts the ice and collects only the resulting interior core meltwater. Here we review the results of bacterial isolations from meltwater generated using this system from the interiors of nonpolar and polar glacial ice cores of different vintage, and from Lake Vostok accretion ice (12,14).

Ice Core Sampling

Ice core exteriors are contaminated during drilling and transport, and a sampling system was designed and constructed to melt ice and collect the resulting

228

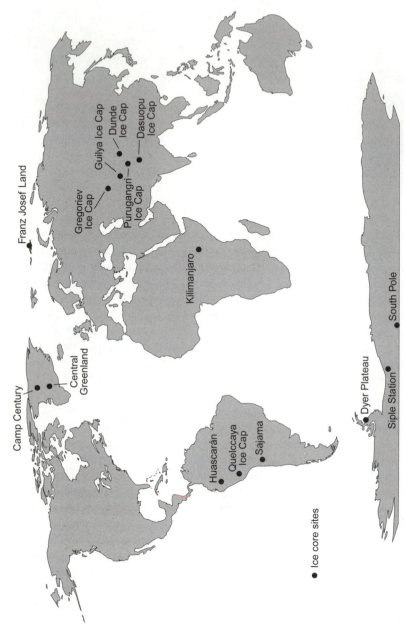

FIGURE 15.1. Sampling sites and ice cores available for study at the Byrd Polar Research Center (BPRC). To date, bacteria have been isolated from ice cores sampled from glaciers at both poles, in the mountain ranges on the subtropical Tibetan plateau, and in the tropical Bolivian Andes. In each case, the nearest major ecosystem, and therefore most likely origin of airborne particulates, is very different.

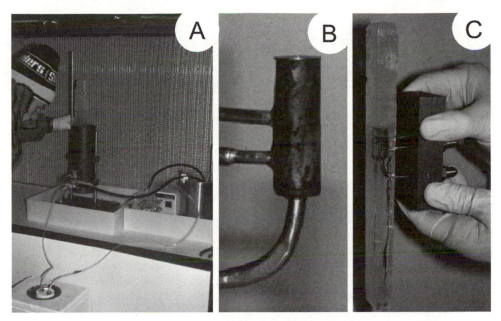

FIGURE 15.2. The ice core sampling system. (A) The sampling system is assembled completely inside a laminar flow hood that is housed within a −10 °C walk-in freezer. All components of the system are autoclaved, dried, and exposed to ethylene oxide for 12 hours before use. An ice core is positioned vertically in the sampler with the cut end of the core contacting (B) the heated sampler head which melts upward (C) through the core and collects the resulting meltwater. In (C), the sampler head is shown disassembled from the main unit to illustrate its movement through the interior of the ice core.

meltwater aseptically only from the inside of an ice core (Figure 15.2) . A thin section is first cut from one end of the core using a dedicated dust-free band-saw, and the newly exposed flat surface is immersed for 2 minutes in 95% ethanol. Exposure to ethanol does not cause the ice core to fracture and, in reconstruction experiments, such an ethanol treatment effectively killed all *Serratia marcesens* cells that were intentionally swabbed onto the saw blade and onto the resulting cut surface of the ice core before the ethanol treatment. However, this treatment may not kill all bacterial endospores, and it certainly would not destroy nucleic acids. Therefore, to monitor for such contamination, the cut surface of each ice core is swabbed after the ethanol treatment before initiating melting. These swabs are used to inoculate growth media and are evaluated for the presence of DNA by polymerase chain reaction (PCR) amplifications using universal 16S rDNA amplification primers. Immediately after the exposure to ethanol, the ice core is positioned vertically in the sampling system with the ethanol-washed surface placed directly in contact with the sampling unit. The sampling unit is heated internally, and as it melts the ice, it moves upward through the ice core. The water generated passes through an orifice in the center of the sampling unit, and is collected aseptically in sterile containers positioned outside the sampling system (Figure 15.2).

During the course of this study, growth occurred only once from a swab-inoculated culture and this isolate had a 16S rDNA sequence consistent with classification as an endospore-forming *Bacillus subtilis* species. If this same species had subsequently been isolated from the interior of the core, it obviously would have been suspected as a potentially introduced contaminant. Taq polymerase preparations from several manufacturers were examined for DNA contamination using a nested PCR approach with the universal primers 27F and 1492R and 515F and 1392R (25). Consistent results were obtained with the LD (low DNA) Amplitaq Gold DNA polymerase (Perkin-Elmer Biosystems), which the manufacturer claims has < 10 copies of bacterial 16S rDNA per 2.5 U of enzyme. No amplicons were generated when material collected on the ethanol-washed surface was used in the PCR, indicating that the level of contamination remaining on the ice core surface is below that detectable by standard PCR procedures. Although there was always a possibility that microbial or nucleic acid contamination would elude decontamination procedures, this concern was addressed by employing quality control measures and drawing conclusions from results obtained from more than one ice core sample.

Bacteria Recovered from Glacial Ice

The numbers and identities of bacteria that form colonies when meltwater is plated directly on solid media have been determined in ice from Sajama (Bolivia), Guliya (China), Greenland, and Antarctica. A range of different media was used during the course of this study, and the largest numbers of colonies were routinely observed after aerobic incubation at 15 to 22 °C on media containing low levels of nutrients, such as R2A and 1% nutrient and tryptic soy agar (Difco). In general, meltwaters from nonpolar, low-latitude, high-altitude glaciers contain greater numbers, as well as more diversity of colony-forming bacteria than meltwaters from polar ice cores. For example, 180 colony-forming units per ml (cfu/ml) were present in meltwater from a 200-year-old sample of Guliya ice, whereas water from a ca. 1800-year-old sample of polar ice from Taylor Dome (Antarctica) contained only ca. 10 cfu/ml. Even fewer cfus were present in meltwater from ice of a similar vintage from the Antarctic Peninsula and from the Summit and Dye 2 sites in Greenland. It is important to note that differences in the amount of annual snowfall and in the subsequent rates of compression mean that equal volumes of meltwater from different cores do not necessarily represent equivalent time periods of microbial deposition. However, these results are consistent with those of Dancer et al. (15) who recovered < 5

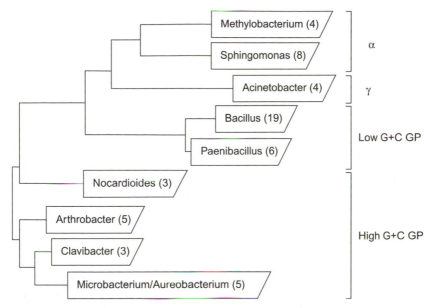

FIGURE 15.3. Bacterial genera represented most frequently by ice core isolates. The number of isolates from both polar and nonpolar ice cores, obtained from each of the bacterial genera shown, is listed in parentheses. The phylogenetic relationships illustrated are based on 16S rDNA sequences. They are not drawn to scale.

cfu/ml from glacial ice from the Canadian high Arctic after enrichment for coliform bacteria, and other reports of recovering even fewer bacteria (< 1 cfu/ml) from meltwaters from polar ice (2, 19). Logically, these differences arise because nonpolar glaciers are closer to major sources of airborne microorganisms such as exposed soils and tropical and subtropical ecosystems. Consistent with this, meltwater from ice from a Taylor Dome site located at the head of the Taylor Valley in the Dry Valley complex of Antarctica contained relatively larger numbers of culturable bacteria (ca. 10 cfu/ml), and microbiological surveys have documented the abundance of bacteria, fungi, and algae in this area despite the very dry and cold climate (6, 30).

Based on their small-subunit ribosomal RNA-encoding sequences (16S rDNAs), most of the ice core isolates are members of the nonsporulating Gram-positive, spore-forming *Bacillus, Paenibacillus*, and *Actinobacteria*, and α- and γ-proteobacterial lineages (Figure 15.3). Many form colored colonies, consistent with pigment production providing protection from solar irradiation during airborne transport and subsequent exposure on the glacier surface. Not surprisingly, a number of the isolates are related to species that have life cycles with radiation- and desiccation-resistant resting stages. Endospore-forming relatives of the genera *Bacillus* and *Paenibacillus* were commonly isolated from nonpolar glacial ices, presumably due to their close proximity to soil ecosystems, and similar but not identical species of *Sphingomonas*, *Methylobacterium, Acinetobacter*, and *Arthrobacter* also were ubiquitous and recovered from both polar and nonpolar locations. While most of the isolates are similar to species frequently found in environmental surveys from around the world, some of the isolates have 16S rDNA sequences most closely related to species recovered previously from Antarctic lake mats (see ref. 6 and chapter 3, this volume), sea ice (see refs. 5, 18, 22 and chapter 4, this volume), and other cold environments (4, 32). The isolation of related microbes from many geographically diverse but predominantly frozen environments argues strongly that these species probably have features that confer resistance to freezing and survival under frozen conditions.

Isolation of Bacteria from Very Old Glacial Ice

An ice core that extends over 300 m below the surface (mbs) to the underlying bedrock was obtained from the Guliya ice cap in Tibet (Figure 15.1), and based on the abundance of ^{36}Cl (half-life = 301,000 years) the ice at the bottom of this core is > 500,000 years old (39). This is the oldest glacial ice recovered to date and provides an opportunity to evaluate microbial survival in ice over an extended period of time (13). Aliquots of meltwater from this ice core from 296 mbs were inoculated into a variety of growth media and, after 30 to 60 days of aerobic incubation at 4 °C, growth was observed in very dilute nutrient and tryptic soy broths. These media were used at 1% of the concentration recommended by the manufacturer (Difco, Inc.). Despite the long period needed for initial growth, and the primary enrichment cultures being grown under oligotrophic conditions at 4 °C, isolates were subsequently obtained from these cultures that grew and formed colonies in two to seven days on nutrient-rich media at 25 °C. Long-dormant cells must eliminate toxic metabolites, such as hydrogen peroxide, superoxide, and free radicals, and repair macromolecular damage that has accumulated before they can grow and divide successfully (16). The results with the very old Guliya ice are consistent with this hypothesis, and indicate that successful recovery is facilitated by providing only a very low level of nutrients initially, sufficient for repair but insufficient to elicit an instant attempt at growth.

Fourteen 16S rDNA sequences, corresponding to nucleotides 27 through 1992 of the *Escherichia coli* 16S rDNA sequence, have been determined from isolates from the very old Guliya ice (Figure 15.4). Based on these data, most of these belong to the same bacterial lineages as the isolates obtained from more recent polar and nonpolar glacial ices, and ca. 50% are members of genera that form endospores known to facilitate long-term survival under nongrowth conditions (8, 40). Light microscopy has revealed that some also have

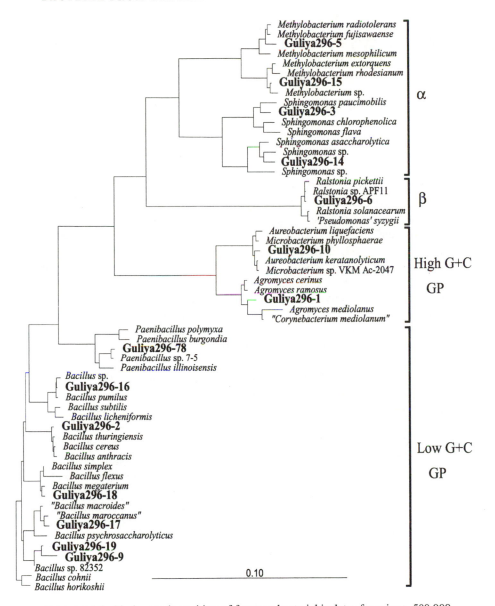

FIGURE 15.4. Phylogenetic position of fourteen bacterial isolates from ice > 500,000 years old from 296 m below surface of the Guliya ice cap. 16S rDNA sequences (ca. 1400 nucleotides) were obtained from the cells from a single colony of each isolate. They were aligned based on secondary structures using the ARB software package (25) and a best fit neighbor-joining tree was constructed. Evolutionary distance is defined as the number of fixed nucleotide changes per position.

thick cell walls and form polysaccharide capsules that presumably also contribute to survival through the physical stresses imposed by freezing, compaction pressure, and thawing (17).

Isolation of Bacteria from Lake Vostok Accretion Ice

More than seventy subglacial lakes have been discovered in Antarctica. The largest, Lake Vostok, has been covered by a layer of glacial ice and isolated from direct surface input for at least 420,000 years (27). Glacial ice melts into Lake Vostok at the northern ice-water interface, and water from Lake Vostok freezes and accumulates as accretion ice directly below the glacial ice over the central and southern regions (21, 23, 33). It seems very likely that viable bacteria are seeded into Lake Vostok as glacial ice melts into the lake. However, whether an active microbial community is established within Lake Vostok remains uncertain, as concerns for contamination have resulted in a moratorium on direct sampling of Lake Vostok water. Ice core drilling also was terminated above the ice-water interface although an ice core was retrieved in which the bottom ca. 150 m are accretion ice and therefore represent a sample of Lake Vostok water. Microbial cells in meltwater from sections of this accretion ice core that originated 3590 and 3603 m below the surface (mbs) have been detected by epifluorescence and scanning electron microscopy (24, 29), and seven bacterial 16S rDNA sequences were amplified from the 3590 meltwater that originated from α- and β-proteobacteria, and from an actinobacteria (29). Evidence for respiration was also obtained by measuring ^{14}C-CO_2 release during incubation at 3 °C and 23 °C after the addition of ^{14}C-acetate or ^{14}C-glucose to meltwater from the 3603 section (24). A section of this core from 3591.965 to 3592.445 mbs, designated as core section 3593, was obtained from the National Ice Core Laboratory (Denver, Colorado), and has been subjected to microbiological investigation (12).

Scanning electron microscopy of materials filtered from core 3593 meltwater revealed the presence of particulates with size and morphology consistent with bacterial cells (Figure 15.5), and four different single-colony isolates were obtained from enrichment cultures inoculated with core 3593 meltwater. Based on their 16S rDNA sequences, these isolates are related to established species of *Brachybacterium, Sphingomonas, Paenibacillus*, and *Methylobacterium* (Figure 15.6, also see photos in chapters 6 and 16). Six bacterial 16S rDNAs also were amplified from core 3593 meltwater with sequences, indicating that they originated from five different bacterial lines of descent. Although not directly comparable to the results of Priscu et al. (29) due to differences in the position of 16S rDNA nucleotides sequenced, bacterial isolates and 16S rDNA sequences originating from the α- and β-proteobacteria and actinobacteria were similarly detected in both studies. It also seems noteworthy that sequence

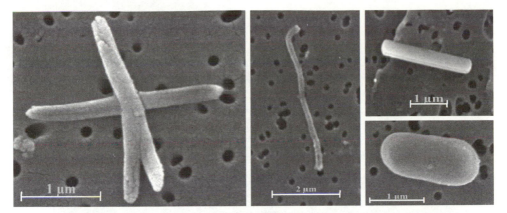

FIGURE 15.5. Scanning electron micrographs of materials filtered from meltwater from Lake Vostok deep ice core section 3593. The particulates shown, apparently bacteria, are retained on the surface of a 0.2 μm isopore (millipore) filter.

pA419 originated from an α-proteobacterium whose nearest 16S rRNA neighbor is an isolate recovered 400 mbs of Lake Baikal (Russia) (4). Only very tenuous extrapolations can be made from 16S rDNA sequences, but the results obtained suggest that Lake Vostok is seeded with bacteria related to those surviving for extended periods in glacial ice, and is probably inhabited by species similar to those found in other permanently cold environments.

Conclusions

Microorganisms recovered from glacial ice are likely to have already endured desiccation, solar irradiation, freezing, a period of frozen dormancy, and thawing. It is not surprising, therefore, that many of the ice core isolates are pigmented and belong to bacterial groups that differentiate into spores that specifically confer resistance to such environmental abuse and facilitate long-term survival under nongrowth conditions. Many also have thick cell walls and polysaccharide capsules and have been demonstrated to be more resistant to repeated cycles of freezing and thawing than standard laboratory bacterial species. Interestingly, closely related bacteria have been recovered from glaciers separated by great distances, suggesting the possibility that some species may indeed have evolved features that help their survival and, conceivably, may even facilitate growth under freezing conditions. Thin films of liquid water may exist between ice crystals, even within apparently solid ice (28), and studies of permafrost (see ref. 31 and chapter 7, this volume), basal glacial ice (34), frozen bacterial suspensions (11), and surface snow (9) have all demonstrated microbial activity under freezing conditions. Evidence for microbial activity

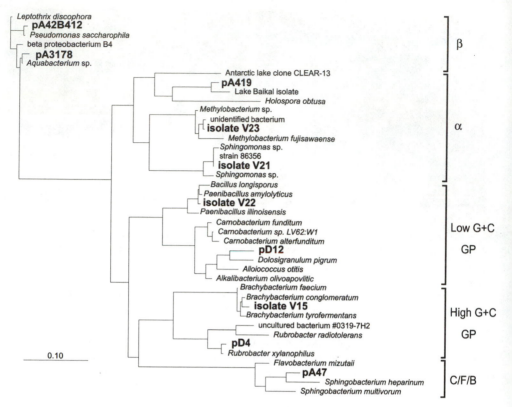

FIGURE 15.6. Phylogenetic analysis of 16S rDNA sequences isolated from bacteria
and directly amplified from meltwater from Lake Vostok core section 3593. Sequences that correspond to nucleotides 515 through 1392 of the *E. coli* 16S rDNA
were obtained, aligned, and used to construct the figure shown as Figure 4 (25).
A best fit tree was created using maximum likelihood with a 771 nucleotide mask
of unambiguously aligned positions using fastDNAml (18).

within glacier ice, as implied by geochemical anomalies in polar and nonpolar
glacial ice cores (7, 35, 36, 37), provides yet another example of the extreme
conditions in which life can exist, extends the known boundary of the biosphere into icy environments, and suggests that *in situ* biological alteration of
gases and ions may skew paleoclimatic interpretations of ice core records.

Ice cores from low-latitude, high-altitude glaciers generally contain more
recoverable bacteria than polar ice cores, presumably because the Andes and
Himalayas are closer to major sources of airborne biological materials. Similarly, polar ice from regions adjacent to the exposed soils and rock surfaces in
Taylor Valley (Antarctica) contains more recoverable bacteria than polar ice
from remote regions. We have established that bacteria remain viable when

frozen in glacial ice for > 500,000 years and, based on other studies of *Bacillus* spore longevity (8,40), this is almost certainly an underestimate. It is also possible that some microorganisms might even maintain some metabolic activity while apparently frozen within ice.

By identifying and counting the microorganisms present in glacial ice of very different ages, we may be able to relate climate change and geography to local airborne microbial populations (see chapter 6). Similarly, by characterizing individual isolates, we can obtain information that contributes to discussions of the possibility that microorganisms might survive frozen in extraterrestrial environments. These isolates should also provide data that are directly relevant to discussions of the prevalence of antibiotic resistance before the advent of antibiotic therapies, and the survival of life through "Snowball Earth" events (20).

Acknowledgments

This research was supported by NSF grant OPP-9714206 awarded through the Life in Extreme Environments Initiative.

Literature Cited

1. Abyzov, S.S., V.Y. Lipenkov, N.E. Bobin, and B.B. Kudryashov. 1982. Microflora of central Antarctic glacier and methods for sterile ice-core sampling for microbiological analyses. Biol. Bull. Academy of Sciences of the USSR 9: 304–349.
2. Abyzov, S.S. 1993. Microorganisms in the Antarctic ice. pp. 265–295. In E.I. Friedmann (ed.), Antarctic Microbiology. Wiley, New York.
3. Abyzov, S.S., L.N. Mitskevich, and M.N. Poglazova. 1998. Microflora of the deep glacier horizons of central Antarctica. Microbiology 67: 547–555.
4. Benson, D.A., I. Karsch-Mizrachi, D.J. Lipman, J. Ostell, B.A. Rapp, and D.L. Wheeler. 2000. GenBank. Nucleic Acids Res. 28: 15–18.
5. Bowman, J.P., S.A. McCammon, M.V. Brown, D.S. Nichols, and T.A. McMeekin. 1997. Diversity of psychrophilic bacteria in Antarctic sea ice. Appl. Environ. Microbiol. 63:3068–3078.
6. Brambilla, E., H. Hippe, A. Hagelstein, B.J. Tindall, and E. Stackebrandt. 2001. 16S rDNA diversity of cultured and uncultured prokaryotes of a mat sample from Lake Fryxell, McMurdo Dry Valleys, Antarctica. Extremophile 5: 23–33.
7. Campen, R.K., T. Sowers, and R.B. Alley. 2003. Evidence of microbial consortia metabolizing within a low-latitude mountain glacier. Geology 31:231–234.
8. Cano, R.J. and M.K. Borucki. 1995. Revival and identification of bacterial spores in 25 to 40 million year old Dominican amber. Science 268: 1060–1064.
9. Carpenter, E.J., S. Lin, and D.G. Capone. 2000. Bacterial activity in South Pole snow. Appl. Environ. Microbiol. 66: 4514–4517.

10. Castello, J.D., S.O. Rogers, W.T. Starmer, C.M. Catranis, L. Ma, G.D. Bachand, Y. Zhao, and J.E. Smith. 1999. Detection of tomato mosaic tobamovirus RNA in ancient glacial ice. Polar Biol. 22: 207–212.

11. Christner, B.C., E. Mosley-Thompson, L.G. Thompson, V. Zagorodnov, K. Sandman, and J.N. Reeve. 2000. Recovery and identification of viable bacteria immured in glacial ice. Icarus 144: 479–485.

12. Christner, B.C., E. Mosley-Thompson, L.G. Thompson, and J.N. Reeve. 2001. Isolation of bacteria and 16S rDNAs from Lake Vostok accretion ice. Environ. Microbiol. 3: 570–577.

13. Christner, B.C. 2002. Incorporation of DNA and protein precursors into macromolecules by bacteria at −15 °C. Appl. Environ. Microbiol. 68:6435–6438.

14. Christner, B.C, E. Mosley-Thompson, L.G. Thompson, and J.N. Reeve. 2003. Bacterial recovery from ancient glacial ice. Environ. Microbiol. 5:433–436.

15. Dancer, S.J., P. Shears, and D.J. Platt. 1997. Isolation and characterization of coliforms from glacial ice and water in Canada's High Arctic. J. Appl. Microbiol. 82: 597–609.

16. Dodd, C.E.R., R.L. Sharman, S.F. Bloomfield, I.R. Booth, and G.S.A.B. Stewart. 1997. Inimical processes: Bacterial self-destruction and sub-lethal injury. Trends in Food Science Technol. 8: 238–241.

17. Fogg, G.E. 1998. pp. 33–47. The Biology of Polar Habitats. Oxford University Press, New York.

18. Gosing, J.J., and J.T. Staley. 1995. Biodiversity of gas vacuolated bacteria from Antarctic sea ice and water. Appl. Environ. Microbiol. 61: 3486–3489.

19. Hardfield, M. H.G. Jones, R. Letarte, and P. Simard. 1992. Seasonal fluctuation patterns of microflora on the Agassiz ice cap, Ellesmere Island, Canadian Arctic. The Musk-ox 39: 119–123.

20. Hoffman, P.F., A.J. Kaufman, G.P. Halverson, and D.P. Schrag. 1998. A Neoproterozoic snowball Earth. Science 281: 1342–1349.

21. Jouzel, J., J.R. Petit, R. Souchez, N.I. Barkov, V.Y. Lifenkov, D. Raymond, M. Stievenard, N.I. Vassiliev, V. Verbeke, and F. Vimeux. 1999. More than 200 meters of lake ice above subglacial Lake Vostok, Antarctica. Science 286: 2138–2141.

22. Junge, K., J.J. Gosink, H.G. Hoppe, and J.T. Staley. 1998. *Arthrobacter, Brachybacterium* and *Planococcus isolates* identified from Antarctic sea ice brine. Description of Plannococcus mcceekinii, sp. nov. system. Appl. Microbiol. 21: 306–314.

23. Kapitsa, A.P., J.K. Ridley, G. Robin, M.J. Siegert, and I.A. Zotikov. 1996. A large deep freshwater lake beneath the ice of central East Antarctica. Nature 381: 684–686.

24. Karl, D.M., D.F. Bird, K. Björkman, T. Houlihan, R. Shackelford, and L. Tupas. 1999. Microorganisms in the accreted ice of Lake Vostok, Antarctica. Science 286: 2144–2147.

25. Lane, D.J. 1991. 16S/23S rRNA sequencing. p. 133. In E. Stakebrandt, and M. Goodfellow (eds.), Nucleic Acid Techniques in Bacterial Systematics. Wiley, New York.

26. Olsen, G.J., J.H. Natusda, R. Hagstrom, and R.Overbeek. 1994. fastDNAml: A tool for construction of phylogenetic trees of DNA sequences using maximum likelihood. Computer Applications in the Biosciences 10: 41–48.

27. Petit, J.R., J. Jouzel, D. Raynaud, N.I. Barkov, J.M. Barnola, I. Basile, M. Benders, J. Chappellaz, M. Davis, G. Delaygue, M. Delmotte, V.M. Dotlyakov, M. Le-

grand, V.Y. Lipendoc, C. Lorius, L. Pepin, C. Ritz, E. Saltzman, and M. Stievenard. 1999. Climate and atmospheric history of the past 420,000 years from the
Vostok ice core, Antarctica. Nature 399: 429–436.

28. Price, P.B. 2000. A habitat for psychrophiles in deep Antarctic ice. Proc. Nat. Acad.
Sci. 97: 1247–1251.

29. Priscu, J.C., C.H. Fritsen, E.E. Adams, S.J. Giovannoni, H.W. Paerl, C.P. McKay,
P.T. Doran, D.A. Gordon, B.D. Lanoil, and J.L. Pinckney. 1998. Perennial Antarctic lake ice: An oasis for life in a polar desert. Science 280: 2095–2098.

30. Priscu, J.C., E.E. Adams, W.B. Lyons, M.A. Voytek, D.W. Mogk, R.L. Brown, C.P.
McKay, C.D. Takacs, K.A. Welch, C.F. Wolf, J.D. Kirschtein, and R. Avci. 1999.
Geomicrobiology of subglacial ice above Lake Vostok, Antarctica. Science 286:
2141–2144.

31. Rivkina, E.M., E.I. Friedmann, C.P. McKay, and D.A. Gilichinsky. 2000. Metabolic
activity of permafrost bacteria below the freezing point. Appl. Environ. Microbiol. 66: 3230–3233.

32. Shi, T., R.H. Reeves, D.A. Gilichinsky, and E.I. Friedmann. 1997. Characterization
of viable bacteria from Siberian permafrost by 16S rDNA sequencing. Microbial.
Ecol. 33: 169–179.

33. Siegert, M.J., R. Kwok, C. Mayer, and B. Hubbard. 2000. Water exchange between
subglacial Lake Vostok and the overlying ice sheet. Nature 403: 643–646.

34. Skidmore, M.L., J.M. Foght, and M.J. Sharp. 2000. Microbial life beneath a high
Arctic glacier. Appl. Environ. Microbiol. 66: 3214–3220.

35. Souchez, R., M. Janssens, M. Lemmens, and B. Stauffer. 1995. Very low oxygen
concentration in basal ice from Summit, Central Greenland. Geophys. Res. Letters 22: 2001–2004

36. Souchez, R., A. Bouzette, H.B. Clausen, S.J. Johnsen, and J. Jouzel. 1998. A
stacked mixing sequence at the base of the Dye 3 core. Geophys. Res. Letters
25: 1943–1946.

37. Sowers, T. 2001. The N_2O record spanning the penultimate deglaciation from the
Vostok ice core. J. Geograph. Res. 106: 31903–31914.

38. Strunk, O., O. Gross, B. Reichel, M. May, S. Hermann, N. Struckmann, B. Nonhoff,
M. Lenke, A. Vilbig, T. Ludwig, A. Bode, K.H. Schleifer, and W. Ludwig. 1998.
ARB: A software environment for sequence data. Department of Microbiology,
Technical University of Munich, Munich, Germany.

39. Thompson, L.G., T. Yao, M.E. Davis, K.A. Henderson, E. Mosley-Thompson,
P.-N. Lin, J. Beer, H. Synal, J. Cole-Dai, and J.F. Bolzan. 1997. Tropical climate
instability: The last glacial cycle from a Qinghai-Tibetan ice core. Science 276:
1821–1825.

40. Vreeland, R.H., W.D. Rosenzweig, and D.W. Powers. 2000. Isolation of a 250
million-year old halotolerant bacterium from a primary salt crystal. Nature 407:
897–900.

41. Willerslev, E., A.J. Hansen, B. Christensen, J.P. Steffensen, and P. Arctander. 1999.
Diversity of Holocene life forms in fossil glacier ice. Proc. Nat. Acad. Sci. 96:
8017–8021.

16

Common Features of Microorganisms in Ancient Layers of the Antarctic Ice Sheet

S.S. Abyzov, M.N. Poglazova, J.N. Mitskevich, and M.V. Ivanov

PREVIOUS microbiological investigations have revealed microorganisms as well as ancient atmospheric dust and air bubbles in the Antarctic ice sheet (1, 2). We characterized the morphological features and estimated the number of microorganisms in deep horizons of the Antarctic ice sheet (up to 2750 m) using direct microscopy. Bacterial cell and atmospheric dust concentrations appeared to be associated. The amount of atmospheric dust reflects climate changes in the geological past (2).

Microbiological data are based on investigations of ice cores from the deepest borehole at Vostok 5G, where the ice is 3750 m deep (4). Drilling has stopped at 3623 m, 150 m above the water surface of subglacial Lake Vostok. The existence of this lake was predicted by Russian scientists more than forty years ago (6, 11), and now has been confirmed by seismic measurements and radio-echo-sounding (7, 10). This lake represents a unique ecosystem where relict microorganisms may survive and function in extreme conditions. Unfortunately, it is impossible to obtain samples from the lake water or its sediments due to an international moratorium on drilling into the lake until special equipment is developed to prevent contamination of its ecosystem.

Nevertheless, during the latest stages of drilling core 5G, we managed to approach the upper layers of the border zone that separates lower ice horizons from the water surface (3530–3611 m). Samples from this part of the border zone have been obtained. Ice at depths of 3538 m and deeper differs significantly from the ice above (see Chapter 17). This ice is formed from the lake water by accretion to the lower glacier surface. This ice is called accretion (i.e. congealed) ice (5).

Our objectives were to investigate the basal layers of the glacial ice and accretion lake ice layers to complete the microbiological examination of the entire 5G ice core from top to bottom. A comparative analysis of the morphology of microorganisms found both in the Antarctic ice sheet and in the accretion lake ice could help us to understand the biological peculiarities of the largest of the subglacial Antarctic lakes.

Materials and Methods

The ice sheet at Vostok station is 3750 m thick (8). The thickness of the accretion ice is about 200 m (see ref. 5, and chapter 17, this volume). Ten ice samples were investigated. The aseptic sampling unit is described in detail in reference 1. This design takes samples from an ice core aseptically by melting out the central part of the core. Its main part is a cast copper or aluminium heater. Its working surface is in the shape of a concave cone, slightly smaller in diameter than the core, which causes the water from the melting core interior to flow through the central hole in the heater along the water-receiving pipe and into the sterile receiver. The heater, tightly covered with a metal lid, and the water-receiving pipe are sterilized in an autoclave. Just before sampling, the end of the core is chopped off to expose a surface area free from contamination. The end of the core is quickly lowered onto the sterile surface of the heater (lid removed), and when the heater is turned on, sampling starts.

Before being sent to the Antarctic, the device was tested for reliability in aseptic sampling from ice cores (1). For this purpose, sterile water was frozen in cylindrical molds under laboratory conditions, and the surface of the resulting ice "core" was coated with a dense suspension culture of *Serratia marcescens* Bizio. The lower end of the core, mounted in the sampling device, then was chopped off as described above, and its central portion sampled. Repeated inoculation of meat-peptone broth (MPB) with the resulting meltwater demonstrated that microorganisms spread on the ice core surface did not penetrate into interior layers, and that the technique described above ensures sterile sampling from the interior of the core.

This sampling method was used in all our microbiological investigations on the Antarctic ice sheet. In this work all samples were obtained aseptically from horizons between 3534 and 3611 m inclusive. The 3534 m sample is from the basement horizon of the glacier. The remaining nine samples are from the accretion ice layer at depths of 3541, 3555, 3565, 3579, 3585, 3592, 3598, 3606, and 3611 m. The last of these samples is only 140 to 150 m above the lake surface.

Approximately 100 to 150 ml of meltwater obtained from each of these samples was filtered through polycarbonate nuclear filters with a pore diameter of 0.23 μm. Microorganisms and other microscopic objects were collected on these filters, which then were stained with the fluorescent dye fluorescamine, which is specific for proteins and other compounds containing amino groups. The filters then were examined using luminescent microscopy (LUMAM-I2) to examine the morphology and to quantify the microorganisms, as described earlier (9). All cells in forty to eighty microscopic fields of view were counted.

Results and Discussion

Prokaryotic and eukaryotic microorganisms were detected in all meltwaters from the basal layers of the glacier. The total number of bacterial cells was slightly less than that from midcore horizons. There were hundreds to thousands of bacterial cells/ml of meltwater from upper glacier samples (2), but only tens to hundreds of bacterial cells/ml in meltwater from accretion ice. Total cell number varied from horizon to horizon (Figure 16.1).

The concentration of microorganisms was not dependent on the depth, and fluctuated within an order of magnitude. A certain periodicity in the alternation of low and high cell concentrations from one horizon to another was observed in the distribution of microorganisms through the entire thickness of the stratum investigated.

The maximum number of cells was observed at the 3534, 3555, 3592, and 3606 m horizons. Nearly a five-fold drop in the number of microorganisms

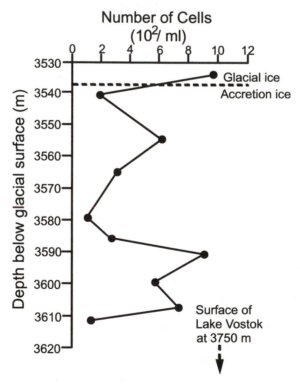

FIGURE 16.1. Concentration of bacterial cells in each of the examined ice horizons extracted from the Vostok 5G core.

was observed at the 3541, 3579, and 3611 m horizons. These variations in the cell number apparently reflect conditions in which the lake water froze: at its shelf where contact with sediments are possible, or in the deep part of the lake (see chapter 17). Rare but small accumulations of mineral particles of moraine origin, which are absent at the horizons below 3606 m, support this supposition (8). Presumably, the observed variation in cell concentration with horizon rather than depth is related to many factors, including heterogeneity of the ice horizons as well as the presence of morainous material in some horizons between 3538 and 3600 m (see ref. 8 and chapter 17, this volume). Specificity and nonstability of conditions in accretion ice did not allow us to identify significant regularities in the distribution of morphological types by horizon. Nevertheless, we observed that horizons differ from each other in the content of different microbial groups. For example, morphological diversity was restricted to micrococci and short rods in those horizons with the lowest concentration of bacterial cells (i.e., 3541, 3579, and 3611 m) (Figure 16.1). Those horizons with low microbial diversity were those that contained a minor mixture of organic particles. However, high microbial diversity was associated with increased total bacterial cell number and with increased concentration of organic particles. Therefore, it appears that the movement of microbes and organic particles into the ice sheet is regulated by the same processes. The association between these two parameters is less obvious in accretion ice than in ice from the main part of the glacier, which is more stable.

Morphological features of microorganisms often reflect their taxonomy. Therefore, we undertook a comparative investigation of the morphological similarities and differences among the microflora from different ice horizons. Ice horizons of different origins contain similar types of bacteria, thus confirming our suspicion that bacteria entered accretion lake ice together with meltwater from the base of the glacier. Many cocci, small rods of different shapes (Figure 16.2 A), and cells and microcolonies of cyanobacteria (Figure 16.2 B) were found in all layers of the accretion ice. Similar cells and colonies were observed in the upper horizons of the glacier. Actinomycete threads (Figure 16.2 C), fungal hyphae and conidia (Figure 16.2 D), and yeast cells (Figure 16.2 E) also were observed. We called these cells first group microorganisms, which dominated and were found in all horizons of the glacier, including accretion ice (Figures 16.2 A–E).

Some microorganisms were detected only in certain horizons of accretion ice. They have rarely been seen in the main part of the glacier. We called these the second group of bacteria. In particular, bacterial species within the genus *Cytophaga* belong to this group (Figure 16.3 A), and were detected in ice horizon 3555 m. Such cells were located on the surfaces of morainous particulates. We suggest that these bacteria were transferred to the lake from the bedrock that underlies the base of the glacier. Many cells similar to the bacteria of the

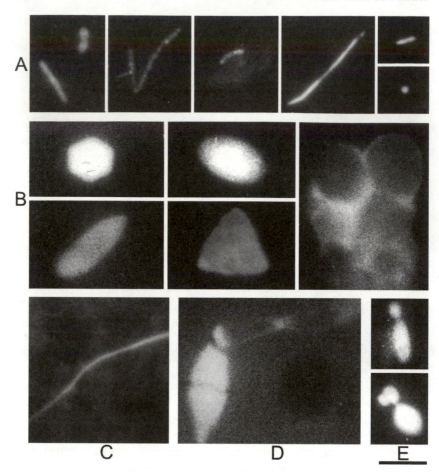

FIGURE 16.2 A–E. Microorganisms detected in the main glacier and in accretion ice by fluorescence microscopy. (A) bacterial cells commonly observed in most ice horizons; (B) cells of cyanobacteria; (C) actinomycete thread observed in ice horizon 3565 m, (D) partially lysed fungal hypha with conidium observed in ice horizon 3585 m, and (E) yeast cells observed in ice horizon 3611 m. Bar = 5 μm in all photos.

genus *Caulobacter* were found at the 3565 m ice horizon (Figure 16.3 B). They were rarely seen in the main bulk of the glacier. These bacteria might have entered this zone by horizontal movement of the glacier from the shelf zone of Lake Vostok. Alternatively, they might have moved to the shelf zone from the lake itself, from snow layers surrounding the ice "dome," or from nearby bordering shallow lakes. Microorganisms observed in the 3611 m horizon were morphologically similar to cyanobacteria, presumably to species of the genus *Pleurocapsa* (Figure 16.3 C). The phenomenon of "increased bright-

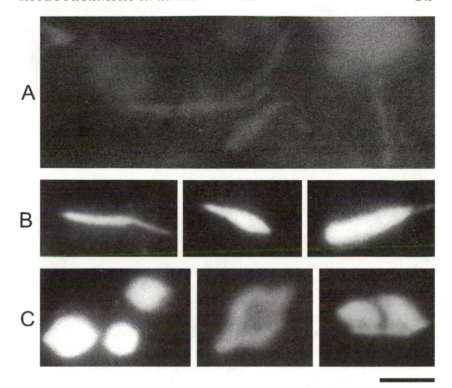

FIGURE 16.3 A–C. Bacterial cells observed in accretion ice horizons by fluorescence microscopy; (A) lysed *Cytophaga*-like bacterial cells observed in horizon 3555m; (B) *Caulobacter*-like bacterial cells observed in horizon 3565 m; (C) autofluorescence of cyanobacteria-like cells (baeocytes) resembling those of the *Pleurocapsa* group observed in horizon 3611 m. Scale bar = 5 μm.

ness," that is, an increase of light emission with time under the influence of blue and ultraviolet light, is specific for these organisms. It was the first time that we observed this phenomenon. A third group of bacteria was found only in the main part of the glacier, and never in accretion lake ice. This group consists of many large rods of different shapes, which may belong to different species of soil bacteria. There were many of them in some horizons (i.e., 161 m, 746 m, 1203 m, and 3299 m, Figure 16.4). We have observed these bacteria in earlier studies (3). We conclude that these bacteria do not reach the accretion ice zone, or if so, they are not preserved there for long. Others, however, have observed rod-shaped cells in accretion ice (see chapters 6 and 15, this volume).

Thus, we conclude that bacteria of the first and third groups are typical for the main part of the glacier, while bacteria of the first and the second groups are typical for the accretion ice.

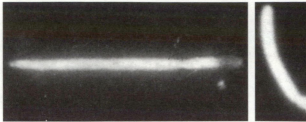

FIGURE 16.4. Large, rod-shaped cells typical of soil bacteria. Scale bar = 5 μm.

Microorganisms were transferred into glacial ice horizons together with atmospheric particulates, and then were conserved there for tens and hundreds of thousands of years. Some of the microorganisms were able to maintain cell integrity and to survive due to their anabiotic state (2). Conservation in the accretion ice horizons was not possible for all microbes because not all microorganisms can survive the periodic freeze-thaw cycles that occur. This supposition may explain the large number of cells that showed weak or nearly absent luminescence after fluorescamine treatment. Decreased luminescence is associated mainly with reduced concentrations of proteinaceous compounds within cells. Approximately 40 to 50% of bacterial cells showing weak luminescence were observed in the main glacier horizons, but this increased to 70 to 80% in accretion ice horizons. There were cocci and rods, cyanobacteria, bacteria of the genus *Cytophaga*, as well as yeast cells and fungal hyphae among the weakly luminescent organisms detected in accretion ice meltwater. Red luminescence typical of cyanobacterial pigments was absent in these cells.

In addition to bacteria, unicellular microalgae (Figure 16.5 A–D and Figure 16.6) and cells similar to the pollen of vascular plants (Figure 16.5 E–H) were detected in all samples of main glacier and accretion ice. Some were brightly luminescent and some were partially or completely lysed. Accretion ice horizons differed in the amount and diversity of these cells.

Diatoms were the predominant microalgae. There were many of them in horizons 3541 m and 3611 m. The total number of microalgae cells detected in accretion ice (without significantly degraded cells) was ten- to 100-fold lower than the number of bacterial cells. Typical microalgae detected in main glacial ice and accretion lake ice horizons are shown in Figure 16.5 A–D. Apparently, microalgae also move from the upper ice horizons (where they were brought in by atmospheric dust) to accretion ice. Representatives of the Coccolithophoridae (Figure 16.5 A), the so-called golden microalgae, were found in nearly all horizons. Their highest numbers were found in horizons 3555, 3579, and 3598 m. Single cells of other microalgae also were observed (Figures 16.5 B–D). Some microalgae were found only in accretion ice. Silica skeletons of diatomic algae never seen before were observed in meltwater

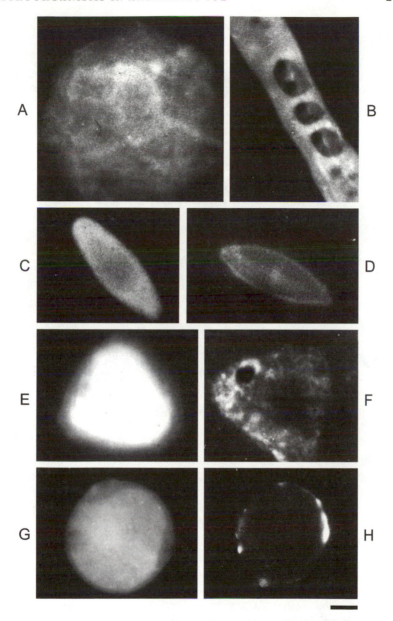

FIGURE 16.5 A–H. Eukaryotic cells commonly observed by fluorescence microscopy in different horizons of the main glacier and in accretion ice; (A) a representative of coccolithophoridea observed in most glacial ice horizons; (B) a fragment of diatom skeleton observed in ice horizons 2902 and 3611 m; (C) and (D) diatoms observed in ice horizons 952m, 1097 m, 2874 m, 2950 m, 3344 m, and 3555 m. The cell in C is brightly fluorescing and probably alive, while the cell in D is weakly fluorescing and probably dead. (E–H) possible vascular plant pollen grains detected in various main glacier and accretion ice horizons. Grains in E and G are brightly fluorescing and probably viable, whereas those in F and H are weakly fluorescing and probably dead. Bar = 5 μm.

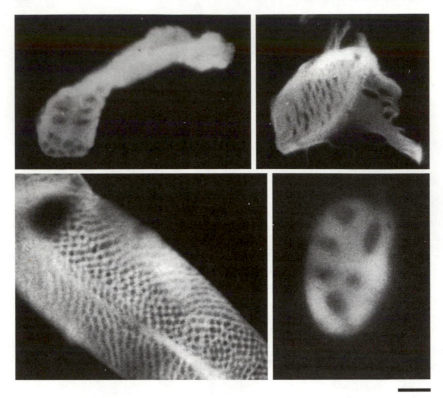

FIGURE 16.6. Skeletal fragments of different diatoms observed in ice horizons 3541–3544 m. Bar = 5 μm.

from horizons 3541 to 3544 m. Because these horizons were enriched with morainous material, we speculate that these organisms entered accretion lake ice as skeletal remains.

Conclusions

We conclude that microorganisms found in accretion ice are of mixed origins. Some of them came from glacial ice horizons, which is confirmed by the similarity of the cells detected. Other microorganisms, however, originated possibly from lateral layers of the glacier together with morainous material. Total cell numbers oscillate in accretion ice, which reflects some periodic changes in the characteristics of the ice horizons in this zone. Based on morphology alone, the microorganisms detected in accretion ice were diverse, but not as diverse as those from the main glacier above. It is theoretically possible that some microorganisms came to accretion ice from the lake itself, as the ice and water surfaces touch each other in some places. This possibility is supported by the presence in the accretion ice of microorganisms rarely found in the main glacier. Up to now, we have failed to find microorganisms with unusual morphology, but we speculate that unusual microorganisms will be detected when deeper layers of accreted ice or the lake itself is sampled.

The microbiological investigations we carried out on the vast zone of the basal stratum of the Antarctic glacier, including accreted ice, demonstrate that bacterial forms and various types of unicellular algae found in this ice are mainly represented by the morphological groups similar to those present in the upper ice sheet horizons. The basal ice horizons contain a mixture of microorganisms that could penetrate into it from the upper glacier layers owing to: (i) the ice melting in the northern part of the lake, (ii) the circulation in the southern part, and (iii) the ice accretion in the southern part, as well as with morainous material from horizontal movements of the glacier. Such a conclusion is logical, but it needs to be confirmed using the methods of molecular microbial identification and taxonomy.

The specific conditions in the accreted ice zone (temperature about −2.4 °C (10), which is considerably higher than in main part of glacier [pressure about 350 atm, at which ice can be melted]) lead us to believe that there are metabolically active and anabiotic microbial cells in this zone, as well as cells of unusual microorganisms, which can function under these conditions.

Literature Cited

1. Abyzov, S.S. 1993. Microorganisms in Antarctic ice. pp. 265–295. In E.I. Friedmann (ed.), Antarctic Microbiology. Wiley, New York.
2. Abyzov, S.S., I.N. Mitskevich, and M.N. Poglazova. 1998. Microflora of the deep glacier horizons of central Antarctica. Microbiology 67: 451–458.
3. Abyzov, S.S., I.N. Mitskevich, M.N. Poglazova, R.B. Hoover, and M.V. Ivanov. 1999. Microorganisms and unicellular algae in the ice sheet of Antarctica. Proc. Spie. 3755: 176–188.
4. Barkov, N.J., and V. Ya Lipenkov. 1995. Changes of the climate in Antarctica during last 220 thousands of years according to the investigations of the ice core from deep borehole at Vostoc station. Problemy Arctiki i Antarctiki, N69:92–107 (in Russian).
5. Jouzel, J., J.K. Petit, K. Soucher, et al. 1999. More than 200 meters of lake ice above subglacial Lake Vostok, Antarctica. Science 286: 2138–2140.
6. Kapitsa, A.P. 1961. Dynamics and morphology of the ice sheet at the central sector of the East Antarctic. Trudy (Transactions) of the Soviet Antarctic Expedition 18: 36–37.
7. Kapitsa, A.P., J.K. Kidley, G. deQ. Robin, M.J. Siegert, and J.A. Zotikov. 1996. A large deep freshwater lake beneath the ice of Central East Antarctica. Nature 381: 684–686.

8. Lipenkov, V. Ya., and N.J. Barkov. 1998. Internal structure of the Antarctic ice sheet as revealed by deep core drilling at Vostok station. pp. 31–32. Lake Vostok Study: Scientific Objectives and Technological Requirements. St. Petersburg, 22–24 March.

9. Poglazova, M.N., and I.N. Mitskevich. 1984. The use of fluorescamine for determination of the total number of microorganisms in sea water by the epifluorescent microscopy. Microbiologia 53: 850–856 (in Russian).

10. Popkov, A.M., G.A. Koudryavtsev, S.K.Verkulich, V.N. Masolov, and V.Y. Lukin. 1998. Seismic studies in vicinity of Vostok Station (Antarctica). pp. 26–27. In Lake Vostok Study: Scientific Objectives and Technological Requirements. International Workshop, St. Petersburg, 24–26 March.

11. Zotikov, Y.A. 1961. Temperature regime of Central Antarctica ice sheet. Inform. Bull. SAE, N 28: 16–22.

17

Comparative Biological Analyses of Accretion Ice from Subglacial Lake Vostok

Robin Bell, Michael Studinger, Anahita Tikku, and John D. Castello

ANCIENT ICE is produced either by the accumulation of snow (meteoric ice) or from the freezing of subglacial water onto the base of an ice sheet (accretion ice). The biological samples from these two types of ancient ice will reflect very different environments, and may be of distinctly different ages. Meteoric ancient ice recovered from deep ice cores, or in locations where the ice flow is convergent, can be over a million years old, but will generally range in age from 0 to 400,000 years old. This meteoric ancient ice will contain biota either native to the surface of major ice sheets or windblown material. Meteoric ice represents the preservation of ice over hundreds of thousands of years, while the accretion ice sampled and imaged to date is linked to ongoing processes—the freezing of lake water onto the base of major ice sheets. Thus, accreted ice derived from subglacial lakes is significantly younger (ca. 40,000 years), but will contain material derived from subglacial lakes and the subglacial water system (4). Some older accretion ice may exist at the base of the East Antarctic ice sheet downflow from the major lakes (4), but the preservation of older accreted ice is unknown. Although the accreted ice is relatively young, these subglacial lake environments may have been isolated from direct exposure to the atmosphere for 15 million years or longer, and the youngest input of material from the surface is 400,000 years old. Many lakes have been identified beneath the thick Antarctic ice sheet from ice-penetrating radar data (24). Most of these lakes, isolated beneath 3 to 4 km of ice, are several kilometers long and located beneath the major ice divides (Figure 17.1). One of these lakes, Lake Vostok, is an order of magnitude larger than the others, and represents the closest analogue to Europa and to a Neoproterozoic subglacial environment.

Lake Vostok

Lake Vostok was identified in 1996 by Russian and British scientists (12), who integrated down-hole seismic data, star sights, airborne ice-penetrating radar, and new spaceborne altimetric observations. This initial discovery has been

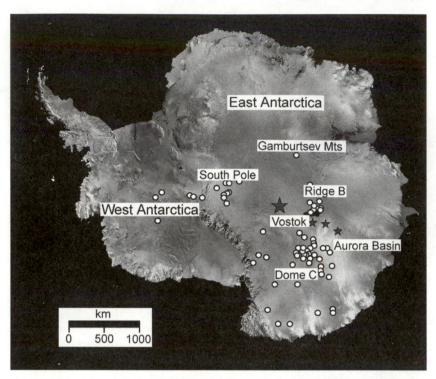

FIGURE 17.1. Advanced Very High Resolution Radiometer (AVHRR) image of Antarctica (9) with location of Lake Vostok, seismic activity, and location of other subglacial lakes (24). Large star south of Lake Vostok is the 5 January 2001 earthquake. Smaller stars are earthquakes in the Vostok region from the most recent compilation of Antarctic seismicity (Reading, in press).

extended by results from the Russian-French-American Vostok ice-coring program, new seismic experiments by the Russian Antarctic Expedition (17, 19), and most recently an aerogeophysical program supported by the U.S. National Science Foundation described here (28).

The horizontal extent of the lake is estimated from the flat surface slope (< 0.01°) observed in the ERS-1 ice surface altimetry (Figure 17.2). The 4-km-thick ice sheet goes afloat as it crosses the lake, just as ice sheets become floating ice shelves at the grounding line. The flat ice surface associated with Lake Vostok extends 280 km in the north-south direction and 50 to 60 km in the east-west direction. Over Lake Vostok the ice surface slopes from 3550 m above sea level in the north to 3480 m in the south. The slope of the ice surface is ten times smaller over Lake Vostok than in surrounding regions. The regional ice flows from an elevated feature known as Ridge B to the west (Figure 17.1) downslope to the east, although the presence of water may significantly alter

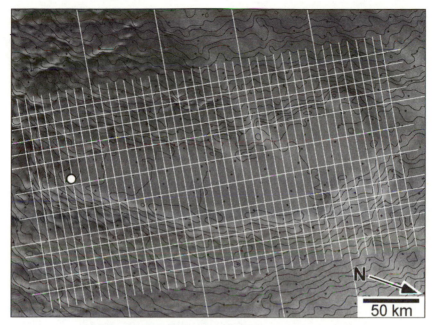

FIGURE 17.2. Grid of aerogeophysical profiles flown during the 2000–2001 field season over Lake Vostok. Basemap is RADARSAT image. Vostok station is noted as a circle. Ice surface elevation is derived from ERS-1 radar altimetry (16). Contour interval is 10 m.

this flow (23). The flow rates across Lake Vostok have been estimated from star sights at Vostok station in 1964 and 1972 (12), and synthetic-aperture radar (InSAR) interferometric methods (14). Recent global positioning system (GPS) results combined with an analysis of ice-penetrating radar yields velocities of 2 to 3 ($\pm$ 0.8) m/yr with a direction of 131° ($\pm$ 4°) against true north (4). The trajectory of the basal ice will be important to the development of the accreted ice and the environments that it represents.

Our early understanding of the 3750–4100 m of ice sealing Lake Vostok was based on limited airborne ice-penetrating radar data acquired by a joint U.S.-U.K.-Denmark program in the 1970s, and from the deep ice core drilled at the Russian Vostok station by an international team of scientists from 1989 to 1998. The early radar data, collected as part of a reconnaissance survey of Antarctica, provides cross-sectional images of the bedrock surrounding the lake, the internal layering within the ice, and the base of the ice over the lake for six flight lines (12). Across the lake the reflection from the base of the ice sheet is strong and very flat, in contrast to the reflections from portions of the ice sheet over bedrock, which is characterized by rugged reflections of varying strength and dominated by reflection hyperbolas. The radar data indicated that the water within the northern half of the lake may be very shallow (ca. 10–30 m), and that several bedrock islands protrude through the lake into the ice sheet. This analysis of the early radar data has suggested a region of rapid melting along the northwestern shoreline of Lake Vostok, and both freezing and melting over the southern reaches of the lake (25).

New Aerogeophysical Data over Lake Vostok

During the 2000–2001 Antarctic field season a series of thirty-five flights were flown over Lake Vostok by a geophysically equipped Twin Otter aircraft (28). This data set provides powerful constraints on the origin and processes that formed the accreted ice over Lake Vostok. The Lake Vostok airborne geophysical survey was designed to cover an area 157.5 km × 330 km with a grid of 45 east-west oriented lines spaced 7.5 km apart. These 157.5 km long east-west lines are constrained by twelve north-south 330 km long lines spaced in general 7.5 km apart (Figure 17.2). Over the lake some of the north-south lines are spaced at 22 km due to limited aircraft time. This high-resolution grid was supplemented by a sequence of lines extending out 200 to 400 km from the main survey area to constrain the regional setting of the lake (28, 29). These long regional lines also will be used to identify any regional subglacial flow into Lake Vostok. Four regional lines were flown both to the east and west of the lake and two long regional lines were flown both to the north and to the south. The aerogeophysical survey was provided by the Support Office for Aerogeophysical Research, a U.S. National Science Foundation Facility (2, 3, 5). The aerogeophysical survey data provide detailed images of the ice thickness, internal layers within the ice sheet, precise altimetry of the ice surface, the regional gravity field, and the regional magnetic field (28). The resulting radar records enable the detailed mapping of the internal layers as the ice flows off the ice divide, traverses the lake, and becomes grounded again to the east. Geodetic global positioning system (GPS) measurements of five locations along a radar profile crossing the lake were used to develop a new velocity vector. The ice thickness maps can be turned into subglacial topography maps in the regions surrounding the lake. We will estimate the lake depth volume and sediment volume using potential field methods.

The Vostok 5G Ice Core

The ice core at Vostok station initially was drilled to recover the record of global climate changes through the past 400,000 years. Near the bottom of the core, beginning at a depth of 3311 m, the ice first shows signs of disruption of the layering by ice dynamics, layers become tilted, and the geochemical indicators become difficult to interpret (8, 18). This layer has been interpreted as ice that was part of the continuous ice column, but has been disrupted by deformation processes between the ice sheet and underlying bedrock. The randomly distributed moraine particles in the base of this section are interpreted as an active basal shear layer. Below this layer, at 3538 m, the ice changes character significantly, with a dramatic increase in crystal size (10–100 cm) and a decrease by two orders of magnitude in the electric conductivity, the stable isotope content, and the gas content. These physical and chemical changes continue to the base of the Vostok ice core at 3623 m. This chemically distinct basal ice is interpreted to represent ice accreted to the base of the ice sheet as it passes over Lake Vostok. The upper 70 m of this basal ice includes numerous sediment inclusions approximately 1 mm in diameter. This inclusion-rich layer is interpreted to be ice accreted during repeated melting and freezing along the lake's margin (11). The distribution of inclusions is not homogeneous through this 70 m. The upper 50 m (3539–3589 m) have an average of 8.6 inclusions/ m of core while the next 6 m (3589–3595 m) are almost inclusion free (< 1 inclusion/m) and the lower 14 m (3595–3609 m) have an average of only 4 inclusions per meter of core. (Figure 17.3) Below the 70 m of ice containing inclusions, the ice is very clear and might have been accreted when the ice sheet moved across Lake Vostok. In this interpretation, the base of the ice sheet consists of a layer of 227 m of disrupted ice, 70 m of ice with varying densities of sediment inclusions, and approximately 150 m of clear accreted ice. Each of these three ice layers has a distinct origin, and will therefore represent samples of distinct environments. The 227 m of disturbed ice formed as a result of basal deformation west of Lake Vostok, the 70 m of inclusion-filled ice formed along the lake shoreline in the region with active grounding, and the 150 m of clear or gem ice formed over the open lake. The origin of the accretion ice recovered in the Vostok core will depend upon the freezing rates (4, 25) and the velocity of the ice sheet. For an accretion rate of 1 cm/year, and an ice sheet velocity of 2 m/year, each meter of accreted ice will have formed over 100 years and will represent an integrated sample of 200 m across the shoreline or the lake. The formation of accreted ice over Lake Vostok has been predicted by a variety of models. Wüest and Carmack (30) developed a simple parameterization that predicted the formation of accretion ice in the south and along the eastern shoreline of Lake Vostok. In general, the circulation within the lake and the distribution of melting and freezing within the lake is predicted to be constrained by the shape of the overlying ice sheet. Regions of thick ice cover, such as the northern lake, should be dominated by melting processes, while regions of thinner ice, such as the southern ice, will be dominated by freezing processes.

The shoreline along the Vostok core flowline is very complex, including a shallow embayment separated from the open lake by a long peninsula. The Vostok core site traverses the grounding line (shoreline) of the embayment,

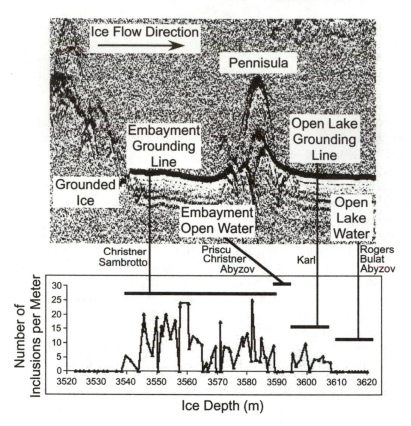

FIGURE 17.3. Location of accreted ice samples. Upper panel: radar profile across the complex western shoreline of Lake Vostok where the accreted ice samples formed. The thick, generally horizontal line is a bright reflector from the surface of the lake. The shallow embayment is in the center, and the open lake on the right. The bedrock, a rough reflector characterized by hyperbolae, is on the left, and the peninsula that separates the embayment from the open lake. The relative water depths based on the returns from below the lake surface are noted at the bottom. Lower panel: Plot of the number of inclusions with depth (Souchez et al.: triangles, and Fitzpatrick: pers. comm. NICL Web site). The location of the accreted ice samples examined for biota is noted between the panels.

the generally shallow embayment, a deeper region of water along the eastern edge of the embayment, the peninsula, and finally the grounding line (shoreline) of the open lake (Figures 17.3 and 17.4). The grounding lines are characteristically 4-km-wide areas along the western shoreline of Lake Vostok where the ice sheet sags into the lake, the water depths are shallow, and accretion of basal ice is active. Similar to the northern regions of the lake, the embayment is generally very shallow, as evidenced by the many diffractors visible beneath the strong reflector from the water, surface. The eastern edge of the embayment may contain deeper water, as suggested by the absence of reflectors from beneath the lake surface close to the peninsula.

Recently, samples of ice from the entire length of the 5G core have been made available for biological analyses. One important issue of concern to ice core biologists and the scientific community at large when evaluating biological studies of ancient ice is the apparent inability to precisely replicate the findings of other laboratories. There are many possible reasons to explain such difficulties, none of which are mutually exclusive. But, one important reason that has not been evaluated to date is that the basal ice, usually considered a vertical ice core, actually represents distinct samples of different physical locations that form a horizontal transect of the shoreline of Lake Vostok. Geographically, the upper samples were frozen onto the ice sheet ca. 33,000 years ago when the Vostok core site was just encountering the lake, while the deeper samples reflect the transition from a shoreline environment to the open lake environment ca. 11,000 years later, or ca. 21,000 years ago.

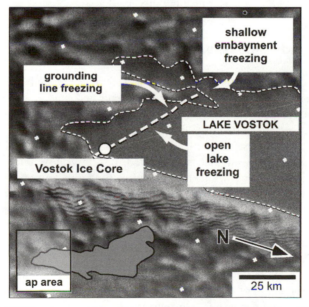

FIGURE 17.4 Different areas of freezing along the Vostok core flow line. RADARSAT image illustrates the flat ice surface over the lake. The heavy dashed line indicates the flow of the basal ice of the Vostok core (4). The thin dashed line is the shoreline or grounding line of the lake determined from reflections in ice-penetrating radar data and the flat ice surface in RADARSAT imagery.

Hypotheses

We predict that some of the variation among results of ice core studies of biota to date is attributed to the fact that different ice core samples represent different portions of the subglacial lake environment, specifically, the shallow embayment and its grounding line, the deep embayment, the open lake grounding line, and the open lake. The accretion ice at the bottom of the Vostok 5G ice core has been well characterized with respect to its origins, and recent although incomplete studies of its biota have been conducted.

These data sets provide us with the opportunity to test several hypotheses. First, ice formed over the two shallow regions will have both higher numbers of inclusions and the highest diversity and numbers of microorganisms, because these represent samples of the lake base. These shallow regions will include the central shallow portion of the embayment and its grounding line and the open lake grounding line. Second, we hypothesize that ice collected from the deeper water regions, which include the eastern deeper portion of the embayment, and the open lake region will contain the fewest numbers and diversity of microorganisms because these are samples of the more dilute water column communities. Third, we hypothesize that not all microorganisms occur uniformly in all four accretion ice regions.

Results and Discussion

The biotic data set from the Lake Vostok accretion ice to date is a first attempt to characterize the biota from this environment, and thus is understandably incomplete. Abyzov has the most detailed data regarding the numbers of cells observed in several horizons of this ice (chapter 16). His data are based on direct microscopic observation. Unfortunately, he did not specifically identify all of the microorganisms that he observed, but only those with distinctive morphologies (chapter 16). Other investigators used direct microscopy, culturing, flow cytometry, and PCR amplification, alone or in combination, to enumerate and to identify the microorganisms in this ice. Unfortunately, direct comparison of these data sets is problematic because the decontamination, isolation, and assay protocols differed, and the accretion ice samples are not identical. Nonetheless, we attempt some comparisons, and discuss these data sets (Figure 17.3, Table 17.1), while keeping these caveats in mind.

The interpretation of accreted ancient ice biota must consider the environment in which the ice was frozen onto the base of the ice sheet. From our analysis of radar data over Lake Vostok, we suggest that the accreted ice samples are derived from at least four distinct environments. The Vostok ice core stratigraphy will contain samples at different levels that represent different subglacial environments. We have identified two shallow water lake environments where accreted ice, dominated by numerous inclusions, will form. These

TABLE 17.1.

Summary of microbial diversity, numbers of cells ml[-1], and number of inclusions m[-1] observed in different horizons of the Vostok 5G deep core accretion ice

Ice Depth (m below surface)	Region of the Accretion Ice Sampled	Inclusion Number (m[-1]) (mean)	Investigator	Type of Microorganisms Detected	Total Numbers of Microorganisms Detected (cells/ml)
3541	Shallow embayment/ grounding line (3540–3584 m)	0–5 (8.6)	Abyzov[1]	Micrococci, short rods, pollen, many diatoms	< 200
3555		15	Abyzov[1]	*Cytophaga* spp. and many Coccolithophoridae	600
3565		5–10	Abyzov[1]	*Caulobacter* spp.	250
3579		5–10	Abyzov[1]	Many micrococci, short rods, and Coccolithophoridae	100
3585		0–5	Abyzov[1]	*Cytophaga* spp.	250
3587		0–5	Sambrotto[3]	Fungi, rod-shaped bacteria, diatoms, pollen grains	Not quantified
3590		0–5	Priscu[4]	*Acidovorax* sp. *Comamonas* sp. *Afipia* sp. *Actinomyces* sp.	2800–36,000
3592		0–5	Abyzov[1]	Many different kinds	900
3593	Shallow open lake grounding line (3593–3608 m)	0–5 (4)	Christner[5]	*Brachybacterium* sp. *Paenibacillus* sp. *Methylobacterium* sp. *Sphingomonas* sp.	Not quantified
3598		5–10	Abyzov[1]	Many Coccolithophoridae	500
3603		5–10	Karl[6]	Gram negative (?) rods, cocci, and vibrios	200–300
3606		5–10	Abyzov[1]	Many different kinds	700
3611	Open lake (3608–3623 m)	0 (0)	Abyzov[1]	Mostly diatoms and cyanobacteria	100
3619		0	Rogers and Bulat[2]	*Serratia liquefaciens Comamonas acidovorans Rhodoturla creatinovora Kocuria* sp. *Friedmanniella antarctica Hydrogenophlus theroluteolus*	Not quantified

[1] Data from Abyzov et al., Chapter 16. Techniques used included light microscopy and SEM. Coccolithophoridae, cyanobacteria, actinobacteria, rod and coccus-shaped bacteria, and fungal conidia and hyphae were detected in all ice horizons examined.

[2] Data from S.O. Rogers and S. Bulat, unpublished. Techniques included culturing and PCR amplification.

[3] Data from Sambrotto and Burckle, Chapter 6. Techniques included culturing and microscopy.

[4] Data from Priscu, et al. (20). Techniques included microscopy and PCR amplification.

[5] Data from Christner, et al. Chapter 15. Techniques included microscopy and PCR amplification.

[6] Data from Karl, et al. (13). Techniques included microscopy, flow cytometry, and ^{14}C uptake.

are the shallow embayment along the western edge of the lake where the water is probably less than 15 m deep, including the embayment grounding line, and the grounding line (shoreline) of the main lake in the west (Figure 17.3). The deeper water environments include the open lake and the region along the eastern edge of the embayment, known as the deep embayment (Figure 17.3).

The Vostok ice core contains ca. 50 m of ice from the shallow embayment grounding line/shoreline region, which extends from 3540 to 3584 m. The deep embayment ice extends from 3584 to 3593 m, the open lake grounding line ice extends from 3593 to 3608 m, and ice below 3608 m represents the open lake environment. By knowing that the ice velocities average ca. 2 m/year, we estimate that all the existing accreted ice samples were accreted to the base of the ice sheet 21,000 to 33,000 years ago (4).

We hypothesized that ice accreted over shallow water will have both higher numbers of inclusions and the highest diversity and numbers of microorganisms than ice accreted over deeper or open lake water because samples from the more shallow regions represent the benthic environment and sediments of the lake.

This hypothesis is only partially correct. For consistency, we utilized only Abyzov's enumeration data for this analysis. The deep embayment (3584–3593 m) contains fewer inclusions (mean = < 1 inclusion/m) than either the shallow open lake grounding line (3593–3608 m, mean = 4 inclusions/m) or the shallow embayment and its grounding line (3540–3584 m, mean = 8.6 inclusions/m) (26), which contained the greatest number of inclusions (Figure 17.3). However, the number of cells detected per ml of meltwater was similar in ice from the deep embayment and the shallow open lake grounding line (mean of 575 cells/ml and 600 cells/ml, respectively), but greater than in the shallow embayment/grounding line ice (mean of 288 cells/ml). Open lake ice (> 3608 m) had no detectable inclusions, yet Abyzov observed approximately 100 cells/ml in meltwater from this ice (chapter 16).

An outstanding issue is the nature of the source environment of the accreted ice. Interpretation of the types, numbers, and diversity of microorganisms within the accreted ice will depend in part upon whether the ice formed over the deep open lake, or in a location with restricted exchange with the deep lake such as the grounding lines at western shoreline or the shallow embayment, which parallels the western shore. Accreted ice from the open lake and the deep embayment should reflect the diversity and nature of the lake biota. In contrast, accreted ice formed along the shoreline/grounding lines should represent samples of the benthic community or sediments of the lake. The accreted ice in the base of the Vostok ice core reflects a traverse of these different environments from the shoreline to the open lake, and thus strict uniformity and reproducibility of ice core studies of biota are probably unlikely.

The Vostok accretion ice contains a diverse range of microbes, including algae, diatoms, bacteria, filamentous fungi, yeasts, and actinobacteria (1, 8). Karl et al. (13), Priscu et al. (20), Christner et al. (chapter 15), Sambrotto and Burckle (chapter 6), Rogers and Bulat (unpublished), and Abyzov et al. (chapter 16) have identified microorganisms within the accreted ice in the Vostok 5G core (Table 17.1). Abyzov observed fungal hyphae and conidia, diatoms, Coccolithophorid algae, coccus and rod-shaped bacteria, cyanobacteria, and pollen grains in all horizons/layers of accreted ice that he examined, regardless of its origins (chapter 16). But in addition to these observations, microorganisms not present in all ice layers/regions were detected and identified. These are detailed below in an attempt to determine if distinct microorganisms occur in specific regions of accretion ice.

Shallow Embayment Grounding Line/Shoreline

Ice accreted over the shallow embayment grounding line (3540–3584 m) had the greatest number of inclusions (mean of 8.6/m, Figure 17.3 and Table 17.1). The inclusions and thus the biota detected here may be associated with subglacial soils and sediments, the embayment water column, or the benthic environment of Lake Vostok. The embayment is isolated from the main lake by a shallow sill, and subsequently there may be little exchange with the open lake. The analyses done to date on the Vostok accretion ice from this region represent the work of Abyzov (Table 17.1 and chapter 16). Microorganisms detected by Abyzov on the basis of their distinctive morphologies include the bacteria *Caulobacter* spp. (α-proteobacteria) and *Cytophaga* spp. (phylum Bacteroidetes) (Table 17.1 and chapter 16, this volume).

Deep Embayment Region

In ice accreted over the deep embayment Abyzov, Priscu, Rogers, and Sambrotto each detected microorganisms (Table 17.1). Sambrotto observed fungi, rod-shaped bacteria, marine and freshwater diatoms, and pollen grains in ice from 3587 m horizon (chapter 6 and Table 17.1). Priscu counted 2800 to 36,000 cells/ml of α- and β-proteobacteria and actinobacteria in ice from the 3590 m horizon (ref. 21 and Table 17.1). Specifically, Priscu amplified DNA sequences closely related to bacteria in the genera *Acidovorax* spp. and *Comamonas* spp. (β-proteobacteria), *Afipia* spp. (α-proteobacteria), and *Actinomyces* spp. (actinobacteria). Rogers, using culture methods, isolated one yeast (*Rhodotorula creatinovora*), two bacteria (*Kocuria* sp. *Friedmanniella antarctica*), and two unidentified bacteria from ice at the 3619 m horizon. Bulat also identified the bacterium *Hydrogenophilus thermoluteolus* from this horizon (Table 17.1). Abyzov observed cells of *Cytophaga* spp.

Shallow Open Lake Grounding Line/Shoreline

The samples from the shallow open lake grounding line are those from 3593 to 3608 m. The water in this region appears to be shallow although not as shallow as the narrow embayment water. These samples contain some inclusions (mean of 4 inclusions/m), although fewer than the grounding line in the shallow embayment (Figure 17.3 and Table 17.1). The samples of Karl et al. (13), Christner et al. (chapter 15), and Abyzov et al. (chapter 16) are from this region. Karl et al. (13) detected vibrios, micrococci, and thin, rod-shaped bacteria that showed some features of Gram-negative bacteria. Abyzov observed many Coccolithophorid algae, and Christner amplified DNA sequences closely related to bacteria in the following genera: *Brachybacterium* sp. (actinobacteria), *Sphingomonas* sp. (α-proteobacteria), *Methylobacterium* sp. (β-proteobacteria), and *Paenibacillus* sp. (Gram-positive endospore-forming rod).

Open Lake Ice

The open lake ice is the inclusion-free ice below 3608 m. This ice will be the youngest ice of the samples recovered in the Vostok core. This ice was accreted from water that has been in the lake for approximately 15,000 years (4). Estimates of lake circulation by Wüest and Carmack (30) suggest the water column should be well mixed at these timescales. Abyzov et al. (chapter 16) and Bulat and Rogers (unpublished) (Table 17.1) evaluated samples from these depths. Abyzov observed mostly diatoms and cyanobacteria in ice at a depth of 3611 m. Bulat and Rogers amplified bacterial DNA most closely related to bacteria in the: β-proteobacteria (i.e., *Comamonas acidovorans*) and γ-proteobacteria (i.e., *Serratia liquefaciens* complex) in ice from the 3619 m horizon.

Is there a relationship between these four specific regions of accretion ice and the types/diversity of microorganisms found in them? Abyzov detected fungi, diatoms, Coccolithophorid algae, cyanobacteria, micrococci, and short rods in all accretion ice horizons that he examined.

Yet some microorganisms were detected in only one or a few specific accretion ice regions. For example, *Caulobacter* spp. were observed by Abyzov only in ice from the shallow embayment grounding line. These bacteria commonly occur in water with very low nutrient levels, and on the surfaces of diatoms, algae, and sediments. They are Gram-negative, aerobic, chemoorganotrophs that colonize surfaces in oligotrophic environments. They do not occur free in a water column, but attached to various surfaces. Bacteria in the genus *Cytophaga* were detected in the shallow embayment as well as in the deep embayment ice. These bacteria are Gram-negative, aerobic, chemoorganotrophic, small, rod-shaped organisms that occur in marine and freshwater environments and sediments, and exhibit gliding motility. They are especially common in low-temperature freshwaters. In addition, *Comamonas* spp. were identified in both deep water environments (the deep embayment and the open lake regions), but not in the shallow environments (Table 17.1).

The specific bacteria identified in the two shallow water environments by Christner and Abyzov (Table 17.1 and chapters 15 and 16, this volume) include species in the genera *Cytophaga, Caulobacter, Sphingomonas, Brachybacterium, Paenibacillus*, and *Methylobacterium*. Speicies of these genera are typical inhabitants of sediments, soils, and freshwater/marine environments..

The microorganisms identified in the two deep water regions by Abyzov, Bulat, Priscu, and Rogers include species in the following genera (ref. 21, and chapter 16 and Table 17.1, this volume): *Rhodotorula* sp., *Serratia liquefaciens, Cytophaga* sp., *Acidovorax* sp., *Comamonas* sp., *Actinomyces* sp., and *Afipia* sp. Some of these we might expect to find in deep, cold, oligotrophic freshwaters, including *Rhodotorula* spp., which are ubiquitous basidiomycetous yeasts found in many different habitats around the world. They are the most common yeasts isolated from cold environments (chapter 12). Bacteria in the genus *Cytophaga* typically are found in aquatic habitats. *Comamonas* and *Acidovorax* are γ-proteobacteria that have been reported from soil and water, as well as from assorted clinical specimens. They may possibly represent contamination at some stage of ice core handling and processing. Bacteria in the *Serratia liquefaciens* complex are members of the Enterobacteriaceae, which are bacteria that occur in the digestive tracts of mammals although they commonly are isolated from soil and water exposed to feces or sewage at some time. Similarly, bacteria in the genera *Afipia* and *Actinomyces* are human and animal pathogens. It seems probable that these organisms represent contaminants picked up during the drilling, shipment, decontamination, and/or assay of the ice cores.

Conclusions

First, there is no apparent direct association between inclusion number, cell number, and accretion ice origin. The numbers and distribution of the biota detected to date in Lake Vostok accretion ice must be related to factors other than just inclusion number. Sediment inclusions may not be the major location of microorganisms in the lake ice. Second, similar microorganisms are present in all accretion ice horizons, regardless of their origin. These include the observations of Abyzov, who has now examined the entire 3623 m length of the Vostok deep core from top to bottom, and are corroborated by others (chapter 16 and references therein). Third, some microorganisms were detected in only one or a few of the four accretion ice regions. Most, but not all, of these microorganisms occur naturally in soils, freshwater and marine environments, and sediments. For example, *Cytophaga* spp. were detected in both the shallow embayment/grounding line and the deep embayment, but not in the other two accretion ice regions. Species of *Caulobacter* were detected only in the shallow embayment/grounding line region. *Acidovorax* sp. was detected only in the deep embayment, whereas *Comamonas* spp. were detected in both the deep embayment and the open lake ice. So there might be some relationship between microbial diversity and the physical environment in which the accretion ice formed.

Fourth, some of the microorganisms detected in ice may be contaminants. The ice core decontamination protocols utilized by some of the investigators were less than completely effective when tested in systematic comparative studies (see chapters 2 and 20 for details). Independent corroboration of the presence of such microorganisms in ice must be conducted to verify these reports. Fifth, although there are no major differences in the diversity of microbial taxa detected in the shallow and deep water environments represented in the accretion ice from Lake Vostok, especially if the potential likely contaminants are excluded from consideration, the distribution of some of the microorganisms reflects the varying habitats inferred from the radar data. Specifically, *Caulobacter* spp., chemoorganotrophs that colonize surfaces, were only observed from the shallow embayment grounding line. *Acidovorax* sp. and *Comamonas* sp., found in soils and waters, were found only in the two deep water habitats. The *Cytophaga*, characterized by gliding motility and common in low-temperature freshwater environments, were found in both shallow and deep water embayment environments. The correlation of these microorganisms with the varying habitats supports the concept that some of the microorganisms identified from the Vostok ice core represent samples of the subice environment, and are not the result of contamination.

Workers have used different methods to decontaminate, detect, and identify the biota within the ice cores. The use of different methodologies biases what has been/can be detected in ice. Nevertheless, generally similar microorganisms have been detected by different workers. Most of the specific microbial identifications are based on comparisons of DNA sequences obtained from meltwater to the most closely related sequence in international sequence databases such as GenBank. Such comparisons do not mean that the DNA from the organism amplified from ice is identical or necessarily even closely related to the organism whose DNA is recorded in the database. Lastly, the microbial biota detected in different accretion ice layers/horizons by the various workers are, in the final analysis, probably more similar than previously believed. This last conclusion supports the contention that many but certainly not all of the microorganisms detected in accretion ice are not the result of contamination, but rather may provide a partial but accurate representation of the subterranean Lake Vostok microbial community. It also underscores the need for a standardized set of decontamination and assay protocols to be used by all those working with ancient ice in order to provide independent corroboration of results (see chapters 2 and 20 for details).

Literature Cited

1. Abyzov, S.S. 1993. Microorganisms in Antarctic ice. pp. 265–295. In E.I. Fried-mann (ed.), Antarctic Microbiology. Wiley, New York.
2. Behrendt, J.C., D.D. Blankenship, C.A. Finn, R.E. Bell, R.E. Sweeney, S.M. Hodge, and J.M. Brozena. 1994. CASERTZ aeromagnetic data reveal late Ceno-zoic flood basalts(?) in the West Antarctic rift system. Geology 22: 527–553.
3. Bell, R.E., V.A. Childers, R.A. Arko, C. Guillemot, D.D. Blankenship, and J.M. Brozena. 1999. Airborne gravity and precise positioning from a multi-instru-mented aircraft. J. Geophys. Res. 104 (B7): 15281–15292.
4. Bell, R. E., M. Studinger, A.A. Tikku, G. Clarke, M.M. Gutner, and C. Meertens. 2002. Origin and fate of Lake Vostok water accreted to the base of the East Antarctic ice sheet. Nature 416: 307–310.
5. Blankenship, D.D., D.L. Morse, C.A. Finn, R.E Bell, M.E. Peters, S.E. Kempf, S.M. Hodge, M. Studinger, J.C. Behrendt, and J.M. Brozena. 2001. Geologic controls on the initiation of rapid basal motion for the West Antarctic ice streams; a geophysical perspective including new airborne radar sounding and laser alti-metry results. Antarctic Res. Ser. 77: 105–121.
6. Crane, K., B. Hecker, and V. Golubev. 1991. Hydrothermal vents in Lake Baikal. Nature 350: 281.
7. Duval, P., V. Lipenkov,, N.I. Barkov, and S. de La Chapelle. 1998. Recrystallization and fabric development in the Vostok ice core. Supplement to EOS Trans., Amer-ican Geophysical Union, Fall Meeting 1998, 79(45): F152.
8. Ellis-Evans, J. C., and D. Wynn-Williams. 1996. A great lake under ice. Nature 381: 644–646.
9. Ferrigno, J. G., J.L. Mullins, J.A. Stapleton, P.S. Chavez, M.G. Velasco, R.S. Wil-liams, G.F. Delinski, and D. Lear. 1996. Satellite image map of Antarctica. United States Geological Survey, Miscellaneous Investigation Map Series 1: 2560.
10. Jean-Baptiste, P., J.R. Petit, V.Ya Lipenkov, D. Raynaud, and N.I. Barkov. 2001. Constraints on hydrothermal processes and water exchange in Lake Vostok from helium isotopes. Nature 411: 460–462.
11. Jouzel, J., J.R. Petit, R. Souchez, N.I. Barkov, V.Ya. Lipenkov, D. Raynaud, M. Stievenard, N.I. Vassiliev, V. Verbeke, and F. Vimeux. 1999. More than 200 m of lake ice above the subglacial Lake Vostok, Central East Antarctica. Science 286: 2138–2141.
12. Kapitsa, A.P., J.K. Ridley, G.d Q., Robin, M.J. Siegert, and I.A. Zotikov. 1996. A large deep freshwater lake beneath the ice of central East Antarctica. Nature 381: 74–76.
13. Karl, D.M., D.F. Bird, K. Björkman, T. Houlihan, R. Shackelford, and L. Tupas. 1999. Microorganisms in the accreted ice of Lake Vostok. Science 286: 2144–2147.
14. Kwok, R., M.J. Siegert, and F.D. Carsey. 2000. Ice motion over Lake Vostok, Ant-arctica: Constraints on inferences regarding the accreted ice. J. Glaciol. 46 (155): 689–694.

15. Leitchenkov, G.L., S.R. Verkulich, and V.N. Masolov. 1998. Tectonic setting of Lake Vostok and possible information contained in its bottom sediments. In Lake Vostok Study: Scientific Objectives and Technological Requirements, International Workshop, 24–26 March, St. Petersburg, Russia, Arctic and Antarctic Research Institute.

16. Liu, H.X., K.C. Jezek, and B.Y. Li. 1999. Development of an Antarctic digital elevation model by integrating cartographic and remotely sensed data: A geographic information system based approach. J. Geophys. Res. 104 (B10): 23199–23213.

17. Masolov, V.N., V.V. Lukin, A.N. Sheremetiev, and S.V. Popov. 2001. Geophysical investigation of the subglacial Lake Vostok in Eastern Antarctica. Doklady Earth Sciences 379A(6): 734–738.

18. Petit, J.R., I. Basile, J. Jouzel, N.I. Barkow, V. Ya Lipenkov, R.N. Vostretsov, N.I. Vasiliev, and C. Rado. 1998. Preliminary investigations and implications from the 3623 m Vostok deep ice core studies. In Lake Vostok Study: Scientific Objectives and Technological Requirements, International Workshop, 24–26 March, St. Petersburg, Russia, Arctic and Antarctic Research Institute.

19. Popkov, A.M., G.A. Kudryavtsev, S.R. Verkulich, V.N. Masolov, and V.V. Lukin. 1998. Seismic studies in the vicinity of Vostok station (Antarctica), In, Lake Vostok Study: Scientific Objectives and Technological Requirements, International Workshop, 24–26 March, 1998, St. Petersburg, Russia, Arctic and Antarctic Research Institute.

20. Priscu, J. C., C.H. Fritsen, E.E. Adams, S.J. Giovannoni, H.W. Paerl, C.P. McKay, P.T. Doran, D.A. Gordon, B.D. Lanoil, and J.L. Pinckney. 1998. Perennial Antarctic lake ice: An oasis for life in a polar desert. Science 280: 2095.

21. Priscu, J., E.E. Adams, W.B. Lyons, M.A. Voytek, D.W. Mogk, R.L. Brown, C.P. McKay, C.D. Takacs, K.A. Welch, C.F. Wolf, J.D. Kirshtein, and R. Avci. 1999. Geomicrobiology of subglacial ice above Lake Vostok, Antarctica. Science 286: 2141–2144.

22. Reading, A. M. 2003. Antarctic seismicity and neotectonics. In J.A. Gamble, D.N.B. Skinner, S. Henrys, and R. Lynch, (eds.). Antarctica at the Close of a Millennium. Proceedings, 8th International Symposium on Antarctic Earth Sciences, Royal Soc. of New Zealand Bull. 35: 479–484.

23. Robin, G. de Q. 1998. Discovery and ice dynamics of the Lake Vostok region. In Lake Vostok Study: Scientific Objectives and Technological Requirements, International Workshop, 24–26 March, St. Petersburg, Russia, Arctic and Antarctic Research Institute.

24. Siegert, M.J., J.A. Dowdeswell, M.R. Gorman, and N.F. McIntyre. 1996. An inventory of Antarctic subglacial lakes. Ant. Sci. 8: 281–286.

25. Siegert, M.J., R. Kwok, C. Mayer, and B. Hubbard. 2000. Water exchange between the subglacial Lake Vostok and the overlying ice sheet. Nature 403: 643–646.

26. Souchez, R., J.R. Petit, J.-L. Tison, J. Jouzel, and V. Verbeke. 2000. Ice formation in subglacial Lake Vostok, Central Antarctica. Earth and Planetary Sci. Letters 181: 528–531.

27. Souchez, R., J.R. Petit, J. Jouzel, M. de Angelis, and J.-L. Tison, 2003. Reassessing Lake Vostok behavior from existing ice core data. Earth and Planetary Sci. Letters 6889: 1–8.

28. Studinger, M., V. Levin, R. Bell, A. Lerner-Lam, W.-X. Du, and A. Tikku. 2001. First results of a short-term seismic experiment on the East Antarctic Plateau at Vostok. EOS Trans., American Geophysical Union, Spring Meeting 2001, 82(20): S273.

29. M. Studinger, R.E. Bell, G.D. Karner, A.A. Tikku, J.W. Holt, D.L. Morse, T.G. Richter, S.D. Kempf, M.E. Peters, D.D. Blankenship, R.E. Sweeney, and V.L. Rystrom. 2003. Ice cover, landscape setting, and geological framework of Lake Vostok, East Antarctica. Earth and Planetary Sci. Letters 205(3–4): 195–210.

30. Wüest, A., and E. Carmack. 2000. A priori estimates of mixing and circulation in the hard-to-reach water body of Lake Vostok. Ocean Modelling 2: 29–43.

18

Search for Microbes and Biogenic Compounds in Polar Ice Using Fluorescence

Ryan Bay, Nathan Bramall, and P. Buford Price

GLACIAL ICE provides a time machine from which can be extracted organisms with known age, whose response to extreme cold and lack of nutrients can be studied. An unsolved problem in microbiology is the length of time, in conditions of extreme temperatures and nutrient deprivation, microorganisms of various types can remain viable. No one has yet carried out a thorough study of the depth dependence of types of microbes, nor of the fractions that are metabolizing, dormant, or dead. There is a great international interest in eventually acquiring samples of water or sediment from a subglacial lake such as Lake Vostok, in which to search for microbes isolated from the surface for millions of years. A better understanding of how microbes arise in and survive hostile environments would have implications for the origin of life (cold versus hot), for the possibility of life on other planets, and for insight into the interface between life and death. Noninvasive yet powerful techniques would be of undeniable use in this fledgling enterprise.

Autofluorescence in Biology

Fluorescence is emission of a longer wavelength photon after excitation by a shorter wavelength photon. Absorption of the incident radiation occurs on a timescale of ca. 10^{-15} seconds, and the molecule typically remains in the excited state on the order of 10^{-8} seconds. Emission continues only during exposure to the exciting radiation, distinguishing the phenomenon from the persistent emission of phosphorescence. Both the excitation and absorption spectra, and possibly the fluorescent decay time-history, are unique characteristics of a given molecule, endowing the phenomenon with great discerning power. An advantage of working with fluorescence in ice is that the lower collisional and photochemical quenching at cold temperatures means a higher fluorescence quantum efficiency.

When used in biology, autofluorescence is that emitted by naturally occurring cellular compounds without introduction of a dye. More common in modern biological sciences is the introduction of specific fluorescent stains,

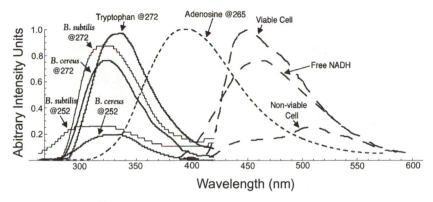

FIGURE 18.1. Fluorescence spectra of some biological autofluorescing compounds, as well as viable and dead microbial cells.

which greatly boost signal and provide diagnostic power, for instance, by isolating certain components of a cell or revealing its metabolic state. Other techniques involve the complexing of an antigen with a target species to produce a unique signature. Staining is impractical in the case of a remote borehole logger, and we hope initially to avoid the use of dyes in any core study because they exacerbate contamination issues. Recent developments in bioagent warning sensors have demonstrated the feasibility of remote detection of biological autofluorescence (6). A pulsed laser irradiates a sample flowthrough volume of air, which then is observed at several wavelengths or over a continuous spectrum. Several groups have achieved sensitivity to single micron-scale biological aerosols using only cellular autofluorescence.

Figure 18.1 gives an overview of a few of the more prominent autofluorescing biogenic compounds as well as fluorescence in living versus dead microbial cells. Tryptophan, which makes up roughly 1% of proteins, fluoresces strongly at neutral pH, with a peak around 330 nm in free form, and blue-shifted about 20 nm when bound in a protein. NADH, an important electron carrier in metabolism, has a moderate fluorescence emission peak at ca. 450 nm, and also is slightly blue-shifted in bound form. The molecule fluoresces only when reduced, not when oxidized ($NADP^+$), so it can be instrumental in tracing cellular metabolic activity. Flavin compounds also have a weak fluorescence, peaking near 550 nm when excited at ca. 460 nm. Polycyclic aromatic hydrocarbons (PAHs) fluoresce at a number of wavelengths, and could be a source of background in searches for viable organisms. Factor 420 (F_{420}), so named due to its excitation at that wavelength, is a clear signal of methanogenic archaea. Certain nucleosides (e.g., adenosine, shown) fluoresce strongly at low pH, and may provide a signature of dead acidophiles whose membranes are no longer able to exclude H^+ ions.

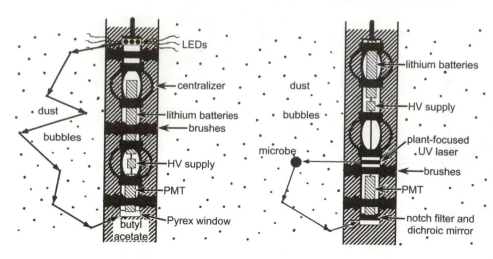

FIGURE 18.2. *Left*: Dust-logger-1, tested in November 2000 in a Siple Dome bore-hole filled with a transparent butyl acetate. Photons emitted from LEDs (370 nm) scatter from dust or bubbles, and some are absorbed by dust. A small fraction returns to the borehole and enters the downward-directed phototube 1 m below, providing a measure of dust concentration. *Right*: Biologger-1 modified to detect biota and biomolecules using filter fluorometry.

Paleoclimatology with Dust-logger-1

We have successfully tested a downhole logger designed to resolve past varia-tions in atmospheric composition and climate in polar ice (2, 4). The dust-logger is a by-product of years of experience in the use of light sources by the AMANDA (Antarctic Muon and Neutrino Detector Array) collaboration to measure optical properties of deep Antarctic ice (1, 3). Emitters and receivers buried at depths of ca. 1 to 2 km in ice and at lateral distances of ca. 0.1 km probe the scattering and absorption of light by air bubbles and dust grains. The dust-logger compresses this instrumentation from a volume of ca. 0.1 km^3 into a compact device that fits into a borehole. A light source shines into the ice surrounding the borehole; photons scatter off bubbles and dust, some are ab-sorbed on dust, and a small fraction scatter back into the borehole and are detected in a downward-looking photomultiplier tube (PMT) below. In ice in which scattering is dominated by bubbles, the absorption from dust impurities is perceived as a drop in signal; in bubble-free ice the scattering from dust increases the light collected.

Dust-logger-1 (Figure 18.2, left) consists of three radial 370 nm Nichia NSHU550E light-emitting diodes (LEDs) 120° apart, 1 m above a downward-looking PMT (Hamamatsu R6094). Centralizers keep the optics steady, and black one-inch-thick nylon brushes slightly wider than the hole seal out scat-

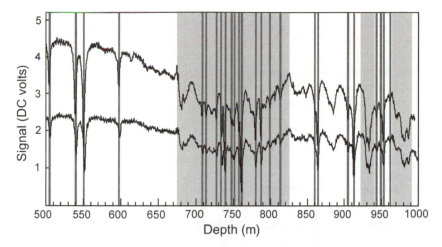

FIGURE 18.3. Raw data taken in Siple Dome hole A. Vertical lines designate sharp spikes due to thin, highly absorbing layers, most of volcanic origin. The shaded regions are long-term climatic changes.

tered light within the borehole and ensure that photons reaching the PMT have sampled the exterior ice (this also can be accomplished with a pulsed source and electronics). The entire unit is constructed from black oxidized stainless-steel pipe. Light is transmitted through Pyrex glass windows sealed with O-rings. The device is tethered and controlled via a four-conductor logging cable. The LEDs are powered from a current supply at the surface through one of the cable pairs. A miniature high-voltage supply for the PMT is powered by on-board lithium-oxyhalide batteries designed for large capacity at cold temperatures. The third wire is used to control the output of the HV supply, and the fourth returns the output current of the PMT anode. The armor of the cable is employed as a fifth conductor for signal and HV-control ground. The PMT output current is dumped across a selectable parallel RC circuit and unity gain amplified and inverted at the surface. This signal voltage is recorded by an ADC connected to a laptop computer. For each voltage measurement, the computer also queries an optical payout encoder on the winch and records depth. This depth is later corrected for cable elongation so that the readings are accurate to ± 2 m.

Figure 18.3 shows the raw signal taken in December 2000 with Dust-logger-1 in the 1000 m Siple Dome Hole A. The top curve is the signal while logging downhole; the bottom curve is taken while logging uphole when the bottom set of brushes are swept downward, obscuring the PMT. The signal drops gradually with depth as scattering from bubbles decreases with growing hydrostatic pressure. Broad depressions in the signal extending over many meters (shaded region) correspond to two long-term climatic changes, the Last Glacial

Maximum between ca. 675 and 825 m, and the 65 kyr peak seen in the dust record at depth ca. 930 m. Monte Carlo simulation verifies that the recurrent, narrow, two-pronged spikes are to be expected when the logger encounters highly absorbing layers localized to ca. 1 cm thick, such as those resulting from nearby volcanic eruptions. Volcanic ash depositions were visually confirmed over portions of the archived ice core, which remained sufficiently intact, and some thirty-five layers are clearly discernible over the length of the borehole.

Fluorescence Photometry

The simplest fluorescence detector is a PMT masked by a dichroic mirror and notch interference filter, which can count photons emitted in specific fluorescence wavebands. Filter fluorometry is relatively simple, and permits the detection of low light levels. It is difficult, however, to separate the fluorescent signal from backgrounds, for example, Mie-scattered and Raman-shifted excitation light, or extraneous fluorescence. Fluorescent dyes greatly boost signal levels, but detection of autofluoresence requires very high signal to noise, and therefore very good wavelength specificity. Gel filters are inadequate; high-quality interference filters are a necessity. Most off-the-shelf filter sets, including those intended for (stained) fluorescent microscopy, allow excessive "bleed-through," that is, excitation light that is transmitted through both the excitation filter and the emission filter and is accounted as signal. Many filters are constructed from materials which themselves fluoresce, and there is a trade-off between the ability of a filter to block UV wavelengths and its tendency to exhibit fluorescence. Achieving sufficient background suppression requires careful optimization of wavelengths and optical components.

Figure 18.4 (top) shows one experimental setup we are using to tune our optics. UV excitation light from a low-pressure mercury lamp is transmitted through a notch filter that selects a prominent line at 254 nm. This shines on a cuvette filled with varying concentrations of *E. coli*. The light emitted from the cuvette is measured with a PMT screened by a dichroic mirror and a 353 nm notch filter to isolate fluorescence from tryptophan. A wave chopper and phase-locked amplifier reduce electronic noise. Figure 18.4 (bottom) shows the resulting approximately linear relationship between cell concentration and signal. In this setup, we were able to detect 10^5 cells per ml before we reached a noise floor, probably due to filter bleedthrough and extraneous fluorescence. We are currently testing improved filters with the hope of detecting 10^3 cells per ml. Figure 18.2 (right) is a schematic of Biologger-1, which is a version of Dust-logger-1 modified to detect biogenic fluorescence in glacial boreholes by filtering the returned light.

Spectrofluorometry

In spectrofluorometry, the detector is a spectrograph, such as a diffraction grating combined with a linear charge-coupled device (CCD) array. It has the advantage that an entire emission spectrum can be captured; signal can be discriminated from noise and different fluorescence signals can be measured simultaneously and distinguished. This added information comes at the expense of increased complexity of the system, and data acquisition must be mediated by a CPU. A cluster of PMTs with individual notch filters is superior in collection power to a bona fide spectrum analyzer. Figure 18.5 is a prototype of Biospectrologger BSL-266. A passively Q-switched 266 nm laser is focused into a plane by a convex conical mirror, and this light is transmitted through a silica window, which may need to be nearly as wide as the borehole if the fluid in it is opaque at this wavelength. Monte Carlo calculation shows that with a source pulse of 10^{17} photons sec^{-1} and 10^3 cells per ml, we can expect a return flux of 10^8 photons sec^{-1} incident on a 3-inch diameter collecting surface.

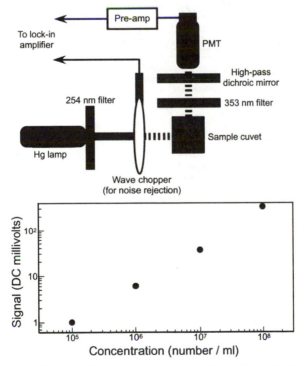

FIGURE 18.4. *Top*: Example of experimental setup used to select optimal wavelengths and optical components. *Bottom*: Signal versus concentration for tryptophan fluorescence from live *E. coli*.

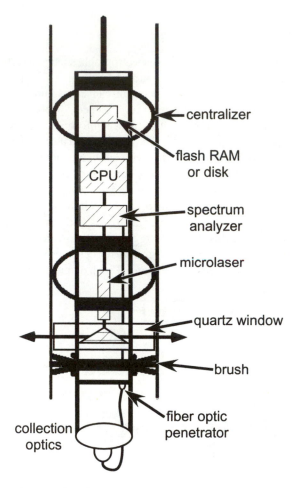

FIGURE 18.5. Prototype Biospectrologger BSL-266.

The light then is focused into a fiberoptic cable that is fed into a miniaturized spectrograph. Data are transmitted serially to the surface in real time or written to onboard flash memory or hard drive for analysis upon retrieval.

Epifluorescence Microscopy in a Cold Box

Price (7) has proposed a habitat for acidophilic microbes in the very low pH liquid veins excluded into the three-crystal boundaries between the grains in deep glacial ice. We have been allotted sections of the Vostok ice core at various depths, which have been visually identified to contain veins. We are cur-

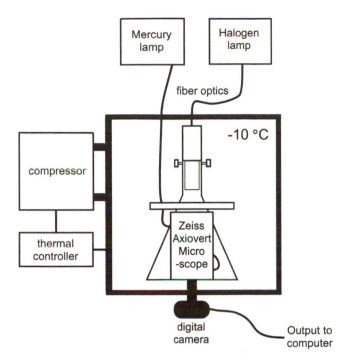

FIGURE 18.6. Schematic of our Zeiss axiovert microscope incorporated into a cold box.

rently instrumenting an epifluorescence microscope to fit into a cold box (Figure 18.6). The excitation (low-pressure mercury) and bright field (halogen) light sources, as well as the CCD camera and acquisition PC, must be kept at room temperature. The scope was constructed with grease that maintains low viscosity at cold temperatures. Our intention is to isolate veins by slicing the core around them into < 1 cm thick sections and using long-working distance objectives to locate organisms by their fluorescence. An advantage of this novel approach is to determine locations of microorganisms in unmelted ice. We are considering boosting signal with stains by placing a drop at the intersection of a vein with the surface and allowing the stain to diffuse along the vein, perhaps with a dye that can identify metabolic activity (e.g., Molecular Probes Live/Dead Viability Kit, 5).

Literature Cited

1. Askebjer, P., et al. 1997. UV and optical light transmission properties in deep ice at the South Pole. Geophys. Res. Letters 24: 1355.

2. Bay, R., P.B. Price, G.D. Clow, and A.J. Gow. 2001. Climate logging with a new rapid optical technique at Siple Dome. Geophys. Res. Letters 28 (24): 4635–4638.

3. He, Y.D., and P.B. Price. 1998. Remote sensing of dust in deep ice at the South Pole. J. Geophys. Res. 103: 17041.

4. Miocinovic, P., P.B. Price, and R. C. Bay. 2001. Rapid optical method for logging dust concentration vs. depth in glacial ice. Appl. Optics 40: 2515–2521.

5. Molecular Probes BacLight Viability Kit. (www.probes.com/servlets/product?item=7007).

6. Pan, Yong-le, S. Holler, R.K. Chang, S.C. Hill, R.G. Pinnick, S. Niles, and J.R. Bottiger. 1999. Single-shot fluorescence spectra of individual micrometer-sized bioaerosols illuminated by a 351 or a 266 nm ultraviolet laser. Optics Letters 24: 116–118.

7. Price, P.B. 2000. A habitat for psychrophiles in deep Antarctic ice. Proc. Nat. Acad. Sci. 97: 1247–1251.

19

Living Cells in Permafrost as Models for Astrobiology Research

Elena A. Vorobyova, V.S. Soina, A.G. Mamukelashvili, A. Bolshakova,
I.V. Yaminsky, and A.L. Mulyukin

ASTROBIOLOGY is a new multidisciplinary branch of science that relies on a synthesis of up-to-date knowledge in the fields of planetology, Earth sciences, ecology, extremophile biology, and the like. Of great importance is the selection of the models and approaches to identify extraterrestrial conditions suitable for life, and to choose the most promising methods for astrobiological research (20). The search for extinct or extant extraterrestrial life must rely on the combined use of several models and direct approaches to the *in situ* analysis of microbial communities. Existing methods and equipment should be modified and tested using the most appropriate terrestrial extreme ecosystems. Then the relevant databases of microorganisms and biomarkers should be created to be used as models for astrobiology.

It is now apparent that almost all terrestrial microorganisms exist in extreme environments, and cope with nutrient unavailability, close to zero or subzero temperatures, high temperatures, reduced oxygen levels, water limitation, high pressure, hypersalinity, and so on. These conditions often are regarded as unfavorable from the human viewpoint, but microbial communities are able to survive under these environmental stresses, and this adaptability is one of the most important factors in maintaining the biodiversity and abundance of microbial populations.

One of the evolutionary strategies developed by microorganisms to survive under these conditions is to enter a resting stage, often associated with the production of spores, cysts, and other specialized dormant cells. However, few microbial species are able to form such specialized resting cells, and the question of whether non-spore-forming bacteria enter the dormant state is still much debated. It has been proposed (4) to consider the dormant stage of nonsporulating bacteria as viable-but-nonculturable (VBNC) cells (1, 10), or as ultramicrobacteria (6). Such cells are unable to produce colonies when plated on solid media under standard conditions, but may possess other attributes of metabolic activity. On the other hand, another way that non-spore-forming bacteria may survive stressful conditions is by the formation of cystlike cells, which display the properties of resting forms (2, 7). We believe that long-term survival of non-spore-forming microorganisms in extreme environments is to a great extent due to their ability to produce dormant cells.

Understanding the physiological state and the mechanisms that permit prolonged survival and maintenance of microbes in various ecosystems is very important for astrobiology because the cells of different physiological states (e.g., viable, VBNC, viable resting forms) can be regarded as target objects for which to search for extraterrestrial life. Microbial micromummies, which are VBNC-fossilized microorganisms, may be a an indicator of past life (15).

Objective and Methods

Our research objective is to use different methodological approaches to the *in situ* analysis of microorganisms in deep permafrost environments. This chapter summarizes our recent research on the use of high-resolution microscopy methods for the detection and study of viable microbial cells in natural ecosystems, particularly in permafrost, and possibly in extraterrestial habitats.

We studied samples of Arctic and Antarctic permafrost that were taken in the Kolyma lowland in Northeast Siberia (67–70 °N, 152–162 °E), and in the Dry Valleys area of the Antarctic (Victoria Land: Miers Valley, Beacon Valley). The upper 100 m of the Late Cenozoic permafrost deposits in the Arctic were formed syngenetically, which means that sedimentation occurs concurrently with the freezing process. Therefore, the age of these Arctic permafrost samples corresponded to the age of the sediments. There is geological evidence that buried permafrost in the Kolyma lowland has remained continuously frozen since its deposition (3, 11, 23). In accordance with geological analyses (isotopic, paleomagnetic, palynological, termo-luminiscence, and bone analyses), the Arctic samples under investigation were dated to the Late Pliocene–Early Pleistocene (0.6–3.0 million years before present [mybp]). The most ancient Antarctic permafrost samples (sampling depth to 20 m) were at least this old (21). According to Sugden et. al. (14), the age of the ash layer buried in permafrost sediments in Beacon Valley was determined as near 8 mybp. Some cores for microbiological research were taken from beneath this layer.

The permafrost was sampled aseptically by the method of slow rotary drilling without fluid. The temperature of the extracted cores was −7 to –10 °C in the Arctic and −20 to –27 °C in the Antarctic. A central (interior) core was extracted from each core aseptically using a sterile knife. These central cores then were treated with ethanol, immediately placed in presterilized aluminum tins, and transported in the frozen state to the lab. These sampling techniques are described in detail (3, 11, 23).

Control procedures were as follows: (i) *In the field*: Permafrost cores were surface-contaminated with *Serratia marcescens* as a biomarker, melted immediately, and bioassayed. *S. marcescens* was not isolated from the central core. (ii) *In the laboratory*: All microbial investigations were carried out in a UV-sterilized box within a UV-irradiated room, and each sample isolation was replicated from three to ten times. The coring and sampling of Antarctic permafrost were described in detail earlier (21).

Several methods were used to detect microorganisms in the permafrost samples, and these are briefly described here. First, direct epifluorescence microscopy using acridine orange to stain prokaryotic cells or calcofluor white to stain eukaryotic cells was used. Second, soil suspensions prepared immediately after melting were inoculated onto solid or enriched media, and the number of colonies that developed were counted. Finally, direct *in situ* observation of microbial cells in native samples and cytomorphological and fine structure analyses were carried out using scanning electron microscopy (SEM), atomic force microscopy (AFM), and transmission electron microscopy (TEM).

An environmental scanning electron microscope (ESEM) (Electroscan Corporation, Wilmington, Massachusetts) equipped with an energy dispersive X-ray spectroscopy (EDS) detector (Oxford Research, Concord, Massachusetts) was used under partial pressure of gas (10 Torr vacuum) in conjunction with scanning electron microscopy to produce high-resolution images. At low temperatures carbon dioxide gas (2–3 Torr) was used to prevent ice formation on the sample. The ESEM operating voltage was 10 to 30 kV to produce magnifications of up to 100,000×. The ESEM was equipped with a 5 axis stage, Kevex detector, and 4Pi analysis EDS data acquisition. The digital images were recorded at 4096×4096 pixel resolution. In addition to the ESEM, a Hitachi S-405A scanning electron microscope (25 kV, magnification up to 150,000×, with a resolution of 70 Å was used. A JEOL transmission electron microscope (100 kV, magnification up to 500,000×, with a resolution of 2 Å) was utilized for investigation of cell fine structure.

Atomic force microscopy (AFM) is a high-resolution technique for biological research, particularly useful for environmental applications. In AFM, a micrometer-sized lever bends as a consequence of the interatomic force between a very sharp tip and a sample. The cantilever deflection is measured with subangstrom resolution by a detector that typically consists of a laser beam reflecting off the cantilever into a segmented photodiode. Thus, forces of the order of 10 pN can be measured. AFM provides the topography of organic and inorganic surfaces at atomic and nanometer scale in air, liquids, and ultra-high vacuum (16).

A multimode scanning probe microscope (AFM, Nanoscope IIIa, Digital Instruments, United States) was used to image bacterial cells *in situ*, as well as in culture. The silicon nitride (Si_3N_4) cantilevers were chosen for the bacterial study. We used cantilevers with different rigid constants (from 0.06 to 48

N/m) (Digital Instruments, United States; MikroMasch, Russia) in contact and force modulation modes. AFM experiments were carried out in air and in liquid (sterile water, meat broth, or soil extract) conditions. For better adhesion of soils and cells we modified substrates (glass, mica) with polyethilenimine, agarose, or polish (nitrocellulose:dibutylphtalate in ethylacetate:butylacetate). The permafrost samples (1 g) were put into sterilized water (10 ml) and shaken for 5 minutes to disperse cells from mineral particles. After sedimentation (5–15 minutes), the upper part of the suspension (0.1 ml) was taken and added to sterilized water (0.5–1.0 ml). Then a droplet of suspension was applied to a substrate. Dried samples were treated with acetic acid to dissolve salt. For pure cultures, a drop of cell suspension (approximately 10^9 cell/ml, depending on size, morphology, adhesion, and other cell features) was placed onto the substrate and dried at 20 to 30 °C. In our research, the height of the observed objects was limited to < 2 microns. FemtoScan Online (Advanced Technologies Center, Moscow, Russia, www.spm.genebee.msu.su/manual/en/index.html) software was used to analyze the images.

Results and Discussion

Microbial Communities in Ancient Permafrost

Permafrost sediments from both polar regions contained large numbers of living cells. Approximately 10^7 to 10^9 bacterial cells/gdw, and 10^5 to 10^6 fungal spores/gdw were counted in frozen sediments after immediate melting. These numbers do not depend on the permafrost sediment origin (Arctic or Antarctic) or age. Others have reported similar bacterial cell numbers in deep subsoil and sea sediments (8, 9, 13). The number of both bacterial and fungal cells in the permafrost sediments was equivalent to that in the mineral horizons of the upper soils (22).

Viable (culturable) bacteria are present in permafrost. Most of those isolated were non-spore-forming species (3, 19) of the following genera: *Micrococcus, Arthrobacter, Rhodococcus, Cellulomonas, Flavobacterium, Aeromonas, Acinetobacter, Pseudomonas, Nitrosospira*, and *Nitrosovibrio*. Spore-forming bacteria also were cultured: *Bacillus, Exiguobacterium*, and *Desulfotomaculum* (17). Even *Cyanobacteria* (18) were isolated in culture. Many of these were present in both Arctic and Antarctic sediments, which were frozen for millions of years. Both the abundance and biodiversity of culturable bacteria were considerably less in Antarctic permafrost. Bacteria were isolated from approximately 1 to 10% of the Arctic permafrost samples, but from only 0.01 to 0.1% (nearly 1% for fungi) of the Antarctic samples. Most of the microorganisms that we observed, however, were nonculturable and could be detected only by direct microscopic investigation. Thus, many of the microorganisms that populate ancient permafrost can be identified only by direct techniques (e.g., PCR-analysis, lipid biomarker analysis).

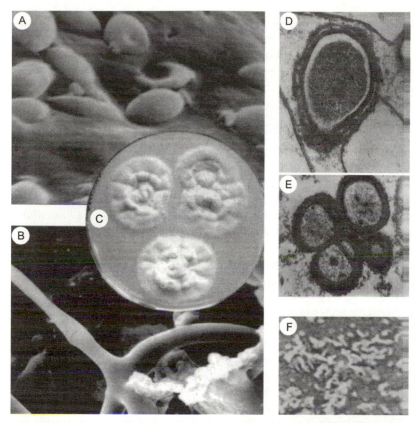

FIGURE 19.1. Fungi in Antarctic permafrost (Taylor Valley, core depth was 16.5 m corresponding to an age of 170,000 years before present). (A) spores, (B) growing mycelium *in situ* (incubated with nutrients for 30 days at 20 °C), (C) colonies on nutrient media. Dormant cells of (D) *Arthrobacter globiformis* and (E) *Micrococcus roseus* from Artic permafrost sediments 1.8 million years old (60,000 x). (F) Biofilm in Antarctic permafrost.

The discrepancy between the total count of microorganisms present in the permafrost sediments and the number of colony-forming units (19) suggests that the permafrost microorganisms there, if all are viable, persist in the VBNC state in permafrost and transition to deep dormancy with time and lowering of temperature (ancient Arctic and Antarctic permafrost).

Numerous fungal conidia were observed in ancient permafrost with the use of epifluorescence microscopy and SEM. Fungal biomass (as direct total count) was ≤ 60 μg/g (absolute dry weight), which is ten-fold greater than the bacterial biomass. Dormant spores and conidia of micromycetes that varied in size and shape were found in permafrost *in situ* (Figure 19.1), whereas in modern soils, fungal mycelium was almost always observed. We suggest that fungi do not multiply in frozen permafrost and, if true, then their age corresponds to the age of the permafrost. Some fungal spores easily germinated, and we observed mycelium within one day of incubation at room temperature under moist conditions (melted samples without added nutrients). But fungal biodiversity on nutrient media was low when the samples were analyzed immediately after melting. In some ancient Arctic permafrost sediments, not one fungal colony was detected. *Penicillium* strains predominated in melted Antarctic permafrost samples. To provide better resuscitation of fungal spores, and to reveal the diversity of micromycetes present, heat (50–100 °C), sonification (15 kHz, 0.3A, 1 minute) shock, and prolonged (1–6 months) incubation of enrichment cultures were used. The appearance of sterile mycelium (white) was common and characteristic for both Arctic and Antarctic samples. Enrichment cultures revealed a high diversity of fungi in permafrost, where the genera *Penicillium, Chrysosporium, Aspergillus*, and *Cladosporium* predominated (5).

Use of High-Resolution Microscopic Techniques for in situ and ex situ Cytological Observation of Cells

Cytological aspects of the structural stability of the different types of bacterial cells isolated from permafrost or *in situ* were studied using direct observation of microbes in frozen and thawed samples. Viability, structure, and growth of the bacteria were examined by studying the effects of temperature fluctuations in media unbalanced with respect to nutrition using scanning electron microscopy (SEM), transmission electron microscopy (TEM), and environmental scanning electron microscopy (ESEM).

TEM of permafrost samples immediately after thawing demonstrated well-preserved bacteria with Gram-positive and Gram-negative cell walls. Bacterial cells did not reveal any signs of lysis, and were small in size (0.2–0.5 μm in diameter). Single cells often occurred in conglomerates surrounded by the fibrillar capsular layer covered with organo-mineral particles. TEM studies *in situ* revealed a limited number of typical spore-forming cells or mature spores.

Some prokaryotic and eukaryotic microorganisms can form cystlike cells that retain viability during long-term storage (as judged by the ability to form colonies after plating), have no detectable respiratory activity, show elevated resistance to heating and other stresses, and are characterized by a specific fine structure. These cystlike cells possess many attributes of resting cells. Such cystlike cells of non-spore-forming bacteria of the genera *Arthrobacter* and *Micrococcus* were isolated from Arctic permafrost sediments (7). Resting cells of the isolates from permafrost sediments had the same ultrastructure (Figure 19.2 A) as that of cells observed *in situ* in the same permafrost sediments (Figure 19.2 B). Therefore, we believe that preservation of some microbial cells in such environments is due to their ability to form dormant cells of this type.

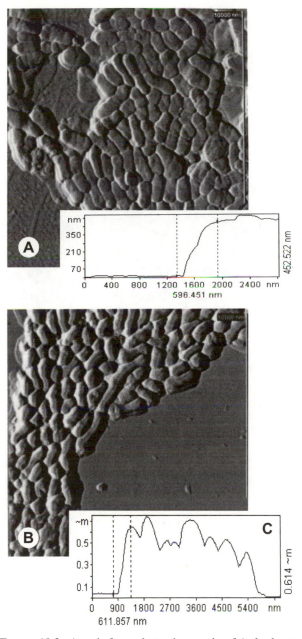

FIGURE 19.2. Atomic force photomicrographs of *Arthrobacter globiformis* from ancient permafrost. (A) reproducing vegetative cells, (B) resting cells (mummies), (C) dimensions of middle cell profile.

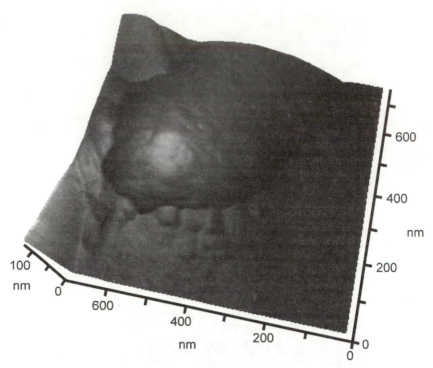

FIGURE 19.3. Atomic force microscopy image of bacterial cell with flagellum.

ESEM allows the study of native cells in conditions close to natural. Using the cooling stage, it is possible to observe cells under changing temperature conditions. ESEM reveals that bacteria frozen in native samples lie in a distinct gel sheath (biofilm), and were found on the surface of microaggregates of organo-mineral particles, but not on the ice inclusions. No destructive features were evident in the bacterial ESEM images when native permafrost samples were frozen at temperatures from −25 to −50 °C. It is sometimes difficult to differentiate between biotic and abiotic objects in complex heterogeneous environments using SEM. Large biotic objects (e.g., yeast and algal cells, conidia or mycelial fragments) are easy to distinguish because of distinct visible features. Biotic objects of micron size often are easy to identify by morphology alone. The problem arises with objects of submicron size. These so-called biofilms (i.e., "biogel") and "cells" of 0.1 to 0.3 μm in size are common in soil sediments (22). Similar-sized objects were found in this work in both Arctic and Antarctic permafrost and ground ice samples. At high resolution they appeared as single "y"-shaped or oval-shaped cells resembling some actinobacteria. The addition of nutrients into the permafrost sample led to extension of the biofilm, suggesting that these biofilms are small cells covered by gel and may be viable.

Atomic force microscopy (AFM) opens up new possibilities for the study of microorganisms in natural environments. This advantage is due to the high resolution in both air and liquid, without the need for specimen coating, high vacuum, and other sample treatments, as well as the small overall dimensions of equipment. The use of AFM enables the examination of cells and cellular surface and subsurface material properties (adhesion, viscosity, etc.) in the native state with a high degree of spatial precision and three-dimensional visualization. Until now there have been no publications concerning the use of AFM in cellular investigations of natural environments. Indeed, AFM techniques should be especially valuable for environmental investigations. The soil and subsurface macrocosms populated by microorganisms represent multiple media composed of different combinations of heterophase systems varying in physico-chemical, granulometric, mineralogical, and other properties. It is necessary to have a technique that can distinguish biotic from abiotic objects. We have demonstrated the use of scanning probe microscopy (AFM) for direct investigation of microorganisms in their native environment (soil, permafrost, and ice) by means of the multimode properties of AFM, such as force mapping, friction mode, lithography, and liquid explorations.

We analyzed bacterial cells from Antarctic permafrost sediments, from ground ice and soils, and in pure cultures. Atomic force microscopy techniques can differentiate actively reproducing from resting bacterial cells based on their size, morphology, and mechanical properties (Figure 19.2). Nonviable cells (micromummies) were easily destroyed under force modulation in contact mode. Lithography mode also was used to differentiate biotic and abiotic objects in heterogeneous environments (mixed cell-mineral suspensions and native samples). Bacterial cells were damaged after the treatment, while mineral particles were unaltered. Round-shaped and rod-shaped cells were observed in situ in all samples investigated. In activated samples (melted sediments incubated for 2 to 4 weeks at 4 and 20 °C), some of the cells had flagella (Figure 19.3), and diatoms (Figure 19.4) as well as micropopulations of cell conglomerates were observed. A great number of cell-like objects ("biofilm," "biogel") with small dimensions (50–100 nm) also were imaged in melted samples incubated for 2 to 4 weeks at 4 and 20 °C).

Conclusions

Exposure to low temperatures even for millions of years did not result in the extinction of life in permafrost. Modern analytical methods now permit the investigation of microbial life in extreme environments. Such methods should play an important role in the search for extraterrestrial life. However, future work in extraterrestrial environments is dependent upon much needed additional work in extreme terrestrial environments to optimize methods and to

FIGURE 19.4. Atomic force microscopy image of diatom in ground ice from Beacon Valley, Antarctica.

create databases for comparison of terrestrial microbial communities in extreme environments to those likely to be found in extraterrestrial ecosystems.

Our studies of deep ancient terrestrial permafrost enable selection of most promising methods for the possible detection of microorganisms beyond the Earth. A combination of methods must be selected for the specific technical task (e.g., robotic missions, space station, sample return mission) and provide for the detection of extant active or anabiotic extraterrestrial life. In our opinion, the following tasks are of priority in the development of methodologies for the detection of extraterrestrial life:

1. Systematic microbiological investigations of natural extreme and ancient Earth habitats (thermal gradients, evaporates, sedimentary and primary rocks, ancient permafrost, ground and continental ice) using identical methodological approaches to create the essential databases.

2. Studies designed to determine how long microorganisms can survive in natural ecosystems, combined with studies of the physiological state of cells and mechanisms of physiological adaptation to stressful environments and specific cell-environment interactions.

3. Development of direct methods to detect biosignatures, biomarkers, living cells, fossils, and microbial communities in ancient and extreme environments. Criteria to distinguish between biotic and abiotic objects need to be developed, and the atlases of images of terrestrial extremophiles *in situ* should be done.

Acknowledgments

Permafrost samples were collected by Dr. David A. Gilichinski et al. (Institute for Physiochemical and Biological Problems in Soil Science, Russian Academy of Sciences, Pushchino, Moscow, Russia). The isolation of fungi from Arctic permafrost was provided in collaboration with T. Vishnivetskaya. We thankfully acknowledge these contributions. This work was supported by the Russian Ministry of Science and Technologies, contract No. 101–10(01)-ï, 2000.

Literature Cited

1. Colwell, R.R., P.R. Brayton, D.J. Crimes, D.B. Roszak, S.A. Hug, and L.M. Palmer. 1985. Viable but nonculturable *Vibrio cholerae* and related pathogens in the environment: Implications for release of genetically engineered microorganisms. Bio/Technol. 3: 817–820.

2. Demkina, E.V., V.S. Soina, and G.I. El-Registan. 2000. Formation of resting forms of *Arthrobacter globiformis* in autolyzing cell suspensions. Microbiology (trans. from Microbiologiya) 69(3): 314–318.

3. Gilichinsky, D., E.A. Vorobyova, L.M. Erokhina, D.G. Fedorov-Davydov, and N.R. Chaikovskaya 1992. Long-term preservation of microbial ecosystems in permafrost. Adv. Space Res. 12(4): 255–263.

4. Kaprelyants, A.S., J.C. Gottshal, and D.B. Kell. 1993. Dormancy in nonsporulating bacteria. FEMS Microbiol. Rev. 104: 271–286.

5. Kochkina, G.A., N.E. Ivanushkina, S.G. Karasev, E. Yu. Gavrish, L.V. Gurina, L.I. Evtushenko, E.A. Vorobyova, E.V. Spirina, D.A. Gilichinsky, and S.M. Ozerskaya. 2001. *Micromycetes* and *Actinobacteria* under long-term natural cryoconservation. Microbiology (trans. from Microbiologiya) (3): 412–420.

6. Morita, R.Y. 1982. Starvation-survival of heterotrophs in the marine environment. Adv. Microbial Ecol. 6: 117–198.

7. Mulyukin, A.L., K.A. Lusta, M.N. Gryaznova, A.N. Kozlova, M.V. Duzha, V.I. Duda, and G.I. El-Registan. 1996. Formation of resting cells by *Bacillus cereus* and *Micrococcus luteus*. Microbiology (trans. from Microbiologiya) 65(6): 782–789.

8. Parkes, R.J., and J.R. Maxwell. 1993. Some like it hot (and oily). Nature 365: 694–695.

9. Pedersen, K. 1993. The deep subterranean biosphere. Earth Sci. Rev. 34: 243–260.

10. Roszak, D.B., and R.R. Colwell. 1987. Metabolic activity of bacterial cells enumerated by direct viable count. Appl. Environ. Microbiol. 53(12): 2889–2893.

11. Shi, T., R.H. Reeves, D.A. Gilichinsky, and E.I. Friedmann. 1997. Characterization of viable bacteria from Siberian permafrost by 16S rDNA sequencing. Microbial Ecol. 33: 169–179.

12. Soina, V.S., E.A. Vorobyova, D.G. Zvyagintsev, and D.A. Gilichinsky. 1995. Preservation of cell structures in permafrost: A model for exobiology. Adv. Space Res. 15(3): 237–242.

13. Stevens, T.O., and J.P. McKinley. 1995. Lithoautotrophic microbial ecosystems in a deep basalt aquifer. Science 270: 450–454.

14. Sugden, D., D. Marchant, N. Potter, R. Souchez, G. Denton, C. Swisher III, and J-L. Tison. 1995. Preservation of Miocene glacier ice in East Antarctica. Nature 376: 412–414.

15. Suzina, N.E., A.L. Mulyukin, N.G. Loiko, A.N. Kozlova, V.V. Dmitriev, A.P. Shorokhova, V.M. Gorlenko, V.I. Duda, and G.I. El-Registan. 2001. Fine structure of mummified cells of microorganisms formed under the influence of a chemical analogue of the anabiosis autoinducer. Microbiology (trans. from Microbiologiya) 70(6): 667–677.

16. Tamayo, J., A.D.L. Humphris, A.M. Malloy, M.J. Miles. 2001. Chemical sensors and biosensors in liquid environment based on microcantilevers with amplified quality factor. Ultramicroscopy 86: 167–173.

17. Vainshtein, M., G. Gogotova, and H. Hippe, 1994. A sulfate-reducing bacterium from the permafrost, pp. 68–73. In D. Gilichinsky (ed.), Viable Microorganisms in Permafrost. Pushchino Science Center, RAS.

18. Vishnivetskaya, T., L. Erokhina, E. Spirina, A. Shatilovich, E. Vorobyova, and D. Gilichinsky. 2001. Ancient viable green algae and cyanobacteria from permafrost. Nova Hedwigia, Beiheft. 123: 427–441.

19. Vorobyova, E., V. Soina, M. Gorlenko, N. Minkovskaya, N. Zalinova, A. Mamukelashvili, D. Gilichinsky, E. Rivkina, and T. Vishnivetskaya. 1997. The deep cold biosphere: Facts and hypothesis. FEMS Microbiol. Rev. 20: 277–290.

20. Wilson, A. (ed.). 1999. Exobiology in the Solar System and the Search for Life on Mars. Report from the ESA Exobiology Team Study 1997–1998. SP–1231, ESA Publ. Division (Ed.), ESTEC, Noordwijk, the Netherlands.

21. Wilson, G.S., D.A. Gilichinsky, D.G. Fyodorov-Davydov, V.E. Ostroumov, V. Sorokovikov, and M.C. Wizevich. 1998. COMRAC drilling, core recovery and handling. In G.S. Wilson and J.A. Barron (eds.), Mount Feather Sirius Group Core Workshop and Collaborative Sample Analysis. Proceedings of a workshop sponsored by the office of Polar Programs, NSF, 19–22 June, 1996. Antarctic Marine Geology Research Facility, Florida State University, Tallahassee, Florida, U.S.A. Rep #14, ISSN: 0896–2472.

22. Zvyagintsev, D.G. 1987. Soil and Microorganisms. Moscow State University, Moscow.

23. Zvyagintsev, D.G., D.A. Gilichinsky, S.A. Blagodatsky, E.A. Vorobyova, et al. 1985. The longevity of microbial life preservation in constantly frozen sedimentary rocks and buried soils. Microbiology (trans. from Mikrobiologiya) 54: 153–163.

20

A Synopsis of the Past, an Evaluation of the Current, and a Glance toward the Future

John D. Castello and Scott. O. Rogers

THIS BOOK addresses a number of important issues. First, it is amply clear that microorganisms can be isolated from ice and permafrost cores from around the world, at depths to nearly 4 km that are up to several hundreds of thousands of years old, as well as from permafrost that is sometimes several millions of years old. Second, remnants of nonviable organisms also are present in these substrates in the form of biological macromolecules such as nucleic acids. Third, the organisms found in these matrices can survive the rigors of freezing and thawing. This is an important evolutionary mechanism, and for pathogens might be vital to their long-term survival while making them more virulent to their hosts.

A major objective of this book is to summarize the evidence to support the theory that viable life is present in the ancient cryosphere, specifically glacial ice and permafrost. A second objective is to support the concept that if life can survive and indeed thrive in such extreme environments on Earth, it is possible that life as we understand it could exist in similar extraterrestrial habitats. These were the objectives of the NSF-sponsored workshop from which this book is derived.

In this final chapter our objectives are to

1. summarize briefly and evaluate the evidence for the presence of life in glacial ice and permafrost.

2. discuss the implications and significance of the research to date

3. identify the more important and immediate research needs, and our recommendations for future research

Many of the contributors to this book are leaders in this relatively new field of scientific research because they have published many papers in leading scientific journals. Collectively, we present a great deal of new and previously published data to support the argument that life is present in ancient ice and permafrost. Specifically, we report the isolation/detection of fungi, bacteria, and even viruses in ice and/or permafrost; the decontamination and detection protocols that were used; the problems that we encountered and solved, and those that still require solutions; some new and potentially useful methods for life detection; the significance and implications of our findings; and finally some immediate needs for future research.

There are many questions yet to be answered, problems that require solutions, and needs that must be addressed. We list some of them here, and revisit them later in this chapter. A major expectation of any scientific research effort that reports an unusual, or at least an unexpected, finding, is that of independent corroboration of the results. This is an immediate need that must be met by those of us who are working in this field of science. This is an important priority because unless it is accomplished we risk disbelief by the general scientific community. Disbelief has been the general reaction to extraordinary reports of viable life in ancient amber and halite crystals in part because of lack of corroboration. Independent corroboration in the strictest sense requires a rigorous set of standardized protocols for the collection of cores to minimize contamination. Decontamination, isolation, and detection of microorganisms and/or their nucleic acids must be performed by two or more independent laboratories working on meltwater from a split ice core. It is not a trivial matter to accomplish these tasks. There is as yet no agreement among us on a standardized set of protocols to achieve this objective. The facilities required to conduct this kind of research in compliance with the rigorous standardized protocols that have been proposed (see chapter 2) will be very expensive or even impossible for many laboratories to implement.

We have yet to establish the "criteria of authenticity" necessary to assure the quality of the research, as well as to satisfy the critics and doubters that the microorganisms or nucleic acids isolated/detected in ancient ice and permafrost are in fact ancient rather than contemporary contaminants. This will not be easy because many of the organisms isolated from ancient substrates are similar or identical to contemporary isolates. Nevertheless, we proposed a set of rigorous criteria at the workshop (see chapter 2), but these have yet to be debated or revised by the ice biology research community.

Another important need is the creation of an organization that includes those who study the biology of the cryosphere, and which meets regularly. Such an organization would provide a forum where important issues are addressed, current research is presented and discussed, collaborations are established, and needs are discussed. At present, such an organization does not exist.

Summary and Evaluation of the Evidence for Life in Ancient Glacial Ice and Permafrost

Many of us have isolated microorganisms from ancient ice and permafrost including ourselves (chapters 2, 11, 12, and 13), Abyzov (chapter 16),

Christner (chapter 15), Priscu et al. (14), Karl et al. (9), Shi et al. (17), Voro-byova (chapter 19), Vishnivetskaya (chapter 10), Rivkina (chapter 7), Faizutdi-nova (chapter 8), Ivanushkina (chapter 9), and Sambrotto (chapter 6). Each of us has worked independently, using different methodologies for decontamina-tion and detection and different ice/permafrost cores in many cases, and yet our results are remarkably similar. Many of the same genera and even species have been isolated. We refer you to the specific chapters for the details, but briefly our results working with microorganisms from glacial ice can be sum-marized as follows:

1. The predominant bacteria detected or isolated from ice appear to be the endo-spore-forming Gram-positive bacteria. These were the predominant forms detected by Abyzov, Christner, and Castello (chapters 16, 15, and 13, respectively) in glacial ice from Lake Vostok, polar and nonpolar glaciers, and Greenland, respectively.

2. In addition to these, various genera and species of actinobacteria, as well as α- and γ-proteobacteria, and other bacterial taxa also have been detected in glacial ice.

3. Curiously, except for one report (19), archaea have not been detected in glacial ice, although they are present in high-latitude oceans, as well as in permafrost in Siberia and Antarctica (refs. 7, 15, and 20, and see chapter 7, this volume).

4. Some of the bacteria, fungi, and archaea from ice and/or permafrost are meta-bolically active *in situ*, while others apparently are in a dormant state.

5. Most of the bacteria and fungi isolated from ice generally do not appear to be true psychrophiles, that is, they are capable of growth at temperatures above 20 °C. But most are cold tolerant (psychrotrophic).

6. So far, based on rDNA sequencing data, while many isolates are very similar to previously described taxa, there are many that are different from any described taxa (see chapter 11).

There are some differences, however, in what has been found by various researchers investigating different horizons of the Vostok 5G core accretion ice. The concentration of bacteria detected by Priscu et al. (15) is greater than what Abyzov detected in Vostok 5G accretion ice (chapter 16). Priscu detected α- and β-proteobacteria (Gram-negative bacteria), and an actinobacterium in horizon 3590 m of this core. Cells numbered 2800 to 36,000/ml. Cultivation or indication of metabolic activity was not achieved. Abyzov did not identify specific microorganisms, but bacterial rods, cocci, actinobacterial threads, cells similar to *Cytophaga* and *Caulobacter*, as well as cells similar to cyanobacteria also were detected in all accretion ice horizons at tens to hundreds per ml. Karl (10), who worked with horizon 3603 of the Vostok 5G core, detected cell wall markers indicative of Gram-negative bacteria, as well as small cocci, rods, and vibrios at about 200 to 300 cells/ml. Uptake studies indicate that some of these cells were viable. Christner et al. (chapter 15) were able to isolate bacteria from horizon 3593 m of the Vostok 5G core accretion ice, as well as to amplify bacteria directly from meltwater. Based on 16S rDNA sequencing, these were identified as *Methylobacterium* (α-proteobacteria), and *Sphingomonas* (ultra-microbactria), *Brachybacterium* (an actinobacterium), and *Paenibacillus* (a Gram positive endospore-forming rod). However, preliminary results by Rogers and S. Bulat (unpublished) based on a Lake Vostok accretion ice core that had been split and assayed by the two independent labs indicate that accretion ice contains very few viable microbes. These differences may have several explanations. It is possible that some of these microorganisms are contemporary contaminants, but which ones? We discuss this possibility later in the chapter. Alternatively, the Vostok accretion ice is highly variable (see chapter 17 for details), and thus strict reproducibility of the results of biological investigation is unlikely and should not be expected (see chapter 17). Third, different protocols were used to detect life. Of course, these possible explanations are not mutually exclusive.

Bacteria similar to those detected in glacial ice and permafrost have been detected in sea ice and lake ice. Priscu (chapter 3) and Nichols (chapter 4) have isolated bacteria from perennially frozen Antarctic lake ice and sea ice surrounding Antarctica, respectively. Priscu detected proteobacteria, planctomycetales, Acidobacterium-Holophaga division, green nonsulfur bacteria, and actinobacteria. Virus-like particles (VLPs) also were detected in abundance (chapter 3). Nichols (chapter 4) also detected α- and γ-proteobacteria, actinobacteria, as well as bacteria of the *Cytophaga-Flavobacterium-Bacteroides* (CFB) group (Phylum Bacteroidetes). These bacteria apparently occur in intimate associations (consortia) with cyanobacteria and microalgae. Similar microorganisms are reported in polar as well as nonpolar glacial ice by Abyzov, Christner, and Castello (see chapters 16, 15, and 13 for details). Are soils and oceans the sources of bacteria detected in ice?

Bacteria have been detected in permafrost too. Vorobyova (chapter 19) has detected bacteria-like particles in Arctic and Antarctic permafrost up to 8 million years old using microscopy techniques. The amount of biomass detected is unrelated to genesis, age, or physico-chemical properties of the permafrost. She observed that fungal biomass was ten-fold greater than bacterial biomass in permafrost. Cystlike cells characteristic of *Arthrobacter* and *Micrococcus* were commonly observed, and may be viable. Vishnivetskaya (chapter 10) also detected viable bacteria in the genera *Microbacterium, Rhodococcus, Microbiospora, Paenibacillus, Agrobacterium, Brevundimonas, Afipia, Aminobacter, Pseudomonas,* and *Xanthomonas* in intimate association with green algae and cyanobacteria isolated from ancient Arctic and Antarctic permafrost. Many of these genera are identical to those found in glacial ice, permafrost, lake ice, and sea ice by others.

Eukaryotes also have been detected in or isolated from glacial ice and permafrost. Fifty-seven taxa of eukaryotes, including green plants, fungi, and protists, were detected in the Hans Tausen ice core in northern Greenland by Willerslev et al. (22). The fungi appear to be the most commonly isolated eukaryotes from ice and permafrost. Faizutdinova (chapter 8) and Starmer (chapter 12) both isolated basidiomycetous yeasts in the genera *Rhodotorula* and *Cryptococcus* from ancient permafrost and from the Greenland ice sheet, respectively. The permafrost yeast isolates were 15,000 to 3 million years old, and some of those from the Greenland ice sheet were at least 100,000 years old. Ma et al. (chapter 11) have isolated fungi from the Greenland and Antarctic ice sheets and Ivanushkina et al. (chapter 9) have isolated very similar fungi from permafrost up to 3 million years old. Species of the genera *Penicillium, Aspergillus, Cladosporium,* and *Ulocladium* are among the most commonly detected filamentous fungi in both ice and permafrost.

Other life forms also have been detected in ice and/or permafrost. We (chapter 13) have detected viruses (a plant virus and bacteriophage) in glacial ice from Greenland, and Priscu et al. (chapter 3) have observed numerous VLPs in ice of perennially frozen Antarctic lakes. So far as we are aware, no one has yet attempted to corroborate these findings, and thus these results await independent corroboration. As mentioned above, Archaea have been reported only once from glacial ice (methanogens from basal ice on Ellesmere Island-19), but methanogens and denitrifiers have been detected in permafrost in Siberia and Antarctica (see refs. 8, 16, 21, 23 and chapter 7, this volume). It is curious that Archaea are present in Antarctic seawater (9), but have not been detected in Antarctic glacial ice.

But Are the Data Reliable?

Independent corroboration in the strictest sense is important, but has been done in only a few cases. In the meantime, how much confidence do we have in the data that have been generated to date? Do the data represent ancient microbial isolates from ice or permafrost, or are these isolates merely contemporary contaminants? To address these questions, it is imperative to know if the specific decontamination protocols used to decontaminate ice and permafrost cores prior to microbiological work with ice are effective.

In a recent publication, we report the results of a study that compared the effectiveness of the protocols most commonly in use by the biological ice community to decontaminate ice and permafrost core surfaces while maintaining the viability/integrity of the microbes and their nucleic acids that are present within the cores (17). The protocols that we tested were based on published protocols (1, 4, 7, 10, 11, 12, 13, 15). The simplest are the ablation protocols including the use of water and mechanical protocols. The other protocols that we tested use chemicals or UV irradiation to decontaminate the outer ice surface, which is then followed by extraction or melting of an interior section of the core. Of the protocols that we tested, treatment with 5% sodium hypochlorite (Clorox) followed by rinsing with sterile water was the most effective. It was the only protocol that removed 100% of the surface contamination without affecting the microorganisms inside the ice core. The results of our study indicate that at least some, but certainly not all, of the organisms reported in publications could be contaminants.

Abyzov (see ref. 1, and chapter 16, this volume) assayed ice core sections from the Vostok 5G deep core for microorganisms using a mechanical ablation decontamination protocol (i.e., removal of the outer end of the core followed immediately by melting of an inner section of the core into a sterile collection vessel). The protocol was tested by creation of sham cores coated with *Serratia marcescens* and looking for growth of this bacterium on media inoculated with interior meltwater. None was detected. Many different viable microorganisms were isolated from different core sections, including yeasts (*Rhodotorula* and *Cryptococcus*), filamentous fungi (*Aspergillus, Penicillium, Phialophora, Mucor, Alysidium,* and *Torula*), non-spore-forming Gram-negative bacteria (*Pseudomonas*), spore-forming Gram-positive bacteria (*Bacillus* spp, *Clostridium* spp. and actinobacteria (*Nocardia, Nocardiopsis*, and *Streptomyces* spp.).

Christner et al. (see refs. 6, 7 and chapter 15, this volume) used a mechanical ablation decontamination protocol similar to that of Abyzov, which involves removal of the end of an ice core section, dipping the cut surface in ethanol, and melting out an interior subcore. In preliminary testing, this protocol effectively killed *S. marcescens* cells swabbed onto the exterior of sham cores and the saw blade. Using this method, they detected various sporulating (*Bacillus* spp., *Paenibacillus* spp.) and nonsporulating Gram-positive bacteria (*Clavibacter* sp.), actinobacteria (*Nocardioides* spp., *Arthrobacter* spp.), ultramicrobacteria (*Sphingomonas* spp.), and α- (*Methylobacterium* spp.) and γ-(*Acinetobacter* spp.) proteobacteria from polar and nonpolar glacial ice cores.

In our study, however, mechanical ablation followed by melting out of an inner core, with or without ethanol treatment, was not an effective decontamination protocol when *Bacillus subtilis* or *Ulocladium atrum* was used as the outer contaminants. Only the innermost melt shells were completely devoid of the externally applied test microorganisms. In addition, treatment with ethanol, use of a heated probe, and drilling all resulted in externally applied microorganisms in the outermost melt shells, but their numbers were reduced or absent by the third meltshell. Thus, there remains a statistical probability that external contaminants may be present even in the innermost part of ice cores so treated (see ref. 17 and chapter 2 this volume for details).

Catranis and Starmer (4) and Castello et al. (see ref. 5 and chapter 13, this volume) used UV irradiation followed by extraction of an inner subcore to investigate ice cores from Greenland for microorganisms and viruses. In later studies Ma et al. (see refs. 11–13 and chapter 11, this volume) utilized a Clorox decontamination protocol as well as the UV irradiation method of Catranis and Starmer to isolate viable fungi from Greenland and Antarctica. Many different filamentous fungi were isolated (e.g., *Penicillium, Aspergillus, Cladosporium,*

Fusarium, Ulocladium, Alternaria, Dactylella, Multiclavula, Tricholoma, Pleurotus, and others). Bacteria closely related to *Bacillus subtilis, Paenibacillus* sp., *Rhodococcus* sp. (actinobacteria), and *Tatumella* (γ-proteobacteria) as well as tomato mosaic virus (ToMV) and bacterial viruses (PBSY group phage) also were detected (see chapter 13).

In our comparative study, UV irradiation was not effective in removal of externally applied *U. atrum*. However, in detailed studies conducted by Ma (11–13) and Castello et al. (see ref. 5 and chapter 13, this volume) UV irradiation at the dosages used on ice effectively killed all microorganisms isolated from ice, including ToMV, with the exception of *U. atrum*, which was highly resistant to UV light.

Both Priscu et al. (15) and Karl et al. (10) used water ablation to decontaminate the Vostok 5G core sections from which they isolated microorganisms. Core sections were cut with ethanol-sterilized saw blades, and the ice section rinsed with ethanol and then nanopure water or just nanopure water until the outer 4 to 10 mm was ablated away (Priscu et al. [15]). Karl et al. (10) rinsed the core section with HPLC grade water, and allowed the rinsed section to melt in a sterile plastic bag. Priscu detected bacteria closely related to the proteobacterial genera *Afipia, Comamonas, Agrobacterium,* as well as *Actinomyces* sp. Twelve amplicons were detected in total, five of which also were detected in the negative control meltwaters, and were discounted. Karl detected evidence of viable Gram-negative bacteria in the same core, but a different horizon. In our comparative study, however, water ablation was ineffective as a decontamination protocol, most probably because of microbial carryover from the exterior to the interior of the core.

Our comparative study of protocols commonly used to decontaminate ice core surfaces prior to microbiological study suggests that results of some prior research using other less than 100% effective decontamination protocols may be suspect. We wish to emphasize several points: (i) Not all decontamination protocols are equally effective against all microorganisms. This finding necessitates testing of all microbial ice isolates against the specific decontamination protocol that was used in its isolation. Unless this is done, it is impossible to determine with 100% certainty which isolate is genuine, and which may represent a contemporary contaminant. (ii) The similarity of microorganisms detected in the same and different ice cores by researchers in different labs using different decontamination and isolation protocols provides a level of independent corroboration of results, although it is still possible that we may all be detecting contaminants. This is why independent replication of results in the strictest sense must be done. (iii) decontamination with Clorox was the most effective of the decontamination protocols that we tested, and should provide the basis for further testing with the ultimate objective of developing a standardized set of decontamination and isolation protocols for the microbiological examination of glacial ice.

Implications and Significance

A great diversity of organisms is entrapped in ice. It is clear that the cryobiosphere is neither simple, nor static, nor sterile. Richard B. Alley refers to
glacial ice as a "Two Mile Time Machine" (3) because there is information
about past climates, geologic events, and biological activity. Thus, it offers a
detailed view into the past. Organisms immured in ice are actually living fossils
that can be used to study speciation, evolution, mutation rates, and population
changes over time. Examples of such organisms include *B. subtilis*, the PBSY-
group phages that infect them, ToMV, certain fungi (*Cladosporium* spp. and
yeasts), and various diatom species. Because each of these organisms has been
detected in ice of many different ages in chronosequence, they can be used to
study the time courses of varied processes, including their evolution. Additionally, it is now possible to extend the study of microbial diversity to geologic
timeframes and to relate microbial diversity to geologic, climatic, and possibly
historic events. For example, it may prove possible to relate historical events
(e.g., mass migrations, extinctions, plagues, etc.) with specific pathogens (e.g.,
influenza pandemics, cholera outbreaks, etc.). The presence of specific types
of microbes in ice may be used to date volcanic eruptions, and perhaps other
geological or climatic events. Abyzov et al. (2) have determined that a correlation exists between the concentration of bacteria and the dust concentration of
ancient atmospheres, and that the amount of dust reflects climate changes in
the geological past. Thus, changes in microbial density and biodiversity can
be measured and related to climate change. We may now be able to address
this and a variety of related questions because ice is a reservoir of ancient
microbial genotypes (see chapter 6).

Some ancient microorganisms are pathogens, possibly gene vectors, and a
few may be sources of new and useful compounds or genes for industry or
medicine (see chapter 9). Because ice can preserve microbes, it may be an
important part of the life cycles for some pathogens. It will be interesting to
determine which pathogens may use this mechanism to survive and hide from
resistant hosts (e.g., influenza, see chapter 14). Also of interest will be whether
there are currently extinct pathogens that may be frozen in glaciers around the
world. This would be valuable from the standpoint of historical outbreaks, but
may be important for potential future outbreaks. In other words, some pathogens thought to be extinct may not be.

There are recent concerns about global warming and its impacts on the
world, and specifically on human populations. As the Earth warms and the ice
melts, an increasing number of microbes are released. If ice does contribute
significantly to human disease, as well as diseases of other organisms, then an
increased rate of melting will lead to an increased risk to the hosts of these
pathogens. The resurrection of ancient genotypes may swamp mechanisms

for resistance and immunity simply by sheer numbers of released microbes; increased rates of genetic recombination and introgression (by viruses injecting new genes into cellular genomes) may also lead to especially virulent pathogens. Although this is all speculation at this point in time, research results to date cannot exclude these possibilities.

Recycling of microbes and their genomes also has population and evolutionary implications. For a given species, the life cycles and population sizes are greater than previously considered. As a practical matter and depending upon the degree of genotype recycling, calculations of evolutionary distances and mutation rates based on contemporary genotypes alone may be inaccurate and misleading. Such distances and rates are valid only within coherent populations, and will be unreliable if ancient genotypes mix with modern ones. Evolutionary theories do not usually consider temporal overlap, especially when the populations are composed of organisms that come from different periods of the geological timescale.

As early as 1973, it had been suggested that Antarctica could be a surrogate for Martian terrain (20). Given current technologies, it will soon be possible to search for microorganisms ancient Martian matrices, comets, ice on the Jovian moon Europa (and others), and lunar ice and other frozen substrates. Work with Earth ice has contributed to the development of the proper protocols and methods. It is essential to know what microorganisms are present in ancient Earth ice for comparative purposes. McKay (14) discussed the importance of studying polar ecosystems to test theories and develop methods to search for extraterrestrial life. Vorobyova et al. suggest that we need a systematic inventory of microorganisms that exist in all extreme environments on Earth detected using similar methods to determine how long microorganisms can survive in such habitats, to develop the methods to detect the biosignatures of life, and to develop and maintain an atlas of images of life forms detected in such extreme habitats (see chapter 19). All will be needed someday for comparison with potential extraterrestrial life forms.

The future of research with ancient ice and permafrost is bright and multifaceted. It will likely lead to many important and valuable outcomes in the near and distant future. These will be in the areas of astrobiology, evolutionary biology, glaciology, geology, microbiology, molecular biology, pharmacology, and others.

Research Recommendations and Needs for the Future

1. To assure reliable data, the contamination concern must be addressed at all levels of handling (drilling, transport, and assay of cores in the lab). Several needs are immediately apparent.

- Strict replication of results in one or more independent laboratories is a priority.
- Rigorous, standardized decontamination protocols are needed. The Clorox decontamination protocol should be further tested as the basis for a standardized protocol to decontaminate ice cores.
- The ice isolates themselves must be tested against the decontamination protocols used to assess their effectiveness.
- Acceptable criteria to evaluate the authenticity of ancient nucleic acids must be developed and adopted.
- Extreme attempts to avoid contamination must be made.
- A systematic comparison of dry and wet drilling methods should be done to determine which is best to minimize contamination of the interior of ice cores.
- Although it is impossible to maintain sterility of the core surface when drilled, additional contamination should be avoided to the greatest extent possible.
- A complete record for each core, including handling, should be maintained. Cores should be clearly labeled and marked. Core sections should be manipulated with utmost attention to sterility.

2. Newer methods to detect life in ancient ice need to be further developed and tested (e.g., AFM, biospectrologging, CT-scan, MRI techniques, among others).

3. Use of quantitative/automated molecular methods to analyze multiple samples is highly desirable (e.g., microarray technology).

4. Investigation should be initiated on DNA repair mechanisms in organisms that survive freezing conditions in ice/permafrost for prolonged times.

5. The issue of microbial metabolism in ice *in situ* must be resolved. Are some microorganisms immured in ice for millennia able to maintain viability, and if so what are the mechanisms involved?

6. The possibility for genome recycling (i.e., temporal gene flow) must be explored.

7. Ice biologists must become formally organized into a group that meets regularly to discuss the issues raised above, to develop and implement protocols, and to promote and coordinate collaborative research efforts among themselves and with other groups/agencies involved in ice/permafrost research as well as NASA.

Literature Cited

1. Abyzov, S.S. 1993. Microorganisms in Antarctic ice. pp. 265–297. In E.I. Friedmann (ed.), Antarctic Microbiology. Wiley, New York.
2. Abyzov, S.S., I.N. Mitskevich, and M.N. Poglazova. 1998. Microflora of the deep glacier horizons of central Antarctica. Microbiology 67: 451–458.
3. Alley, R.B. 2000. The Two Mile Time Machine: Ice Cores, Abrupt Climate Change, and our Future. Princeton University Press, Princeton.

4. Catranis, C.M., and W.T. Starmer. 1991. Microorganisms entrapped in glacial ice. Ant. J. U.S. 26: 234–236.

5. Castello, J.D., S.O. Rogers, W.T. Starmer, C. Catranis, L. Ma, G. Bachand, Y. Zhao, and J.E. Smith. 1999. Detection of tomato mosaic tobamovirus RNA in ancient glacial ice. Polar Biol. 22: 207–212.

6. Christner, B C., E. Mosley-Thompson, L G. Thompson, V. Zagorodnov, K. Sandman, and J.N. Reeve. 2000. Recovery and identification of viable bacteria immured in glacial ice. Icarus 144: 479–485.

7. Christner, B.C., E. Mosley-Thompson, L.G. Thompson, and J.N. Reeve. 2001. Isolation of bacteria and 16S rDNAs from Lake Vostok accretion ice. Environ. Microbiol. 3: 570–577.

8. Friedmann, E.I. et al. 1996. Viable bacteria, methane and high ice content in Antarctic permafrost: Relevance to Mars. Eighth ISSM meeting, 11th International Conference on the Origin of Life, Orleans, 5–12 July, Abstract 5–1, p. 60.

9. Garrison, D.L., C.W. Sullivan, and S.F. Ackley. 1986. Sea ice microbial communities in Antarctica. Bioscience 36: 243–250.

10. Karl, D.M., D.F. Bird, K. Björkman, T. Houlihan, R. Shackleford, and L. Tupas. 1999. Microorganisms in the accreted ice of Lake Vostok, Antarctica. Science 286: 2144–2147.

11. Ma, L.-J. 2000. Ancient Fungi Entrapped in Glaciers. Ph.D. dissertation, SUNY-College of Environmental Science and Forestry, Syracuse.

12. Ma, L.-J., C.M. Catranis, W.T. Starmer, and S.O. Rogers. 1999. Revival and characterization of fungi from ancient polar ice. Mycologist 13: 70–73.

13. Ma, L-J., S.O. Rogers, C.M. Catranis, and W.T. Starmer. 2000. Detection and characterization of ancient fungi entrapped in glacial ice. Mycologia 92: 286–295.

14. McKay, C.P. 1993. Relevance of Antarctic microbial ecosystems to exobiology. pp. 593–601. In E.I. Friedmann, Antarctic Microbiology. Wiley, New York.

15. Priscu, J., et al. 1999. Geomicrobiology of subglacial ice above Lake Vostok, Antarctica. Science 286: 2141–2144.

16. Rivkina, E.M., D. Gilichinsky, S. Wagener, J. Tiedje, and J. McGrath. 1998. Biogeochemical activity of anaerobic microorganisms from buried permafrost sediments. Geomicrobiology 15: 187–193.

17. Rogers, S.O., V. Theraisnathan, L.J. Ma, Y. Zhao, G. Zhang, S.-G. Shin, J.D. Castello, and W.T. Starmer. 2004. Comparisons of protocols to decontaminate environmental ice samples for biological and molecular examinations. Appl. Environ. Microbiol. 70: 2540–2544.

18. Shi, T., R.H. Reeves, D.A. Gilichinsky, and E.I. Friedmann. 1997. Characterization of viable bacteria from Siberian permafrost by 16S rDNA sequencing. Microbial Ecol. 33: 169–179.

19. Skidmore, M.L., J.M. Foght, and M.J. Sharp. 2000. Microbial life beneath a high Arctic glacier. Appl. Environ. Microbiol. 66: 3214–3220.

20. Vishniac, W.V., and S.E. Mainzer. 1973. Antarctica as a Martian model. pp. 3–31. In P.H.S. Sneath (ed.), Life Sciences and Space Research XI. Akademie Verlag, Berlin.

21. Vorobyova, E., et al. 1997. The deep cold biosphere: Facts and hypotheses. FEMS Microbiol. Rev. 20: 277–290.

22. Willerslev, E., A.J. Hansen, B. Christensen, J.P. Steffensen, and P. Arctanger. 1999. Diversity of Holocene life forms in fossil glacier ice. Proc. Nat. Acad. Sci. 96: 8017–8021.
23. Zvyagintsev, D.G., D.A. Gilichinsky, G.M. Khlebnikova, D.G. Fedorov-Davydov, and N.N. Kurdryavtseva. 1990. Comparative characteristics of microbial cenoses from permafrost rocks of different age and genesis. Microbiology 59: 332–338.

Index

Aberdeen, V. 181

Abyzov, S.S., 1, 4, 55, 70, 102, 206, 240, 258, 259, 260, 261, 262, 263, 290, 291, 292, 294, 296

Adams, E.E., 43

Aeolian transport, 95, 100, 101

Africa, 5, 94, 191, 221

Alaska, 92, 106, 108, 178, 218, 220

Amundsen-Scott Station, 72

Amundsen Sea, 71

Amundsen Sea Low Pressure System, 83, 87

anabiosis, 18, 107, 175, 288

Antarctica, v, ix, x, xi, xii, 5, 6, 15, 16, 18, 20, 22, 23, 29, 36, 43–52, 66–72, 75, 76, 78, 79, 80, 81, 82, 85, 86, 88–95, 97, 100–105, 108, 116, 117, 118, 122, 126, 133, 138, 140, 145, 156, 157, 159, 161, 162, 175, 177, 179, 180, 191, 206, 223, 230, 231, 234, 236, 237, 238, 239, 249, 250, 252, 253, 265, 266, 267, 286, 288, 291, 292, 293, 294, 297, 299; Antarctic peninsula, 72, 80, 81, 85, 230; Dome C, 70, 71, 72, 79, 83, 85, 86, 101; Dronning Maud Land, 100; East Antarctic ice sheet (EAIS), 73, 74, 83, 84, 100; Fisher Massif, 71, 79, 85, 86, 91; Lake Bonney, ix, xiii, 23, 24, 26–30, 32–37, 39, 40, 44, 46–49; Lake Fryxell, ix, xiii, 29–32, 35, 38, 41, 43, 45, 47, 145, 237; Lake Hoare, xiii, 34, 36, 38, 40, 43, 46; Lake Vanda, 35, 145; Lake Vida, ix, xiii, 35; Lake Vostok, vii, xi, 3, 19, 20, 32, 43, 45, 48, 58, 59, 63, 95, 98, 100, 104, 105, 160, 177, 179, 227, 234, 235, 236, 238–240, 244, 249–258, 261, 263–268, 291, 292, 299; Law Dome, 51, 83; McMurdo Dry Valleys, ix, xiii, 29, 32, 40–48, 72, 81, 237; McMurdo Sound, xiii, 29, 30, 46, 56, 66, 67, 68, 145, 156; McMurdo Station, 145; Miers Valley, 144, 145, 278; Mt. Crean, 71, 79, 80, 85, 86, 145; Mt. Feather, 71, 79, 80, 85, 86, 89, 144–146, 288; Pagodroma Formation, 80; Ross Ice Shelf, 29, 81; Schirmacher Oasis, 72; Siple Dome, xii, 71, 73, 74, 75, 82, 83, 84, 86, 270, 271, 276; South Pole, 22, 43, 70–72, 74, 75, 81,

83, 84, 86, 91–93, 97, 110, 116, 177, 228, 237, 275, 276; Table Mountain, 71, 79, 80, 85, 86; Taylor Dome, 71, 73, 74, 75, 79, 82, 83, 84, 86, 91, 92, 95, 96, 230, 231; Taylor Valley, 23, 24, 32, 44, 91, 129, 144, 146, 156, 231, 236, 281; Transantarctic Mountains, 80, 83, 85; Victoria Land, 45, 141, 144, 278; Victoria Land, northern, 72; Victoria Land, southern, 23, 24, 44, 89, 92, 156, 157; Vostok Station, 6, 47, 160, 162, 241, 250, 253, 254, 266; Vostok 5G core, xi, xii, 95, 101, 102, 161, 162, 240, 242, 254, 258, 259, 291, 294, 295; West Antarctic ice sheet (WAIS), 73, 74, 84

antibiotics, 138, 174, 175, 188, 201, 237

archaea, 1, 46, 107, 196, 269, 291, 293; methanogens, 107, 110, 111, 113, 269, 293

Asia, 5, 160, 174, 176, 218, 221

astrobiology, vii, 2, 88, 116, 277, 278, 297

Australia, 2, 85, 139

bacteria, vi, ix, xi; Acidobacterium-Holophaga division, 32, 292; acidophilic, 274; actinobacteria, 32, 55, 96, 106, 125, 127, 138, 203, 206, 231, 234, 259, 260, 261, 262, 284, 287, 291, 292, 294, 295; anaerobic, 143; bacterioplankton, 56, 64, 67; Bacteroidetes, 261, 292; chemolithotrophic, 115; Cytophaga-Flavobacterium-(Flexibacter)-Bacteroides (CFB) group, 54, 55, 62, 63, 292; denitrifying, 108, 110; endospore-forming, 127, 203, 206, 231, 262; epiphytic, 57; eubacteria, 199; Gram-negative, 62, 148, 203, 205, 259, 262, 282, 291, 294, 295; Gram-positive, 54, 55, 148, 203, 231, 262, 282, 291, 292, 294; green nonsulfur division, 32, 292; marine, 57, 58, 64–67; micrococci, 243, 259, 262; nitrifying, 108, 117; oligotrophic, 58; Planctomycetales, 32, 292; proteobacteria, 32, 53, 54, 55, 65, 96, 127, 148, 197, 206, 231, 234, 261–263, 291, 292, 294, 295;l thermoactinomycete, 6, 19; thermophilic, 6, 18, 44

bacteria, genera of: Acidovorax, 259, 261, 263; Acinetobacter, 231, 232, 280, 294;